D1261333

Chemical Oceanography

VOLUME 4

2ND EDITION

Chemical Oceanography

Edited by
J. P. RILEY
Department of Oceanography, The University of Liverpool, England

and

G. SKIRROW
Department of Inorganic, Physical and Industrial Chemistry, The University of Liverpool, England

VOLUME 4

2ND EDITION

1975

ACADEMIC PRESS
LONDON NEW YORK SAN FRANCISCO
A Subsidiary of Harcourt Brace Jovanovich, Publishers

ACADEMIC PRESS INC. (LONDON) LTD.
24/28 Oval Road,
London NW1

United States Edition published by
ACADEMIC PRESS INC.
111 Fifth Avenue
New York, New York 10003

Library of Congress Catalog Card Number: 74–5679
ISBN: 0–12–588604–7

Printed in Great Britain by

PAGE BROS (NORWICH) LTD
NORWICH

Contributors to Volume 4

E. BOOTH, *275 Haslingen Old Road, Rossendale, Lancashire, England*
W. F. MCILHENNY, *The Dow Chemical Company, Freeport, Texas, U.S.A.*
YUZURU SHIMIZU, *College of Pharmacy, University of Rhode Island, Kingston, Rhode Island, U.S.A.*
M. WHITFIELD, *Marine Biological Laboratory, Plymouth, Devon, England*
HEBER W. YOUNGEN, JR., *College of Pharmacy, University of Rhode Island, Kingston, Rhode Island, U.S.A.*

Preface to the Second Edition

Rapid progress has occurred in all branches of Chemical Oceanography since the publication of the first edition of this book a decade ago. Particularly noteworthy has been the tendency to treat the subject in a much more quantitative fashion; this has become possible because of our much improved understanding of the physical chemistry of sea water systems in terms of ionic and molecular theories. For these reasons chapters dealing with sea water as an electrolyte system, with speciation and with aspects of colloid chemistry are now to be considered as essential in any up-to-date treatment of the subject. Fields of research which were little more than embryonic only ten years ago, for example sea surface chemistry, have now expanded so much that they merit separate consideration. Since the previous edition, there has arisen a general awareness of the potential threat to the sea caused by man's activities, in particular its use as a "rubbish bin" and a receptacle for toxic wastes. Although it was inevitable that there should be some over-reaction to this, there is real cause for concern. Clearly, it is desirable to have available reasoned discussions of this topic and also an examination of the role of the sea as a potential source of raw materials in view of the imminent exhaustion of many high grade ores; these subjects are treated in the second, third and fourth volumes.

Most branches of marine chemistry make use of analytical techniques; the number and range of these has increased dramatically over recent years. Consequently, it has been necessary to expand greatly and restructure the sections dealing with analytical methodology. These developments are extending increasingly into the very important and rapidly developing area of organic chemistry.

Rapid advances which have taken place in geochemistry, particularly those that have stemmed from the Deep Sea Drilling Project, have made it necessary to devote a whole volume to topics in sedimentary geochemistry.

Both the range and accuracy of the physical constants available have increased since the first edition and a selection of tabulated values of these constants are to be found at the end of each of the first four volumes.

No attempt has been made to discuss Physical Oceanography except where a grasp of the physical concepts is necessary for a better understanding of the chemistry. For a treatment of the physical processes occurring in the sea the reader is referred to the numerous excellent texts now available on physical oceanography. Likewise, since the distribution of salinity in the sea is of greater relevance to the physical oceanographer and is well discussed in these texts, it will not be considered in the present volumes.

This series is not intended to serve as a practical handbook of Marine Chemistry, and if practical details are required the original references given in the text should be consulted. In passing, it should be mentioned that, although those practical aspects of sea water chemistry which are of interest to biologists are reasonably adequately covered in the "Manual of Sea Water Analysis" by Strickland and Parsons, there is an urgent need for a more general laboratory manual.

The editors are most grateful to the various authors for their helpful co-operation which has greatly facilitated the preparation of this book. They would particularly like to thank Messrs R. F. C. Mantoura and A. Dickson for their willing assistance with the arduous task of proof reading; without their aid many errors would have escaped detection. They would also like to acknowledge the courtesy of the various copyright holders, both authors and publishers, for permission to use tables, figures, and photographs. In conclusion, they wish to thank Academic Press, and in particular Mr. E. A. S. Cotton, for their efficiency and ready co-operation which has much lightened the task of preparing this book for publication.

Liverpool J. P. RILEY
November, 1974 G. SKIRROW

CONTENTS

Chapter 22 *by* E. BOOTH

Seaweeds in Industry

Chapter 23 *by* HEBER W. YOUNGKEN, JR and YUZURU SHIMIZU

Marine Drugs: Chemical and Pharmacological Aspects

Appendix *compiled by* J. P. RILEY

Tables of Physical and Chemical Constants Relevant to Marine Chemistry

Contents of Volume 1

Contents of Volume 2

Contents of Volume 3

Symbols and units used in the text

A list of the more important symbols used in the text is given below. It is not exhaustive and inevitably there is some duplication of usage since some symbols have different accepted usages in two or more disciplines. The generally accepted symbols have been altered only when there is a possibility of ambiguity.

Concentration. There are several systems in common use for expressing concentration. The more important of these are the molarity scale (g molecules l^{-1} of solution = mol l^{-1}) usually designated by C_i, the molality scale (g molecules kg^{-1} of solvent* = mol kg^{-1}) designated by m_i and the mole fraction scale usually denoted by x_i, which is of more fundamental significance in physical chemistry. In each instance the subscript i indicates the solute species; when i is an ion the charge is not included in the subscript unless confusion is likely to arise. Some other means of indicating the concentration are also to be found in the text, these include: g or mg kg^{-1} of solution (for major components), μg or ng l^{-1} or kg^{-1} of solution (for trace elements and nutrients) and μg-at l^{-1} of solution (for nutrients). Factors for conversion of μg to μg-at are to be found in Appendix Tables 4 and 5.

Activity. When an activity or activity coefficient is associated with a species the symbols a_i and γ_i are used respectively regardless of the method of expressing concentration, where the subscript i has the significance indicated above. Further qualifying symbols may be added as superscripts and/or subscripts as circumstances demand. It is important to realize that the numerical values of the activity and activity coefficient depend on the standard state chosen. It should also be noted that since activity is a relative quantity it is dimensionless.

UNITS

Where practicable SI units (and the associated notations) have been adopted in the text except where their usage goes contrary to established oceanographic practice.

* A common practice is to regard sea water as the solvent for minor elements.

LENGTH

Å	= Ångstrom	= 10^{-10} m
nm	= nanometre	= 10^{-9} m
μm	= micrometre	= 10^{-6} m
mm	= millimetre	= 10^{-3} m
cm	= centimetre	= 10^{-2} m
m	= metre	
km	= kilometre	= 10^{3} m
mi	= nautical mile (6080 ft)	= 1·85 km

WEIGHT

pg	= picogram	= 10^{-12} g
ng	= nanogram	= 10^{-9} g
μg	= microgram	= 10^{-6} g
mg	= milligram	= 10^{-3} g
g	= gram	
kg	= kilogram	= 10^{3} g
ton	= metric ton	= 10^{6} g

VOLUME

μl	= microlitre	10^{-6} l
ml	= millilitre	10^{-3} l
l	= litre	
dm^3	= litre	

CONCENTRATION

ppm	= parts per million (μg g^{-1} or mg l^{-1})
ppb	= parts per billion (ng g^{-1} or μg l^{-1}
μg-at l^{-1}	= μg atoms l^{-1} = (μg/atomic weight) l^{-1}

ELECTRICAL

V	= volt	
mV	= millivolt	10^{-3} V
μV	= microvolt	10^{-6} V
A	= ampere	
mA	= milliampere	10^{-3} A
μA	= microampere	10^{-6} A
Ω	= ohm	

TIME

s	= second
ms	= millisecond
min	= minute
h	= hour
d	= day
yr	= year

ENERGY AND FORCE

J	= Joule	= 0·2390 cal
N	= Newton	= 10^5 dynes
W	= Watt	

LIGHT FLUX

klux	= kilolux

General Symbols

For Chapter 21 see list of symbols, pp 211–214

A	Helmholtz free energy
AOU	apparent oxygen utilization
C_o	initial concentration
C_t	concentration at time t
C_x	molar concentration of species x; charges of the species x are omitted when they are obvious
CA	carbonate alkalinity (meq 1^{-1})
Cl	chlorinity (g kg^{-1} = ‰)
D_x	diffusion coefficient of x
E	E.M.F. of a cell
E°	standard of potential
E_h	redox potential
E_x	overall cell potential
E_x°	standard potential of x
F	Faraday equivalent of electric charge
G	Gibbs free energy
H	enthalpy

I	ionic strength
i	current
i_p	peak current
K	equilibrium constant
K_{ij}	selectivity coefficient of electrode for i to interfering ion j
M_x	molarity of component x
m_x	molality of component x
P_G	partial pressure of gas G in solution
R	gas constant
S	entropy
S	salinity ($g\,kg^{-1}$ = ‰)
T	temperature in K
t	temperature in °C
t	time
V	volt
VP	saturation water vapour pressure
z_i	ionic charge of ion i

Greek Symbols

β	Bunsen coefficient of a gas
γ_x	activity coefficient of species x
Δ	change of (as in ΔG)
ε_j	junction potential
ε_R	potential of reference electrode
ε_x°	standard potential of working electrode
η	viscosity
λ	wavelength
λ_i	equivalent conductivity of ion i
$\boldsymbol{\mu}_i$	chemical potential of component i

Chapter 20

The Electroanalytical Chemistry of Sea Water

M. WHITFIELD

Marine Biological Laboratory, Plymouth PL1 2PB, Devon, England

20.1. INTRODUCTION

Electrochemical techniques are basically simple. The working unit in all cases is a cell in which two conducting electrodes make electrical contact with the sample solution. This cell acts as a transducer which produces modified current or voltage signals in response to excitations applied to the electrodes from a controlled source. The art of electroanalytical measurement lies in the manipulation of the conditions within the cell, the structure of the electrodes and the nature of the input so that the signals transmitted by the cell are directly related to the activity or concentration of specific solution components. The basic instrumental requirements are a stable power supply capable of applying voltage or current loadings to the cell in a variety of modes (Meites, 1965; Lingane, 1958; Delahay, 1954; Reilley and Murray, 1963) and a voltmeter or a recording device to follow the resulting current and voltage fluctuations. If these fluctuations are very rapid then an oscilloscope may be required.

In recent years these procedures have assumed a more important role in the chemical analysis of sea water. The reasons for this are partly technological. The revolution in electronics over the past decade has resulted in an increase in the reliability and functional simplicity of electrochemical instruments. In particular it has become a simple matter to rationalize the equipment for a wide variety of electrochemical techniques into a small number of individual circuit blocks (Springer, 1970; Schroeder, 1972). This has resulted in compact multifunction instruments with a high precision and reliability so that a range of electroanalytical procedures may be selected by the flick of a switch (Ewing, 1969; Flato, 1972; Bond and Canterford, 1972a). In addition, the basic building blocks can be purchased separately so that an analyst with only a rudimentary knowledge of electronics can assemble an electrochemical system to suit his particular needs (Morgenthaler, 1968; McKee, 1970).

Another consequence of the development in electronics has been the increasing sophistication of the electroanalytical procedures themselves with the introduction of new cell excitation modes and signal processing procedures which may give the basic methods greater discrimination and sensitivity. The development of microelectronic circuitry will soon result in the design of reliable field instrumentation with low power requirements and rugged dependability for most of the electroanalytical procedures considered here. Techniques for the direct conversion of charge to a digital number are already available (Goldsworthy and Clem, 1972). These will enable measurements of a high fidelity to be recorded and stored *in situ* in a directly accessible digital form even under adverse conditions.

The emphasis of marine chemistry itself has shifted noticeably in the last ten years. The oceans are now being treated as an interlinked chemical system (Sillén, 1967a; Broecker, 1971; MacIntyre, 1970) and the questions that are being asked about the nature of this system require a knowledge not only of the total concentrations of the various elements but also of their chemical forms (Andersen, 1973). This information requires more subtle analytical procedures, and it is becoming clear that electrochemical methods, where they are applicable, are likely to provide the most direct insight into the problems of chemical speciation. Since electrochemical procedures are so sensitive to the chemical forms of the components being determined, calibration procedures must be performed with care.

Electrochemical techniques also provide particularly simple means for the *in situ* measurement of chemical parameters. Such measurements are becoming more important as the significance of fine structure in the distribution of physical and chemical properties in the oceans becomes more apparent, and as society demands more detailed information about the complex chemical processes occurring in estuaries and in-shore waters. Electrochemical sensors are already available for the continuous measurement of more than half a dozen chemical parameters in sea water (pH, $[S^{2-}]$, Eh, $[F^-]$, $[O_2]$, $[Zn^{2+}]$, salinity, carbonate saturation) and many more are currently being developed. A number of reviews have considered the use of electroanalytical methods for the determination of trace constituents (Laitinen, 1967; Maienthal and Taylor, 1968; Minear and Murray, 1973). The present discussion will focus on analytical applications to marine chemistry.

20.2. Classification of Electroanalytical Procedures

The great variety of electrochemical procedures can be conveniently classified according to the experimental conditions established within the cell (Fig. 20.1) and the parameters measured (Reilley and Murray, 1963; Lingane, 1958; Taylor *et al.*, 1965; von Fraunhofer and Banks, 1972).

20.2.1. Zero current flow

If the electrodes are allowed to come to equilibrium with the solution and the resulting potential difference between the electrodes is measured in the absence of current flow then the cell potential can be related to the solution composition by equilibrium thermodynamics (*potentiometry*; Rossotti, 1969). The exchange current densities at the electrode surfaces must be high for the equilibrium condition to be attained within a reasonable time. By suitably manipulating the molecular architecture of the electrode material it is possible to arrange for the potential generated at each electrode surface

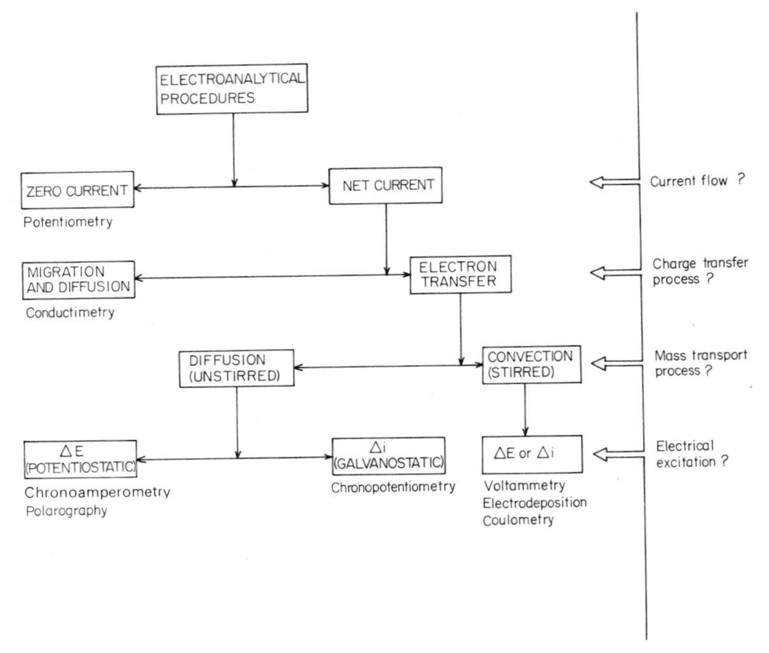

FIG. 20.1. Classification of electroanalytical procedures according to the scheme suggested by Reilley and Murray (1963) using the four criteria listed on the right of the figure. ΔE indicates a voltage excitation and Δi a current excitation.

to be dominated by the activity of a specific ionic species in solution (see Durst, 1969a; Moody, 1973; Covington, 1974). In establishing a suitable cell for analytical purposes one electrode (the working electrode) is exposed directly to the solution and generates a potential in response to changes in the *activity* of the component of interest. The second electrode (the reference, or counter electrode (Covington, 1969; Ives and Janz, 1961)) is usually a chloride selective electrode which is encased in a protective glass envelope containing a solution of fixed chloride concentration. Electrical contact between the reference solution and the sample solution is made by establishing a controlled leakage of solution from the reference compartment via a liquid junction. If the response of the working electrode is completely controlled by a single component (X) the overall cell potential (E_x) is related to the concentration of X (C_x) by the Nernst equation,

$$E_x = (\varepsilon_x^0 + \varepsilon_R + \varepsilon_J + \varepsilon_\gamma) + kT \log C_x \tag{20.1}$$

where $k = R \ln 10/nF$, R is the gas constant, T is the absolute temperature,

F is the Faraday and n is the stoichiometric number of electrons exchanged in the charge transfer process (negative for an anion, positive for a cation). This equation is generally valid for concentrations down to 10^{-6} M but its validity may extend to much lower concentrations if the ion X is effectively buffered in the solution.

The standard potential of the working electrode (ε_x^0) is a constant at any particular temperature and pressure. The potential generated by the reference electrode (ε_R) will also depend only on the temperature and pressure once the composition of the reference solution has been fixed. The potential generated at the liquid junction (ε_J) cannot be estimated precisely (Covington, 1969) and it depends on the detailed composition of the sample solution. ε_γ is the proportion of the cell response associated with changes in the activity coefficient of X (γ_x) and it is a measure of the deviation of the solution properties from ideal behaviour (Whitfield, 1975). ε_γ is given by

$$\varepsilon_\gamma = kT \log \gamma_x \tag{20.2}$$

Since they depend on the detailed composition of the solution, both ε_J and ε_γ complicate the direct use of ion-selective electrodes for solution analysis.

A further limitation on the use of direct potentiometry is the slope factor in equation 20.1 (kT) which restricts the cell response to about 60 mV for each ten-fold change in the concentration of monovalent ions. The sensitivity for divalent ions is only one half of this value.

Since the working electrodes usually have a rather complicated internal structure (Durst, 1969a; Whitfield, 1971a) they exhibit day-to-day drifts in potential of several millivolts, and even in the laboratory with frequent calibration it is seldom possible to attain a precision better than 0·4% (equivalent to 0·1 mV) for monovalent ions. Under field conditions the precision is typically closer to 10 or 15% (equivalent to 3 or 4 mV; Ross, 1969). Direct potentiometry is therefore not a very precise procedure although it has the merit of showing equal precision over its whole range because of the logarithmic form of equation 20.1.

A final problem arises because it is rarely possible to construct a working electrode that is exclusively selective to a particular ion, although the electrodes for fluoride, sulphide and hydrogen ions come close to this criterion. Consequently other ions in the solution may contribute to the cell potential, and their individual contributions are not always easy to disentangle. Detailed reviews of the structure and behaviour of the wide variety of commercially available electrodes have been presented (Durst, 1969; Whitfield, 1971a; Moody and Thomas, 1971; Koryta, 1972; Warner, 1972a; Covington, 1974).

If the total concentration rather than the activity of a particular ion is required then the most precise results can be obtained by using ion-selective electrodes to follow the course of a titration. In marine work both standard addition of the selected ion and removal of the selected ion by precipitation or complex formation have been used. Such procedures are capable of a precision better than 0·1% using mathematical techniques for locating the end point.

20.2.2. NET CURRENT FLOW FOCUSING ON BULK MIGRATION

If a potential is applied to an electrochemical cell, then reactions will occur at the electrodes. These reactions will alter the concentrations of electroactive species near the electrode surface with the result that the chemical conditions there are compatible with the new cell potential (Charlot *et al.*, 1962, p. 17). Electrons will be released into the solution at one electrode (the cathode) and removed from the solution at the other (the anode). Two major processes are responsible for the transfer of charge from the cathode to the anode within the solution (Fig. 20.1). The first process is the *migration* of ions through the bulk of the solution under the influence of the applied potential gradient, and the second is the *diffusion* of dischargeable (or electroactive) species across the concentration gradient that is established at the electrode–solution interface (Charlot *et al.*, 1962, pp. 18–21).

In the simplest case the working and counter electrodes will behave merely as inert electron-exchange sites, and will not themselves be chemically involved in the discharge process. To try and achieve this the working electrodes are usually constructed from the noble metals (rhodium, iridium, palladium, platinum, gold or mercury) or from carbon (graphite, glassy carbon).

If inert electrodes with large surface areas (e.g. platinized platinum electrodes) are used, the discharge process is so rapid that the potential drop across the electrode–solution interface is approximately constant. The current flowing through the cell in response to the applied potential can then be related directly to the migration of ions in the bulk of the solution. If an alternating potential is applied, the build up of electrolysis products at the electrodes is prevented and the migration current can then be accurately measured. By suitable manipulation of the ac circuitry it is possible to design a cell on the transformer principle with inductive coupling between the solution and the imposed voltage thus eliminating the electrodes from the solution entirely. For example, conductometric analysis (Loveland, 1963; Pungor, 1965) is useful for estimating the total ionic content of a solution, or for following the course of reactions that yield or consume ions with an unusual mobility in solution (e.g. H^+ or OH^-: Lee, 1969; Szekielda, 1969; Erhardt, 1969). The migration process is, however, common to all charged components, and the conductometric technique is not a good method for looking at specific solution components. The conductance (K) of a cell is related to the solution composition by the equation

$$K = b \sum_i C_i \lambda_i z_i \tag{20.3}$$

where b is a constant which depends on the geometry of the cell, C_i is the concentration of the ion (i), λ_i is its equivalent conductance and z_i is its charge. At present, there is no satisfactory theoretical procedure for calculating λ_i in electrolyte mixtures (see for example Fuoss, 1968).

Conductometric procedures are widely used to determine the salinity or total salt content of sea water both in the laboratory and *in situ* (Cox *et al.*, 1967). The detailed design of the conductivity meters (or salinometers) used has been described in Chapter 6. The resulting salinity values, together with the *in situ* pressure and temperature, are used to analyse the very small density differences that provide the driving force behind much of the water circulation in the ocean depths. The utility of this procedure depends rather heavily on the concept of constant composition of sea salts since the salinities are assigned by comparison with a standard sea water. Measurements of the relative contributions of the major electrolyte components to the overall conductivity of sea water (Connors and Weyl, 1968; Connors and Park, 1967) indicate that the observed variations in sea water composition are not likely to introduce serious errors into the density calculations. However, there are occasions when shifts in the carbon dioxide system accompanying photosynthesis (Park and Curl, 1968; Park *et al.*, 1964) or carbonate dissolution (Park, 1964) may exert a significant influence. The detailed interpretation of conductometric salinity data and their relationship to chlorinity etc. will not be considered further here (see Wallace (1973) for a detailed discussion).

20.2.3. NET CURRENT FLOW FOCUSING ON ELECTRON TRANSFER

Greater selectivity can be obtained by focusing attention on the diffusive transport of charge across the electrode–solution interface since the rate of this process and its dependence on applied potential are intimately connected with the concentration and nature of the electroactive species in solution.

To ensure that the contribution of the electroactive species to the cell characteristics is restricted to the electrode–solution interface, a large excess of an inert electrolyte is usually added to the solution. This *supporting electrolyte* does not discharge at the working electrode to any marked degree over the potential range of interest, but it accounts almost entirely for the bulk migration of charge between the two electrodes. It also reduces the resistance of the cell solution so that there is a negligible ohmic potential (or

"iR") drop across the cell. The condensed diffusion layer that results at the electrode–solution interface simplifies the mathematical description of the system. In sea water the major ionic components act as a very effective supporting electrolyte. When current is passed through the cell, reactions will occur at *both* electrodes in response to the applied potential. The interpretation of the resulting signals from the cell is greatly simplified if the response of one of the electrodes (the working electrode) is dominant. Consequently, the other electrode in the cell (the counter electrode) should be designed so that the electron exchange process at its surface is rapid and so that its potential is independent of current density over the normal range encountered. It is rather difficult to produce stable, non-polarizable electrodes of this kind for use over a wide potential range. Consequently, a third electrode (the reference electrode) is usually introduced into the cell. This electrode is identical in construction to the reference electrodes used in potentiometry. The input to the cell is applied between the working and counter electrodes and the potential of the working electrode is measured with respect to the reference electrode which is undisturbed by the passage of current through the cell. This potential may be used in turn in a feedback loop to control the potential applied to the cell.

If the solution in the cell is efficiently *stirred*, then the thickness of the diffusion zone (δ) at the electrode surface will be independent of the applied potential, and ions at the outer edge of the diffusion zone will be replaced by convection as rapidly as they are discharged. If there is only a single electroactive species in solution (X^{2+} say) then the current passing through the cell at a particular potential will depend on the rate at which X^{2+} ions reach the electrode, and on the ease with which they are discharged there. For each electroactive species there is an optimum discharge potential (or half-wave potential, $E_{\frac{1}{2}}$) which will depend on the exact composition of the supporting electrolyte and on the material used to construct the working electrode. If a linearly varying potential is applied across the cell the current flowing will gradually increase as more and more X^{2+} ions reach the electrode surface with sufficient energy to be discharged. In the immediate vicinity of $E_{\frac{1}{2}}$, there will be a sudden, stepwise increase in the current flowing (Fig. 20.2). In this region the electrode is effectively depolarized since quite large shifts in the current density result from only small changes in the potential applied across the cell. Beyond the half wave potential the current flow is actually limited by the characteristics of the diffusion process in the immediate vicinity of the electrode surface and by the nature of the discharge process.

If the discharge process proceeds so rapidly that the reactants and products of the discharge reaction are continuously at thermodynamic equilibrium (reversible process) then the diffusion current on the plateau (i_D, Fig. 20.2) is

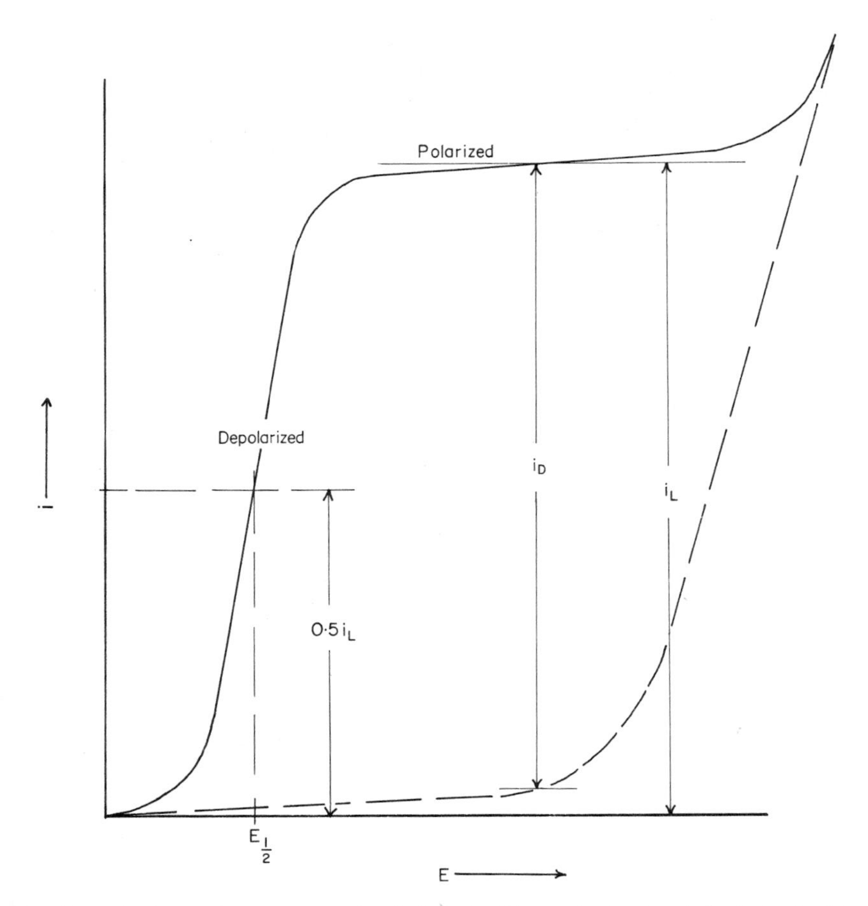

FIG. 20.2. Idealized voltammogram of a single reducible species (or depolarizer) obtained in a stirred solution (cf. Headridge 1969, p. 18). By convention the applied potential (E) is plotted with an increasingly negative value going from left to right. The cathodic current (i) corresponding to a reduction is positive.

i_L is the limiting current and $E_{\frac{1}{2}}$ is the half-wave potential. i_D is the diffusion current and i_R ($= i_L - i_D$) is the residual current obtained in the supporting electrolyte in the absence of the depolarizer (dashed line).

related to the concentration of the electroactive species (C_x) by the equation (Reilley and Murray, 1963),

$$i_D = nFAD_xC_x/\delta \tag{20.4}$$

where A is the area of the electrode and D_x is the diffusion coefficient of the component X^{2+}. D_x is dependent on the exact composition of the electrolyte solution and is very sensitive to changes in temperature. δ is critically dependent on the nature of the flow processes induced by the particular stirring regime. The steady state represented by the limiting current is, in practice, only observed when the concentration of the depolarizer is low. In concentrated solutions very high current densities would be required to keep pace with the rapid rate of recruitment of the X^{2+} ions to the diffusion layer. The experimental limiting current (i_L, Fig. 20.2) will also include contributions from the migration current (i_M) and from the residual current (i_R) associated with the discharge of other components in the solution and of the water molecules themselves. These contributions must be estimated by employing some calibration procedure before the concentration of component X^{2+} can be estimated from equation (20.4). The theoretical treatment is a little more difficult if the discharge reaction is too slow to maintain an equilibrium distribution of reactants and products at the electrode surface (i.e. if the reaction is irreversible). However, it can be shown (Reilley and Murray, 1963) that even under these circumstances the limiting current is proportional to the concentration of the electroactive component.

While the limiting current is characteristic of the concentration of the electroactive component, the half-wave potential ($E_{\frac{1}{2}}$, Fig. 20.2) of the current–potential step is characteristic of the nature of the discharging species. If the discharge process is reversible, the equation may be written as

$$E_{\frac{1}{2}} = E_x^0 + kT \log (D_x'/D_x) \qquad (20.5)$$

where D_x is the diffusion coefficient of the product of the discharge process in the particular supporting electrolyte. E_x^0 is the standard potential of the discharge reaction (de Bethune and Swendeman-Loud, 1964).

For an irreversible process $E_{\frac{1}{2}}$ is an arbitrary, though convenient, way of specifying the position of a wave, and its value depends on the conditions under which the voltammogram is measured. The voltammogram (Fig. 20.2) therefore provides a means for identifying (from $E_{\frac{1}{2}}$) and estimating the concentration (from i_D) of electroactive species. If the technique is used analytically, the voltage range must be scanned rapidly so that there is no appreciable depletion of the bulk concentration of the electroactive species. If several depolarizing species are present in the same solution, the resulting voltammogram resembles a series of steps, provided that the half-wave potentials of the individual discharge reactions are adequately separated. The degree of separation required depends on the electronic sophistication of the measuring technique.

To obtain maximum current flow through the cell the working electrode must have a large surface area and δ must be kept as small as possible, by efficiently stirring the solution or by rapidly rotating the working electrode (equation (20.4)). Many applications of voltammetry in stirred solutions have taken advantage of the high efficiency of the resulting electrolytic process to effect quantitative removal of the electroactive species from solution. The cell potential is usually fixed somewhere on the limiting current plateau so as to combine maximum current flow with maximum selectivity. A three electrode cell is employed to increase the efficiency of the process (Lingane, 1958) and to prevent errors caused by the polarization of the reference electrode. For a reversible process the concentration of the selected component at a time t (C_t) may be related to the initial concentration (C_0) by the equation (Browning, 1969, p. 122),

$$C_t = C_0 . 10^{-bt}, \tag{20.6}$$

where

$$b = 0{\cdot}43 D_x A/\delta V \tag{20.7}$$

and V is the volume of the solution. The electrolytic process is therefore speeded up by using electrodes with a large surface area in small volumes of efficiently stirred solution. The use of a working electrode with a large surface area also reduces the possibility of unwanted side reactions that might otherwise be important at high current densities. The amount of material removed from solution can be determined by weighing the electrode before and after the electrolysis (*electrogravimetry*, Lingane, 1958) if the product forms an insoluble, adherent and chemically stable deposit on the working electrode.

A sensitive and selective procedure for monitoring the quantitative removal of electroactive components from the solution is provided by primary or *constant-potential coulometry* (Rechnitz, 1963; Milner and Phillips, 1967; Abresch and Claasen, 1964). The number of coulombs of electricity (Q_∞) consumed during the removal of electroactive components is related to their concentration by the equation (Meites, 1965, p. 523)

$$Q_\infty = \int_0^\infty i\,dt = nFVC_0 = nFN^0, \tag{20.8}$$

where N^0 is the number of moles of electroactive agent present at the start of the electrolysis. This procedure is not restricted to the analysis of species that give insoluble deposits, but it does require that the electrode process should be quantitative and proceed with a 100% current efficiency. The accuracy of the method is limited by the accuracy of the current-time integration (equation (20.8)) and by the correction required to allow for the residual (non-Faradaic) current (Fig. 20.2).

The reverse of electrodeposition is used in secondary or *constant-current coulometry* (Milner and Phillips, 1967; Abresch and Claasen, 1964) where advantage is taken of the very high precision of current and time measurements to *generate* precisely known amounts of reactants in solution by electrolysis. These components can be used as reagents in titrimetric procedures whose endpoints may be estimated electrometrically or by other means. It is somewhat surprising that this technique has found little application in marine chemistry. More than fifty reactions have been documented with a 100% current efficiency, see Rechnitz (1963), Charlot *et al.* (1962) and Meites (1963). The reagents can be generated externally for addition to the sample via a burette, or they can be generated within the solution itself. Hydrazine can be used as a depolarizer for the counter electrode to prevent the discharge of chlorine in sea water (Lingane, 1958). The precision of coulometric technique is such that it has been used to confirm the purity of primary standard materials (e.g. Marisenko and Taylor, 1967, 1968; Yoshimori and Tanaka, 1971, 1973).

The procedure of *electrodeposition* can provide a valuable preconcentration step by removing metals from solution for analysis by other methods. Some measure of separation is possible if the current waves of the various electroactive components are sufficiently well separated, and this often simplifies subsequent analysis. Because of the logarithmic relationship shown in equation (20.6) the quantitative (or stoichiometric) removal of electroactive components is a time-consuming business. Consequently most analytical uses of electrodeposition depend on the removal of a fixed and reproducible portion of the electroactive component by electrolysis in cells with a uniform and highly reproducible stirring regime (non-stoichiometric deposition). All the parameters defined in equations (20.6) and (20.7) must be kept constant throughout the analysis of both the sample solutions and the standards. The quantitative removal of trace components by prolonged electrolysis can be used to advantage to clean up reagent solutions used for trace analysis.

Rapid and convenient analytical procedures may be developed if the discharge process is monitored in a *static or quiescent solution.* If the electrolysis is continued for some time under these circumstances, the solution around the working electrode becomes depleted in the electroactive species so that the outer edge of the diffusion zone moves further and further away from the electrode surface. In the initial stages when the zone is thin and well defined, the well developed theories of diffusion can be used to analyse the data since there is no mass convective transport to confuse the picture. However, as the diffusion zone becomes progressively thicker, convective processes become more important and the current–voltage curves become less reproducible. In quiescent solutions, it is therefore usual to apply the potential or current excitation to the cell rapidly and to measure the associated disturbances as a function of time. This procedure also helps to minimize the extent of polarization of the counter electrode. Under these circumstances a fair proportion of the initial current flowing will be used to charge the capacitance associated with the charge gradient across the electrolyte–solution interface. The presence of the inert supporting electrolyte helps to compress this capacitance close to the electrode surface so that it is not very sensitive to changes in voltage during the excitation. A *small* working electrode is used to minimize the depletion of the electroactive substance during the excitation, and thus to minimize the bulk migration of the electroactive species.

If a linear potential ramp is rapidly applied to the working electrode (*chronoamperometry*) the current-potential curve will show a peak at a potential that is related to the half-wave potential measured on the voltammogram. The current rise associated with the reversible discharge of the electroactive species near the half-wave potential is counterbalanced by the progressive depletion of the electroactive species near the electrode surface. For a reversible reaction,

$$E_{\text{peak}} = E_{\frac{1}{2}} - 0{\cdot}5kT \log D'_x/D_x - 1{\cdot}1kT \tag{20.9}$$

(Reilley and Murray, 1963). The peak current (i_p) is related to the rate of the voltage scan (V volts s^{-1}) by the Randles–Sevick equation (e.g. Randles, 1948),

$$i_p\,(25^\circ\text{C}) = 2{\cdot}72 \times 10^5\, n^{\frac{3}{2}} A D_x^{\frac{1}{2}} C_x V^{\frac{1}{2}}. \tag{20.10}$$

The linear relationship between peak current and C_x is maintained even if the electrode reaction is irreversible (Reilley and Murray, 1963).

The reverse of this technique is commonly used in *anodic stripping voltammetry* for the measurement of trace concentrations of electroactive species (Barendrecht, 1967; Shain, 1963). In this technique the components of interest are deposited on to a stationary mercury electrode by non-stoichiometric deposition from a well stirred solution. The trace constituents are effectively concentrated at this stage. The stirring is then stopped and the electroactive components are removed by rapidly applying a potential ramp in the reverse direction and stripping the metals off the electrode into the solution by an anodic process. In the stripping step the sample current is considerably enhanced without a concomitant increase in the background (or residual) current. Such procedures are capable of determining constituents with concentrations as low as 10^{-11} M in electrolyte solutions.

Analytically, the greatest attention has been paid to the use of chronoamperometry at a dropping mercury electrode (*polarography*: Milner, 1957; Meites, 1965; Heyrovsky and Kuta, 1966). The electrode consists of a short length of fine bore ($<0{\cdot}05$ mm) glass capillary tubing which is attached to a mercury reservoir. A succession of fine mercury droplets forms at the base of the capillary at a rate controlled by the height of the mercury reservoir and the diameter of the capillary. If a linear potential ramp is now applied between the dropping mercury electrode (DME) and the counter electrode in a quiescent solution, a current–voltage profile is obtained that is analogous to the voltammogram observed in a stirred solution (Fig. 20.2) except that the discrete contributions of each individual drop are apparent, giving the curve a characteristic "saw tooth" appearance. As each drop grows the current flow increases as a result of the increase in surface area and of the necessity to charge the capacitance that is established across the electrode–solution interface. These effects are superimposed on the current rise associated with the voltage ramp. Since the properties of successive mercury drops are accurately reproducible at a given applied potential, the overall effect is to yield a mercury micro-electrode with a continually renewed surface. The voltammograms are therefore highly reproducible and are less sensitive to mechanical vibration than are those from other procedures in quiescent solution. Each drop has a lifetime of only a few seconds so that the diffusion layer remains thin, and when the drop breaks free it stirs the adjacent solution sufficiently to prevent local depletion of the electroactive species. The associated solution disturbance has died down to acceptable proportions before the removal of the next drop. The result is the development of a voltammogram without the theoretical and experimental uncertainties associated with the need for efficient and reproducible solution flow. Because the solution is quiescent and a micro-working electrode is used the current flow is such that there is negligible depletion of the electroactive substance from the solution (Meites, 1965, p. 511). The main limitation of the technique is that the mercury electrode is relatively easily oxidized so that it cannot be used to study electrode reactions more anodic than +0·4 V versus the saturated calomel electrode. Solid electrodes (e.g. platinum) may be used at more anodic potentials to study the oxidation of anions and organic compounds in solution (Adams, 1969). The diffusion equations relating to the surface of an expanding sphere are naturally quite complicated. The simplest approximation was made by Ilkovič who considered linear diffusion at the expanding spherical surface to give the average diffusion current over the life of a drop (i_{av}) at 25 °C as

$$i_{av} = 607{\cdot}0 n m^{\frac{2}{3}} D_x^{\frac{1}{2}} t_d^{\frac{1}{6}} C_x \qquad (20.11)$$

where t_d is the drop time and m is the average mercury flow rate through the capillary. The half-wave potential of the voltammogram is given by

$$E_{\frac{1}{2}} = \mathrm{E}^0 + 0{\cdot}5kT \ln (D_x'/D_x) \qquad (20.12)$$

Several electronic techniques are available for minimizing the influence of the current artefacts associated with the growth of the mercury drop. These have been extensively reviewed (Breyer and Bauer, 1963; Brinkman, 1968; Schmidt and von Stackelberg, 1963; Kambara, 1966; Bond and Canterford, 1972a) and details will be introduced where necessary in the subsequent discussion.

The complementary technique of following the voltage changes accompanying the imposition of a controlled current between the electrodes in a quiescent solution (*chronopotentiometry*) has not been so widely used in analytical applications, although its selectivity is comparable with that of normal d.c. polarography and it is more rapid and sensitive (Reilly *et al.*, 1955). When the current is imposed the most readily reduced component will be discharged at the electrode until, after a certain time (the transition time, τ), its surface concentration falls to zero. At this point the potential will shift rapidly until it reaches a level at which a second component can be discharged in order to maintain the impressed current. At a constant impressed current the voltage-time plot will therefore appear as a series of steps. At each successive step a further electroactive component will contribute to the transport of current across the electrode-solution inteface. The transition time to each consecutive step therefore depends not only on the concentration of the dominant depolarizer at that time, but also on the concentration of all preceding depolarizers. This successive dependence makes it necessary to calibrate the procedure by standard addition techniques. The transition time is related to the concentration of the electroactive species by the Sand equation,

$$\tau^{\frac{1}{2}} = \pi^{\frac{1}{2}} nFAD_x^{\frac{1}{2}} C_x / 2i \qquad (20.13)$$

where i is the value of the impressed current. The transition time must be sufficiently short (less than 2 minutes) to prevent interference from convective transport phenomena and sufficiently long (greater than 10^{-3} s) to minimize distortion of the curves by the initial capacitance charging current. One of the convenient aspects of chronopotentiometry is the possibility of adjusting τ to a convenient time interval by altering the impressed current (i). For reversible reactions the half-wave potential occurs when $t = \tau/4$ and it is related to the diffusion coefficients of the reactants and the products by equation 20.12. The problems involved in the development of chronopotentiometry as an accurate and sensitive method of analysis for concentrations as small as 10^{-5} M have been discussed by Bos and Van Dalen (1973).

The classification of electroanalytical techniques is summarised in Fig. 20.1. The concentration ranges of the various procedures are shown in Fig. 20.3 together with, for comparison, a logarithmic display of the composition

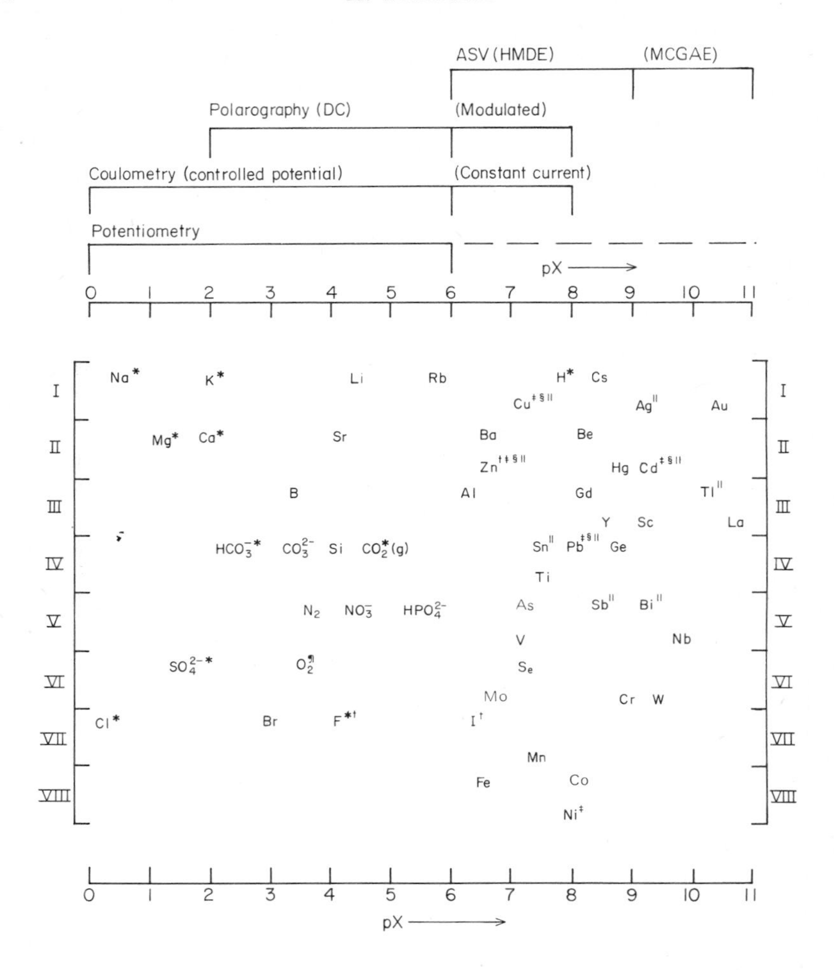

FIG. 20.3. Comparison of the concentration ranges of common electroanalytical methods (Laitinen, 1967; Allen *et al.*, 1970) with the distribution of elements found in natural oceanic seawater (Carpenter, 1972). Concentrations are shown on a logarithmic scale across the diagram ($pX = -\log_{10}[X]$). The elements are arranged vertically according to their location in the periodic table. Group numbers are given at the side of the diagram. For the sake of clarity the actinide and lanthanide elements have been omitted. Reported analyses are indicated as follows (cf. Table 20.15)

- * Potentiometry
- † Polarography (direct)
- ‡ Polarography (with pre-concentration)
- § Anodic stripping voltammetry (ASV) with the dropping mercury electrode (HMDE)
- || ASV with the mercury coated graphite electrode (MCGE)
- ¶ Voltammetry.

of sea water. These figures will serve as a general guide to the subsequent detailed discussion of the application of electroanalytical techniques to marine chemistry.

20.3. Zero Current Flow—Potentiometry

20.3.1. Direct Potentiometry

The development of a wide range of ion-selective electrodes (Durst, 1969a; Whitfield, 1971a; Warner, 1972a; Moody and Thomas, 1971; Moody, 1973; Covington, 1974) has provided fresh impetus to the application of potentiometric methods to marine chemistry. The measurement procedures employed in direct potentiometry are simple and rapid, and rugged battery-powered voltmeters are available for field measurements.

The most striking property of ion-selective electrodes, as opposed to, say, atomic absorption procedures, is that they actually respond to ionic activities rather than to total concentrations. This is inconvenient if one is trying to elucidate the mass budget of a particular constituent since the proportion of the component locked up by complex formation or even that proportion involved in strong ionic interactions will not be sensed—the reading will be low. However if one is interested in the *mechanisms* involved in the transport of that constituent through the marine system, then the activity measurements are an essential prerequisite for any detailed physico-chemical calculations. In addition, by chemically manipulating the sample and measuring the resulting changes in activity it is often possible to obtain information about the nature and strength of the various interactions.

No satisfactory procedure has yet been developed for calibrating electrodes for the determination of ion activities in complex electrolyte mixtures such as sea water. The basis of the calibration procedure is the assumption that individual ion-activities are known in certain solutions of precisely defined composition. When the calibration solutions contain only a single electrolyte, the single-ion activities are usually calculated from the experimental mean-ion activity value by some conventional procedure (Whitfield, 1973, 1975; Bates and Alfenaar, 1969). In oceanographic and estuarine work, the background electrolyte is a rather complex mixture of high ionic strength. Since little is known about the mean-ion activities in such solutions (Whitfield, 1975) the calibration procedure for ionic activity measurements is restricted in a number of ways. It is rarely possible to prepare a calibration solution containing a known activity of the selected ion *and* a composition approaching that of the natural system. Single electrolyte solutions or dilute buffers are therefore used to calibrate the electrodes with the result that the accuracy of the measurements is affected by the introduction of an unknown and variable liquid junction potential (Section 20.2.1). The precision is also reduced because the electrode potential drifts for some time after the electrodes are immersed in the sample as the working electrode adjusts to its new environment. Other difficulties arise if the electrodes are not perfectly selective since the selectivity coefficients themselves depend on the ratio of the activities of the selected and the interfering ions.

The resolution of these difficulties depends on the adoption of some conventional procedure for assigning single-ion activity coefficients. Such a procedure has been developed for the calibration of hydrogen-ion selective electrodes in dilute solutions (Bates, 1973) and some consideration has been given to the extension of these ideas to solutions of high ionic strength (Bates, 1973; Baumann, 1973; Light and Fletcher, 1967). There are indications that a parallel procedure, based on the concept of ionic hydration, may be adopted for the calibration of electrodes selective to other ions (Bates *et al.* 1970; Robinson *et al.*, 1971). When these ideas have been extended in a rational manner to electrolyte mixtures (Robinson and Bates, 1973; Bagg and Rechnitz, 1973a) and to solutions of high ionic strength (Bagg and Rechnitz, 1973b) a simple and self-consistent approach to the use of ion selective electrodes for the measurement of activities in the marine environment will be possible. In the absence of such a coherent treatment each set of measurements must be treated independently and great care must be taken to ascertain the mode of electrode calibration before the data are used for detailed mechanistic calculations. Most applications discussed here will consequently be concerned with the use of ion-selective electrodes for concentration measurements.

Electrodes are available for the potentiometric determination of most of the inorganic components with concentrations greater than 10^{-6} M in normal sea water (Fig. 20.3). For many of these components the use of the potentiometric method is restricted by the selectivity characteristics of the available electrodes. The full equation describing the variation in cell potential in the presence of interfering ions may be written as (Moody and Thomas, 1971, 1972)

$$E = \text{const.} \pm kT \ln[a_i + \Sigma_j K_{ij}(a_j)^{n/y}] \tag{20.14}$$

where the subscript i refers to the selected ion and j to the interfering ion, y is the charge on the interfering ion and K_{ij} is the selectivity coefficient. Procedures for estimating K_{ij} have been considered in some detail (Moody and Thomas, 1971, 1972), but at present there is no consensus of opinion about the most satisfactory method. Selectivity studies involving the simultaneous presence of more than one interfering ion have so far only been empirical (see e.g. Whitfield and Leyendekkers, 1969b). For present purposes it is sufficient to consider K_{ij} as defining the point at which the selected and interfering ions make equal contributions to the cell potential in a mixed electrolyte solution. The selectivity coefficients for liquid ion-exchange electrodes vary with both the solution composition and the ionic strength (Leyendekkers and Whitfield, 1971; Srinivasan and Rechnitz, 1969; Whitfield and Leyendekkers, 1970) so that it is unrealistic to assess the detailed behaviour of these electrodes in a wide range of solutions using the single-valued selectivity "constants" often provided in the manufacturers' literature. However, a *rough* screening procedure can be devised for assessing the applicability of any electrode to a particular problem by defining a deviation function, pX, so that,

$$pX = \log a_i - [\log K_{ij} + (n/y) \log a_j] \tag{20.15}$$

If pX is greater than 2 then the error in a_i contributed by the interfering ion will be less than 1 %; if pX is between 1 and 2 the error will be less than 10 %, and if it is less than 1 the interference will be greater than 10 %.

Using tabulated selectivity data (Moody and Thomas. 1971; Whitfield, 1971a; Warner, 1972a) and calculated activities (Whitfield, 1975) it is possible to estimate pX values for a range of ion-selective electrodes when used in sea water (Tables 20.1 and 20.2, compare Warner, 1972a). Data are only given for those electrodes that show a good selectivity for the ion of interest. Electrodes with a suitable selectivity are available for sensing hydrogen, sodium, potassium, calcium, chloride and fluoride activities in natural sea water. Some of the commercial electrodes also show good selectivities towards other ions. For example, the Orion calcium electrode is three times more sensitive to zinc than it is to calcium, and the Corning potassium electrode is ten times more sensitive to rubidium than it is to potassium. These adverse selectivities are rarely important with natural waters.

Several electrodes have been developed that are not considered in Tables 20.1 and 20.2. The tetrafluoroborate electrode has been used for the determination of boron in soils (Carlson and Paul, 1968, 1969). The boron must first be extracted from the sample and then converted to BF_4^- by the action of hydrofluoric acid. The procedure is not as convenient as available photometric methods for the determination of the element in sea water (Dyrssen *et al.* 1972) and the potentiometric method is not sufficiently sensitive to follow the relatively small changes in the boron:chlorinity ratio that are normally observed in the sea (Riley and Chester, 1971). Potentiometric oxygen electrodes based on a semiconducting perovskite (strontium-doped $LaCoO_3$; Tseung and Bevan, 1973) and on a sodium tungsten bronze (Na_xWO_3:

TABLE 20.1

Selectivity characteristics of representative electrodes for the measurement of cationic components with concentration greater than 10^{-6} M in sea water.

Ion	Electrode		K_{ij}[a]				
	Identification	Type[b]	H^+	Na^+	K^+	Mg^{2+}	Ca^{2+}
Na^+	NAS 11–18[c]	G	—	—	3×10^{-3} (4.19)[a]	—	—
K^+[d]	Beckman 39622	L	2×10^{-4} (9·72)	5×10^{-5} (2·63)	—	2×10^{-5} (3·49)	2×10^{-5} (3.86)
	Corning 476132	L	—	0·012 (0·25)	—	3×10^{-3} (1·31)	5×10^{-3} (1·45)
	Philips 560-K	L	5×10^{-5} (10·32)	$2·5 \times 10^{-4}$ (1·93)	—	2×10^{-4} (2·49)	2×10^{-4} (2·86)
	Orion 92–19	L	0·01 (8·02)	7×10^{-4} (1·49)	—	2×10^{-4} (2·49)	2×10^{-4} (2·86)
Mg^{2+}[d]	Beckman 39614	L	—	0·013 (0·98)	0·13 (3·31)	—	1·05 (0·84)
	Orion 92–32	L	—	0·015 (0·91)	0·15 (3·25)	—	1·00 (0·86)
Ca^{2+}[d]	Orion 92–20	L	10^{-7} (12·52)	$1·6 \times 10^{-3}$ (1·14)	10^{-4} (5·69)	0·014 (1·11)	—
	Beckman 39608	S	$1·5 \times 10^{-4}$ (9·35)	0·029 (−0·12)	0·034 (3·15)	0·34 (−0·27)	—
	Corning 476041	L	—	0·029 (−0·12)	0·034 (3·15)	0·34 (−0·27)	—
	Ruzicka[e]	S	$1·6 \times 10^{-4}$ (9·32)	$6·31 \times 10^{-6}$ (3·55)	$1·99 \times 10^{-6}$ (7·38)	$2·51 \times 10^{-4}$ (2·85)	—
	Ammann[f]	S	$4·1 \times 10^{-2}$ (6·91)	$5·7 \times 10^{-3}$ (0·59)	$7·3 \times 10^{-2}$ (2·82)	3×10^{-5} (3·78)	—
	Griffiths[g]	S	—	$2·8 \times 10^{-3}$ (0·90)	$1·6 \times 10^{-4}$ (5·48)	$2·4 \times 10^{-2}$ (0·88)	—

TABLE 20.1—*continued*

[a] Selectivity coefficient quoted by manufacturer. The number in parentheses is the deviation function pX (equation (20.15)). pX was calculated using the following activities (molal) for sea water at 25°C, 1 atm. pressure and 35‰ salinity: $a_{Na^+} = 0{\cdot}306$, $a_{K^+} = 6{\cdot}57 \times 10^{-3}$; $a_{Mg^{2+}} = 0{\cdot}0115$, $a_{Ca^{2+}} = 2{\cdot}09 \times 10^{-3}$, $a_{H^+} = 6{\cdot}3 \times 10^{-9}$.

[b] G, glass electrode; L, liquid ion-exchange electrode; S, solid electrode with PVC or collodion matrix.

[c] Specification of glass formulation (Eisenman, 1967).

[d] Data from Moody and Thomas (1971), pp. 21, 123.

[e] Ruzicka *et al.* (1973). All solid-state electrode.

[f] Ammann *et al.* (1972). Selectivity coefficients in mixtures readily calculated from simple theory.

[g] Griffiths *et al.* (1972). Orion 92–20 exchanger set in PVC matrix.

TABLE 20.2

Selectivity characteristics of representative electrodes for the measurement of anionic components with concentrations greater than 10^{-6} M in sea water.

Ion	Electrode: Identification	Electrode: Type[b]	K_{ij}[a]: Cl^-	Br^-	I^-	SO_4^{2-}	NO_3^-	CO_3^{2-}	HCO_3^-
Cl^-	Corning 476131	L	—	2·5 (2·42)	15 (4·88)	—	2·5 (3·96)	—	—
	Orion[c] 92–17	L	—	3·73 (2·25)	26·7 (4·63)	0·14 (1·77)	5·89 (3·58)	—	0·19 (3.22)
	Orion 94–17	IS	—	333[d] (0·30)	2×10^6 (−0·25)	—	—	—	—
	Beckman 39604	IS	—	333 (0·30)	2×10^6 (−0·25)	—	—	—	—
	Coleman 3-802	IS	—	204 (0·51)	10^6 (0·06)	—	—	—	—
Br^-	Beckman[f] 39602	IS	$2·5 \times 10^{-3}$ [g] (−0·22)	—	6×10^3 [h] (−0·46)	—	—	—	—
	Philips IS 550-Br	IS	6×10^{-3} (−0·60)	—	20 (1·94)	—	—	—	—
	Pungor	IS	5×10^{-3} (−0·52)	—	130 (1·13)	—	—	—	—
I^-	Orion[i] 94-53	IS	10^{-6} [j] (−0·06)	2×10^{-4} [k] (0·46)	—	—	—	—	—
	Philips IS 550-I	IS	$6·6 \times 10^{-6}$ (−0·45)	$6·5 \times 10^{-5}$ (2·19)	—	—	—	—	—
	Pungor	IS	$5·9 \times 10^{-6}$ (−0·83)	$4·8 \times 10^{-3}$ (−0·92)	—	—	—	—	—
NO_3^-	Orion 92-07	L	6×10^{-3} (−2·13)	0·1 (−0·54)	20 (0·40)	6×10^{-4} (−0·22)	—	6×10^{-3} (0·08)	0·02 (−0·15)
	Corning 476134	L	4×10^{-3} (−1·96)	0·01 (0·46)	25 (0·30)	10^{-3} (−0·44)	—	—	10^{-3} (1·15)
	Beckman 39618	L	0·02 (−2·66)	0·28 (−0·99)	5·6 (0·95)	10^{-5} (1·56)	—	$1·9 \times 10^{-4}$ (1·58)	—

TABLE 20.2—*continued*

Ion	Electrode		K_{ij} [a]						
	Identification	Type[b]	Cl^-	Br^-	I^-	SO_4^{2-}	NO_3^-	CO_3^{2-}	HCO_3^-
	Davies (Orion)[l]	S	4×10^{-3} (−1·96)	—	16 (0·50)	3×10^{-4} (0·09)	—	—	—
	Davies (Corning)[m]	S	5×10^{-3} (−2·05)	—	17 (0·47)	10^{-5} (1·56)	—	—	—
CO_3^{2-}	Ag_2CO_3 based[n]	IS	$2·5 \times 10^{-8}$ (2·74)	3×10^{13} (−15·53)	$1·18 \times 10^{21}$ (−19·88)	—	—	—	—
	$PbCO_3$ based[o]	IS	$2·06 \times 10^{-9}$ (3·82)	$8·46 \times 10^{-10}$ (7·02)	$4·65 \times 10^{-6}$ (6·52)	—	—	—	—
	Herman and Rechnitz (1974)	L	$1·85 \times 10^{-4}$ (−0·68)	—	—	$1·49 \times 10^{-4}$ (1·25)	—	—	—
F^-	All manufacturers	IS			$K_{F^-, OH^-} = 0·1$				
S^{2-}	All manufacturers	IS			No recorded interferences in natural systems				

[a] Selectivity coefficient quoted by manufacturer (Moody and Thomas, 1971, pp. 23, 26–27). The number in parentheses is the deviation function pX (equation (20.15)). pX was calculated using the following activities (molal) for sea water at 25°C, 1 atm. pressure and 35‰ salinity: Cl^-, 0·355; Br^-, $5·41 \times 10^{-4}$; I^-, $3·12 \times 10^{-7}$; SO_4^{2-}, $1·84 \times 10^{-3}$; NO_3^-, $1·57 \times 10^{-5}$; HCO_3^-, $1·12 \times 10^{-3}$; CO_3^{2-}, $4·84 \times 10^{-6}$.

[b] L, liquid ion exchange; IS, insoluble salt membrane; S, PVC matrix.

[c] Moody and Thomas (1971), p. 23.

[d] The selectivity coefficient of an insoluble salt membrane is defined as $K_{x,y} = K^s_{mx}/K^s_{my}$ where K^s_{mx} is the solubility product of the salt used to prepare the X selective electrode and K^s_{my} is the solubility product of the corresponding salt formed by the interfering ion Y. From Meites (1963) $K_{Cl, Br} = 346$ for an AgCl membrane.

[e] See note d from Meites (1963) $K_{Cl, I} = 2·2 \times 10^6$ for an AgCl membrane.

[f] Also for Orion 94–35 and Coleman 3–801 electrodes.

[g] From Meites (1963) $K_{Br, Cl} = 2·89 \times 10^{-3}$, see note d.

[h] From Meites (1963) $K_{Br, I} = 6·27 \times 10^3$, see note d.

[i] Also for Coleman 3–802 electrodes.

[j] From Meites (1963) $K_{I, Cl} = 4·61 \times 10^{-7}$, see note d.

[k] From Meites (1963) $K_{I, Br} = 1·60 \times 10^{-4}$, see note d.

[l] Davies *et al.* (1972) using Orion liquid ion-exchanger.

[m] Davies *et al.* (1972) using Corning liquid ion-exchanger.

[n] Calculated selectivity coefficients (Note d) assuming $K_{CO_3^{2-}, Y} = K_{Ag_2CO_3}/K^2_{AgY}$ (Walker *et al.*, 1927).

[o] Calculations on a hypothetical lead-based electrode assuming $K_{CO_3^{2-}, Y} = K_{PbCO_3}/K_{PbY_2}$.

Hahn *et al.* 1973) have been reported. Their application is generally restricted to alkaline solutions (pH 11 to 12).

Sulphate sensitive electrodes have been prepared by incorporating barium sulphate into a silicone rubber matrix (Pungor and Havas, 1966). These electrodes have rather a poor selectivity towards sulphate ions (Rechnitz *et al.*, 1967) requiring a forty-fold excess of sulphate over chloride. Hot pressed membranes incorporating lead sulphate in a mixed sulphide matrix (Ag_2S 32 %, PbS 31 %, Cu_2S, 5 %: Rechnitz *et al.* 1972) appear to yield more stable and selective electrodes.

Phosphate selective electrodes can be prepared by incorporating insoluble manganese or iron phosphates into a silicone rubber matrix (Pungor *et al.* 1966; Pungor and Toth, 1964; Guilbault and Brignac, 1969). The performance of such electrodes is not very satisfactory, possibly because of the slow rates of dissociation of the metal phosphates (Rechnitz *et al.*, 1967). A more useful electrode has been prepared (Shu and Guilbault, 1972) by incorporating a coordination complex (the phosphate salt of a polymerized silver thiourea) into a silver sulphide membrane. At present this electrode shows equal selectivity for chloride ions and dibasic phosphate ions (HPO_4^{2-}). Considerable improvements in selectivity may be expected as the specificity of coordination complexes is more fully exploited. Fig. 20.4 summarizes the applicability of potentiometric techniques to the analysis of open ocean waters and provides a guide to the subsequent discussion.

20.3.1.1. *The glass electrode—determination of hydrogen ions*

The glass electrode has been used to measure the pH of sea water for over forty years. The procedure currently adopted (Strickland and Parsons, 1972) is to calibrate the electrode couple in dilute, NBS standard phosphate buffers ("equimolal", pH 6·865 or "blood buffer", pH 7·413 at 25°C: Bates, 1973). The electrodes are then immersed in a sea water sample at the same temperature and the pH is read off the calibrated scale after a suitable equilibration time (3–5 minutes) has elapsed. The slow equilibration can be attributed almost entirely to the gradual adjustment of the glass electrode surface to its new ionic environment (Pytkowicz *et al.*, 1966).

Typical of the more precise measurements are those made by Culberson for the 1969 Geosecs Intercalibration study (Takahashi *et al.*, 1970). He used a Beckman 46850 micro blood pH electrode to minimize errors due to the loss or absorption of carbon dioxide. The samples were all measured at 25°C and the cell was calibrated with 0·05 M potassium hydrogen phthalate (quoted as pH = 4·007 at 25°C) and a solution containing 0·008695 M potassium dihydrogen phosphate and 0·03043 M disodium hydrogen phosphate (quoted as pH = 7·411 at 25°C). The slope of the glass electrode

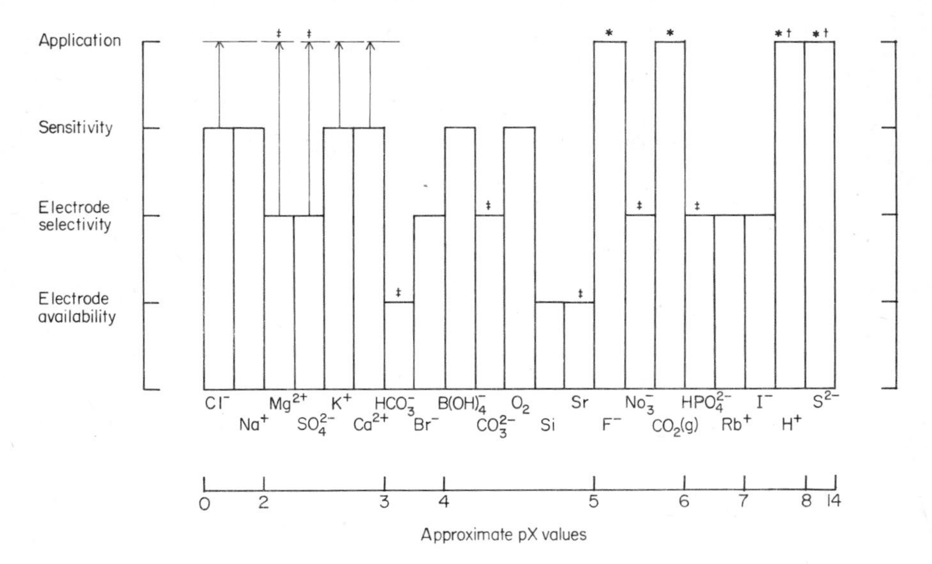

FIG. 20.4. The use of potentiometric techniques for the analysis of sea water constituents. The relevant components (Fig. 20.3) are arranged in order of decreasing concentration. Three horizontal barriers are identified on the left of the figure that may prevent the application of the potentiometric method. The column representing a particular element stops at the barrier that is currently preventing the potentiometric determination of that element in sea water. A vertical arrow indicates that barriers may be overcome by the use of titrimetry rather than direct potentiometry.

† These ions are effectively buffered in sea water and can therefore be estimated at very low free ion concentrations.

* Electrode used *in-situ.*

‡ Electrodes with adequate selectivity would find immediate application.

response was 58·46 mV/pH units (98·8% of theoretical) and the sea water pH values were calculated from the equation

$$pH_{sw} = 7{\cdot}411 + (E_{sw} - 65{\cdot}5)/58{\cdot}46 \qquad (20.16)$$

The cell gave a potential of 65·5 mV ($\pm$0·24 mV) in the pH 7·411 buffer. The precision of the sea water measurements calculated by combining the standard deviations for duplicate measurements on 29 samples and for one triplicate set was $\pm$0·003 pH units. For comparison of pH values measured between different casts the uncertainty was $\pm$0·005 pH units (cf. Pytkowicz *et al.*, 1966).

Hansson (1973a) has suggested that "pH-values determined by this procedure are no measure of the activity of H^+ ions on the activity scale based on pure water or any other solvent but are just numbers determined by a specific procedure". This may seem rather a harsh judgement, but it has ample justification. The difficulties centre around the liquid junction that is used to make contact between the sample solution and the standard chloride solutions surrounding the reference electrode. The potential generated here depends on the composition of the reference and sample solutions, and on the way in which the liquid junction is constructed (Covington, 1966, 1969). The standard buffer solutions normally used are sufficiently dilute ($<0{\cdot}05$ M) for theoretical procedures to be used to calculate conventional single-ion activity coefficients (Bates, 1973). Since the ionic strength of sea water is usually nearer 0·7 M, the cell potential includes a contribution of several millivolts from the liquid junction potential. This contribution will manifest itself as a shift in the hydrogen ion activity from the true value. Hawley and Pytkowicz (1973) have estimated that a liquid junction error of 3 mV (0·05 pH units) is involved when a saturated calomel–glass electrode pair is transferred from a dilute buffer into a NaCl–$NaHCO_3$ solution with an ionic strength similar to that of sea water.

Although Ben-Yaakov (1972, 1973) has made ingenious use of the liquid junction potential to measure the mobilities of ions in sea water, most efforts have been directed towards minimizing its effect on potentiometric measurements. Smith and Hood (1964) prepared secondary standard buffers with a background electrolyte corresponding to that of sea water. This device increases the precision of pH measurements by removing the major source of electrode drift, but it does not enhance the accuracy of the measurements since the liquid junction contribution is still present in the standardization of the secondary buffer against the NBS reference.

The liquid junction can be removed from the solution altogether if a second membrane electrode, specific to a component with a constant activity is used instead of the conventional reference electrode (Merril, 1961; Garrels, 1967; Wilde and Rogers, 1970b). Wilde and Rogers (1970a) have designed a useful *in situ* amplifier for measuring the potential difference betwen two such high resistance electrodes. Precise pH measurements can be made in sea water using a sodium selective glass counter electrode (Garrels. 1967, Wilde and Rogers, 1970b). pH *changes* will be accurately recorded provided that the accompanying variations in solution pH are not sufficient to alter the sodium ion activity. However, the procedure does not overcome the basic calibration problem since cells with a liquid junction must be used initially to calibrate the counter electrode itself (Wilde and Rogers, 1970b; Merril, 1961). The technique should prove useful under field conditions where large and erratic liquid junction potentials may be expected (e.g. in the presence of suspended sediment).

More general procedures for accommodating the liquid junction potential have been considered in some detail (see for example Light and Fletcher,

1967; Baumann, 1973; Bates, 1973) and two distinct approaches may be traced in marine chemistry.

The first approach, which has a long and somewhat confused history (Lyman, 1972; Edmond, 1972), is to accept that the pH values obtained by the conventional method are just empirical numbers. In order to accommodate these numbers into useful oceanographic parameters it is necessary to define "apparent" stability constants to describe the distribution of protolytic species as a function of conventional pH values. For example, for the second ionization constant of carbonic acid,

$$HCO_3^- \rightleftharpoons CO_3^{2-} + H^+$$

the "apparent" constant K'_{2C} is given by

$$\log K'_{2C} = \log [CO_3^{2-}] - \log [HCO_3^-] - pH_{conv}.$$

So long as exactly the same procedures have been used to measure the conventional pH and to determined the apparent constant (Pytkowicz, 1969) this equation can be accurately used to determined the ratio of carbonate and bicarbonate ion concentrations. A different value of K'_{2C} is required for each solution composition, temperature and pressure. This empirical approach has proved useful for measuring the changes in chemical conditions of particular water masses with time, and it has been of considerable help in unravelling the rather complex chemistry of the carbon dioxide system in sea water. However if we are able to explain, rather than simply to describe, the chemistry of protolytic species in sea water it is important that pH values should be available that are defined on some thermodynamically acceptable scale rather than specified in terms of particular experimental procedures.

In an attempt to give oceanographic pH measurements a less empirical basis, Hansson (1973a) has developed a series of pH scales based on synthetic sea water as the ionic medium (Sillén, 1967b). Using this activity scale the activity coefficient of the hydrogen ion approaches unity as the hydrogen ion concentration approaches zero in sea water of a specified composition. Sillén (1967b) has shown that, for any species with a total concentration of less than 10% of the concentration of the medium electrolyte, the activity and the concentration are the same (i.e. the activity coefficient is unity). Consequently, buffers used to define pH scales on this convention are readily prepared by using solutions with a known hydrogen ion *concentration*.

Hansson (1973a) prepared buffers containing known molalities of "tris" (B = 2-amino-2-hydroxymethylpropane-1,3-diol) and "tris" hydrochloride (B.HCl) in synthetic sea water solutions. The buffers were standardized over a range of salinities and temperatures by means of a precise potentiometric titration, using an Ag–AgCl reference electrode in sea water saturated with silver chloride. Each salinity corresponds to a different ionic medium, and hence each salinity has its own pH scale that is defined over an appropriate temperature range. For work in the normal salinity range (30–35‰) the 35‰ buffer can be used for field measurements. Suitable corrections can be made to pH values measured in the lower half of the salinity range by calibrating the electrodes in the 35‰ buffer and then taking a reading in the 30‰ buffer at the same temperature. The difference between the observed and tabulated values (Hansson, 1973a) provides the necessary correction factor. Detailed instructions for the preparation and calibration of the buffers are given by Hansson (1973a).

These buffer scales are thermodynamically well-defined so that the pH values obtained have a clear and exact meaning. As our understanding of interactions in ionic solutions develops it will be possible to calculate values for ionic activity coefficients in sea water relative to the widely used physico-chemical standard at infinite dilution. Theoretical models based on this approach will then be able to incorporate pH data based on Hansson's pH scales directly, whereas the precise meaning of pH values based on the NBS scale will remain obscure. A number of experimental advantages also follow from the adoption of the ionic medium scale. The precision of measurement is enhanced, and the time required for the electrodes to equilibrate with the sample is greatly reduced since the calibrating and sample solutions have closely similar compositions. In addition, pH measurements in the sea water buffers are not as sensitive as measurements in dilute buffers to changes in the geometry of the liquid junction (Hansson, 1973a). Consequently, measurements from a wide range of cell configurations can be compared directly.

A few cautionary words are also appropriate about the limitations of the sea water pH scales. The buffer scales as defined are valid only for the ionic media specified. Each new salinity represents a new ionic medium with its own pH scale, and in practice the buffer used must have a total concentration of medium salts within ten percent of that present in the sample (Almgren *et al.*, 1974). The utility of these scales might therefore be restricted in estuarine situations where the salinity is very variable. The pH values ascribed are also only valid for the particular sea water recipe used by Hansson (1973a). Attempts to prepare buffers in natural sea water media would introduce deviations from the tabulated pH values since protolytic species (e.g. carbonate, borate, fluoride) are present in the natural sample but not in the recipe used in the definition of the pH scales. However, it is not a difficult matter to calibrate buffer solutions in other media using the titration procedure described in detail by Hansson (1973a).

Almgren *et al.* (1974) have used the new pH buffer scales at sea and give some useful practical advice. They prepared their standard buffer by dissolving 10 mmol of "tris" in 50·00 ml of solution A (Table 20·3) and adding solution B (Table 20.3) to make the volume up accurately to 1000 ml. Since the "tris"

TABLE 20.3

Solutions used to prepare[a] standard sea water buffers on the sea water ionic medium scale for 35‰ salinity[b] (Hansson, 1973; Almgren et al., 1974).

Component	Solution A ($mmol\,l^{-1}$)	Solution B ($mmol\,kg^{-1}$)
NaCl	319·20	412
KCl	10·23	10
Na_2SO_4	28·65	28
$MgCl_2.6H_2O$	55·26	54
$CaCl_2.2H_2O$	10·23	10
HCl	100·00	—

[a] The preparative technique is described in the text.
[b] Buffers for other salinities are prepared by altering the concentrations of the first five ingredients in proportion. The hydrochloric acid concentration is unaltered. The pH of these buffer solutions is given by the equation (Almgren *et al.*, 1974).

$$pH_s = (4{\cdot}5\,S‰ + 2559{\cdot}7)/T°K - 0{\cdot}0139\,S‰ - 0{\cdot}5523.$$

buffer has rather a large temperature coefficient ($dpH_s/dT = 0{\cdot}031$ units per degree at 25°C and 35‰ salinity) the solutions must be thermostatted to within ±0·1°C. Samples are transferred to glass or polythene bottles as rapidly as possible with the minimum of aeration. A 20 ml sample is transferred to a 25 ml beaker which is placed in the thermostat alongside a second 25 ml beaker containing 20 ml of the appropriate buffer. The glass and reference electrodes may be mounted in a rubber stopper that neatly fits the beakers. The electrodes are carefully wiped with soft tissue when they are transferred from one solution to another. They should on no account be rinsed in distilled water but should be stored in fresh sea water between readings. When not in use the glass electrode may be stored dry after rinsing with distilled water. After being stored in this way the glass electrode should be conditioned by soaking in fresh sea water for 24 hrs before it is used. The buffer solutions should be renewed after three to five transfers from the sea water samples. Almgren *et al.* (1974) have suggested that an accuracy of ±0·003 units may be attained if care is taken to prevent nucleation of CO_2 bubbles on the sides of the sample bottles and on the walls of the measuring vessel.

The results of an intercalibration experiment by Almgren *et al.* (1974a) on mid-Pacific waters (Table 20.4) indicate the extent of disagreement between measurements made on the NBS scale and on the sea water scale. The *in situ* pH values calculated from the two conventions differ by as much as 0·1 pH units. The procedure used for making pH measurements on the dilute buffer scale was closely similar to that described by Strickland and Parson (1972) and did not incorporate the refinements contributed by Culberson (see Takahashi *et al.*, 1970).

TABLE 20.4

Results of an intercomparison[a] between pH measurements made on the NBS scale and on the mol kg^{-1} sea water (M_w) scale.

Depth (m)	Temperature °C		Salinity (‰)	pH (NBS scale)		pH (M_w scale)	
	in-situ	labora-tory		observed	calculated *in situ*[b]	observed	calculated in situ[b]
0·4	23·10	22·0	33·980	8·27	8·28	8·179	8·180
50	21·02	21·5	34·357	8·29	8·30	8·206	8·238
148	14·36	21·0	33·892	8·13	8·20	8·029	8·160
246	11·54	20·5	34·587	7·71	7·80	7·625	7·787
394	8·96	20·0	34·588	7·65	7·76	7·551	7·747
590	6·78	20·0	34·522	7·62	7·75	7·548	7·775
790	5·36	19·7	34·506	7·62	7·76	7·531	7·777
990	4·52	20·0	34·601	7·61	7·76	7·545	7·805
1195	3·80	19·6	34·583	7·68	7·84	7·614	7·892
1500	2·92	19·6	34·618	7·67	7·83	7·608	7·898
2035	2·16	19·3	34·682	7·75	7·92	7·693	8·002

[a] Station 652, East of the Hawaiian Islands 18° 09 N, 121° 54′ W, March 1st 1973. Reproduced with permission from Almgren *et al.* (1974a).
[b] Calculated for temperature variations only.

The procedure for calculating the *in situ* pH is quite complex (Ben-Yaakov, 1970; Gieskes, 1970) since it requires a knowledge of the dissociation constants and the total concentrations of all the contributing protolytic species over the appropriate temperature, pressure and salinity ranges. The constants must be appropriate to the pH scale used in the surface measurement. Great care must be taken in the manipulation of the sample to ensure that there is no degassing, and no precipitation, nor dissolution of suspended carbonates.

Because of these problems considerable effort has been put into the design of probes for measuring the actual pH at depth. Since these probes clearly illustrate the problems common to all *in situ* electrochemical devices they will be considered in some detail. The most obvious problems are the mechanical difficulties involved in encapsulating the electrodes and their leads. In relatively shallow in-shore waters (down to a hundred meters or so) most glass electrodes are sufficiently robust to be used without pressure compensation, and *in situ* assemblies are readily prepared by locating conventional electrodes in suitable water-tight housings (Allegier *et al.*, 1941; Kramer, 1961; Manheim, 1961; Whitfield, 1971a). However, some care is required in the design of the reference electrode since it must respond readily to changes in ambient temperature, and continuous *outflow* of reference solution is required through the liquid junction to maintain a steady potential. Silver/silver chloride or thallium amalgam/thallium chloride reference electrodes are recommended since they have better temperature characteristics than does the more commonly used calomel electrode (Whitfield, 1971a; Warner, 1972a). A 3·5 M potassium chloride filling solution is to be preferred since the use of saturated potassium chloride solutions results in sluggish and erratic behaviour in the field, particularly at low temperatures (Kramer, 1961). In shallow waters there are distinct advantages in separating the working and reference electrodes and allowing the latter to float at the surface (Manheim, 1961). This greatly simplifies the calibration of the cell over a range of temperatures and removes the problems associated with pressure equilibration across the liquid junction. A compact probe based on a small cartridge-type glass electrode has been developed to monitor pH, Eh and sulphide activity in in-shore waters (Whitfield, 1971b). The Eh and sulphide electrodes are based on platinum and silver foils respectively which are cemented to the conical face of an araldite casting. The miniature glass electrode is mounted axially in the casting and protrudes from the apex of the cone. The glass electrode is mercury filled and has a cylindrical pH sensing surface. The system gives readings reproducible to within $\pm 0{\cdot}02$ pH units and is currently being modified for work in deep well waters for which its robustness and small size (2 cm by 9 cm overall) are particularly useful.

Conti and his co-workers (Conti *et al.*, 1971; Conti, 1972) have developed an *in situ* system that is particularly suited for detailed surveys in relatively shallow in-shore waters (down to 30 m). The electrodes and the recording system are incorporated in a towed vehicle which is capable of working either at a fixed depth, or at a fixed distance, from the bottom (bottom contouring). The vehicle, which is made of acrylic plastic, is shaped rather like an aeroplane with fixed wings on the fuselage and controllable elevators on the tail section. These, together with a pressure transducer and a servomechanism, provide the necessary depth control (Conti *et al.*, 1971). The fuselage contains the control electronics, the probe scanning and signal conditioning equipment for the electrochemical sensors, and the recording unit. The recording unit is a cheap and ingenious assembly that consists basically of a digital voltmeter, an event counter and an electronic watch mounted together and photographed at predetermined intervals by a modified cine-camera, using incident light coming through the transparent walls of the fuselage (Conti, 1972). The electrode signals are switched in sequence through a suitable amplifier to the digital voltmeter. The readout system is somewhat cumbersome because of the delays involved in developing the film and manually transferring the data to punched tape or cards. In addition, the storage capability is restricted to 3200 data points (cf. the 11 000 data points in the system of Ben-Yaakov and Kaplan (1971a)). However, the system is cheap and effective. The sensors themselves are mounted in pods on the body of the fuselage just aft of the wings. The electrodes are sealed into an acrylic plastic housing by O-ring compression seals (Fig. 20.5) and pressure equilibration

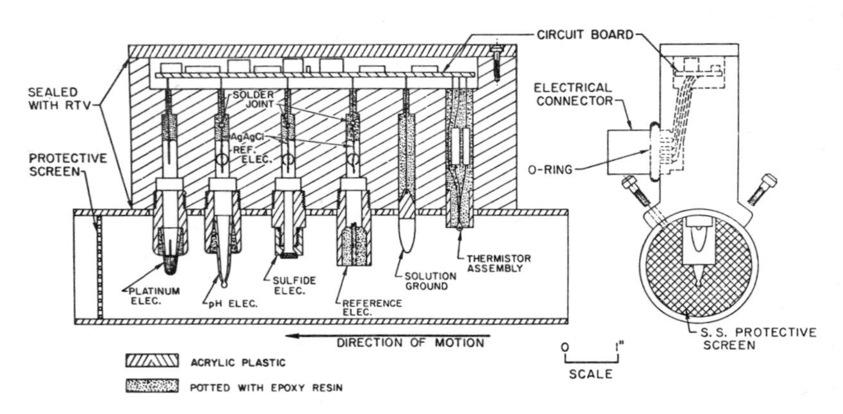

FIG. 20.5. Assembly of potentiometric sensors for *in-situ* measurements in shallow waters. The instrument pod is mounted on the towed under-water vehicle described in the text. Reproduced with permission from Conti (1972).

is achieved *via* a small neoprene disc that is sealed into the wall of the electrode chamber. The vehicle is light and easily handled and gives stable and reliable performance in a variety of conditions (Conti, 1972).

A diver operated electrode system (Conti and Wilde, 1972) offers further flexibility in sampling shallow waters. A commercial glass electrode is cemented into a plexiglass "wand" with silicone rubber cement. Pressure equilibrium is unnecessary at the shallow depths encountered (30 m maximum). Other ion-selective electrodes can be constructed by drilling holes perpendicularly into the face of a flat plexiglass wedge (Fig. 20.6). Up to eight electrodes can be accommodated in a stout wedge 25 cm × 10 cm × 2·5 cm. The individual electrodes are connected by 50 cm of cable to a plexiglass box which houses the electronics. The electrode signals are fed via a selector

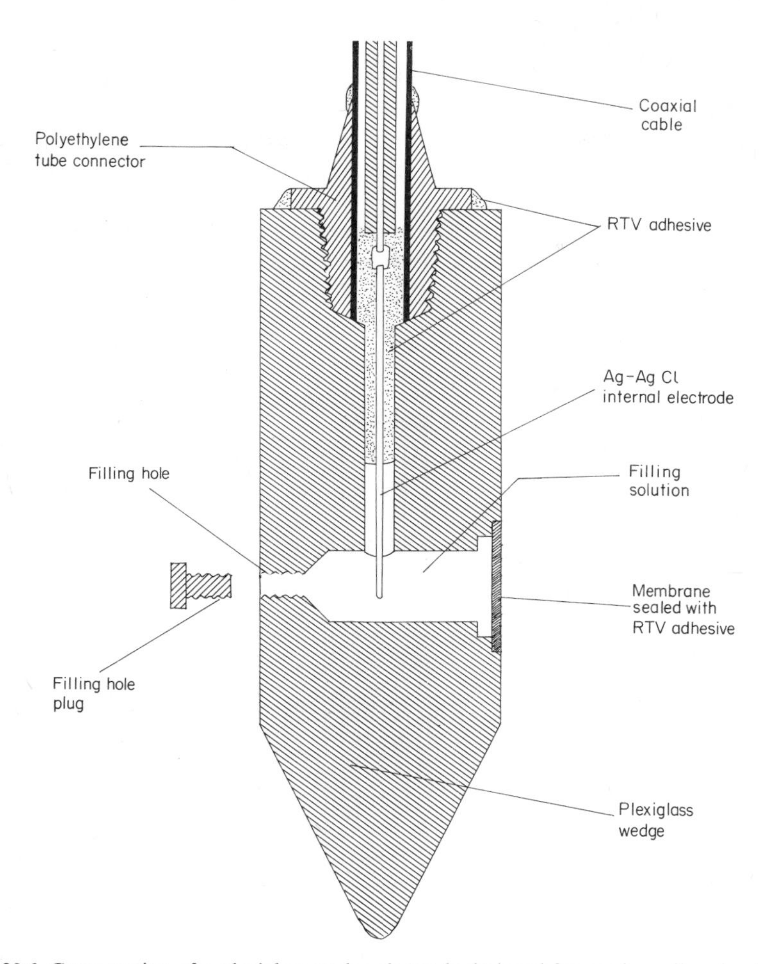

FIG. 20.6. Cross-section of a plexiglass wedge electrode designed for use by a diver in shallow water. Several electrodes can be mounted side-by-side in this way in a single wedge. Reproduced with permission from Conti and Wilde (1972).

switch to a truly differential amplifier that is designed to accept two high impedance electrodes (Wilde and Rogers, 1970a). This enables a high impedance electrode (e.g. a sodium selective glass electrode) to be used as a reference thereby eliminating the problems associated with sulphide precipitation, junction blockage and electrostatic interference often experienced when conventional liquid junction electrodes are used in sediments or resuspended muds (Pommer, 1967; Whitfield, 1969, 1971a). To take a reading the diver inserts the probe into the sediment, switches to the appropriate electrode and manipulates a potentiometer to give a null reading on a meter visible through the walls of the box. The reading on the turns–counting dial is then recorded on a slate. The system has a low power consumption and can run on two Mallory cells for several weeks. Again. the procedure is cheap, simple and effective.

If pH measurements are to be made in deep oceanic waters then the problems of pressure equilibration must be overcome. In an elegant design (Fig. 20.7) Distèche applied the principle of pressure compensation to remove

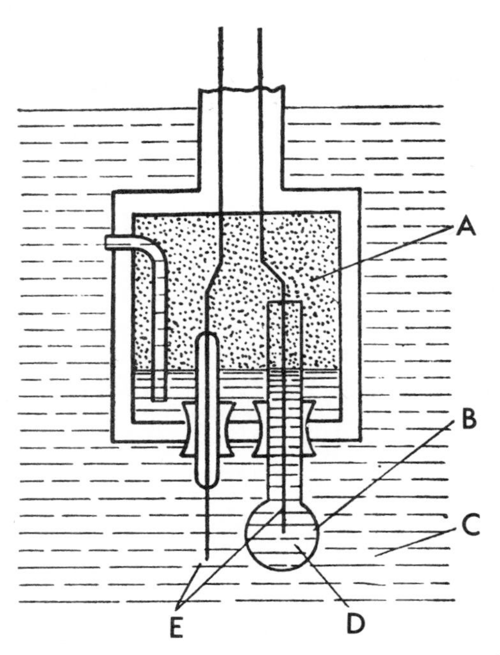

FIG. 20.7. Principle of pressure compensation used by Distèche (1959, 1964) for the development of a deep-sea pH probe.

A. Silicone oil B. Glass electrode
C. Sea water D. 0·1 M HCl
E. Silver–silver chloride electrodes.
Reproduced with permission from Distèche (1959).

the stress from the glass electrode at high ambient pressures (Distèche, 1959; 1964; Distèche and Dubuisson, 1960). A layer of silicone oil was used both for pressure transmission and electrical insulation, and problems of pressure equilibration at the reference electrode were obviated by using a silver–silver chloride reference electrode immersed directly in the sea water. Although bromide and iodide ions will both interfere with the performance of this electrode (Table 20.2) they do not appear to cause any *erratic* changes in potential so that its performance as a reference electrode is not noticeably impaired (Peterson and Groover, 1972). This equipment was primarily designed for operation from a deep-diving bathyscaphe so that the necessary recording equipment could be stored on board in a relatively "normal" ambient environment. This cell has been used down to 2500 m (approximately 250 atmospheres pressure). Cells of a closely similar design have been used in the laboratory to measure a number of parameters of oceanographic interest at much higher pressures.

Ben-Yaakov and Kaplan (1968a, b, 1971a) have developed a self-contained instrument package that can be used from a surface ship. Their glass electrode design is similar to that of Kunkler *et al.* (1967) in that pressure equilibration is achieved by replacing a section of the electrode stem by a length of silicone rubber or Tygon tubing (Fig. 20.8). This flexible wall equilibrates the internal solution to changes in the ambient pressure. A specially formulated pH glass is used that gives a resistance as low as 10 MΩ at 0°C. The electrode design is rather complex, but the reproducibility of the data obtained ($\pm 0{\cdot}02$ pH units: Ben-Yaakov and Kaplan, 1968c, 1969, 1971b) and the ability of the cell to distinguish fine structure in pH profiles (Ben-Yaakov and Kaplan, 1973b) indicate that it is very effective. The reference electrode is mounted *in situ* and consists of a silver–silver chloride electrode immersed in 2·7 M potassium chloride solution. The reference solution is pressurized by a spring-loaded piston (Ben-Yaakov and Kaplan, 1968a) or by a distended rubber bulb that is filled via a spring loaded clamp which also holds the liquid junction in place (Ben-Yaakov and Ruth, 1974; see also Kunkler *et al.*, 1967). A nylon wick (Ben-Yaakov and Ruth, 1974) or a plug of porous glass rod (Kunkler *et al.*, 1967) can be used for the liquid junction. The glass and reference electrodes are secured to the casing of the instrument package via Swagelok fillings (Crawford Fitting Co., Solon, Ohio) which are also used to provide pressure tight electrical leads.

The instrument package itself is mounted in a stainless steel pressure vessel and contains amplification and recording systems. In the most recent version (Ben-Yaakov and Ruth, 1974), a high input impedance, low input bias amplifier has been constructed to give a truly differential input. Since the reference electrode need not be connected to ground with this configuration, a galvanic connection can be made between the electronics ground and the instrument case giving improved noise rejection. The cell emf together with pressure and temperature data are recorded *in situ* on a miniature entertainment type magnetic tape recorded (Ben-Yaakov and Kaplan, 1971a).

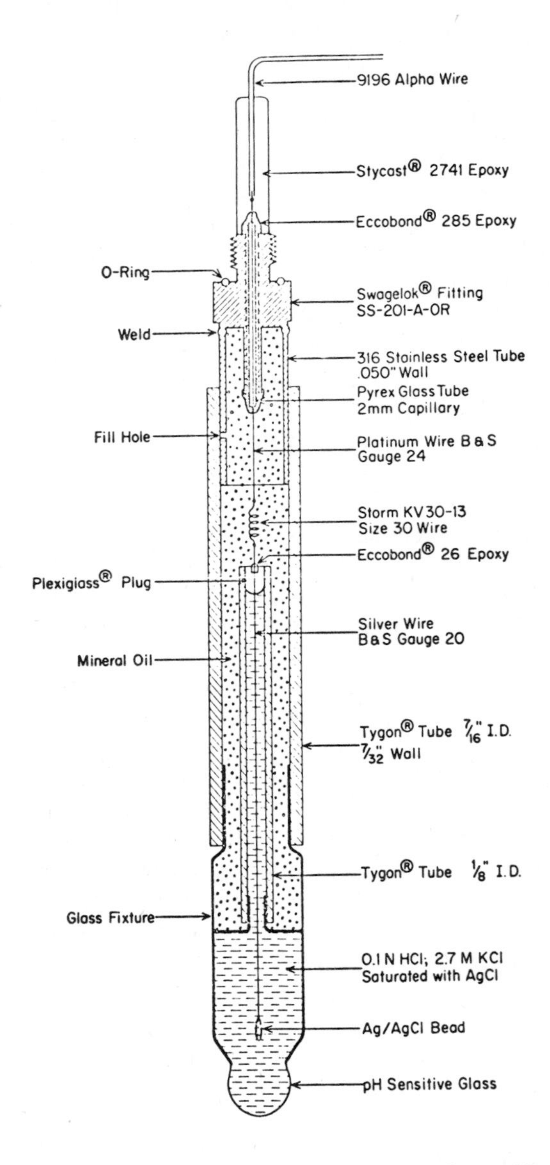

FIG. 20.8. Pressure compensated glass electrode used by Ben-Yaakov and Ruth (1974) for *in situ* pH measurements. The later versions use a combination of silicone rubber tubing and castor oil to increase the life of the probe. Reproduced with permission from Ben-Yaakov and Ruth (1974).

The relatively poor performance of this type of recorder requires careful design of the tape read-out system (Ben-Yaakov and Kaplan, 1971a). Once the tape has been transcribed the data are available for immediate computation. The main advantage of such a system is that the small power demands of the tape recorder and scanning circuitry enable the probe to run for eight hours on power from a small battery pack. The low cost of the recording unit is also important in oceanographic work in which risks of loss or breakage are high.

Invariably *in situ* pH assemblies have been calibrated on the NBS dilute buffer scale so that the comments made about surface pH measurements also apply here. The variations in ambient temperature and pressure experienced by the *in situ* electrodes must also be taken into account in the calibration procedure. Temperature effects have been considered in some detail (Whitfield, 1971a; Negus and Light, 1972; Covington, 1974). Two cell configurations, the thermal and the isothermal, are in common use (Fig. 20.9). They correspond respectively to the use of reference electrodes mounted at the surface (Manheim, 1961; Whitfield, 1971b) and with the glass electrode

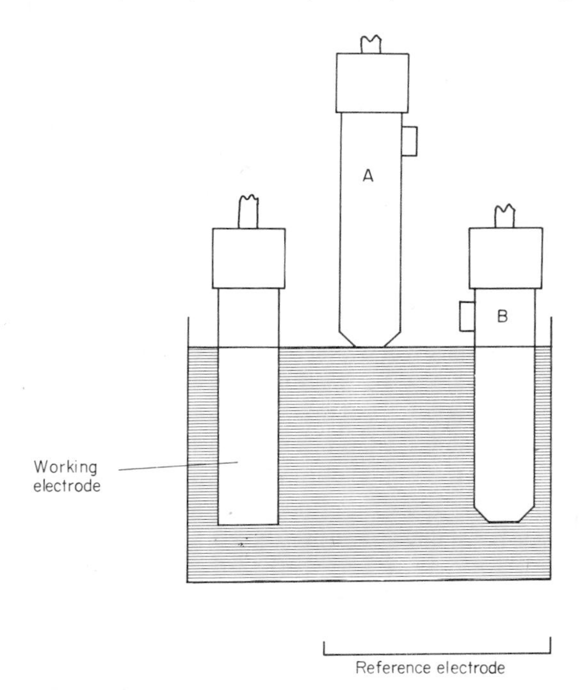

FIG. 20.9. Cell configurations used in the calibration of ion-selective electrodes over a range of temperature. A. Thermal cell. B. Isothermal cell.

at depth (Distèche, 1959; Ben-Yaakov and Kaplan, 1968a, b). The equation relating the cell potential to the pH may be simplified to

$$E = E' - kT\,pH, \tag{20.17}$$

when the cell is being used to measure the pH at a particular temperature. Values of kT as a function of temperature are readily calculated (see Bates, 1973 for tabulated values), and the major problem in electrode calibration is to estimate the temperature coefficient of E'. This temperature coefficient may be estimated using the isopotential principle (Jackson, 1948). pH measurements in a series of buffers over a suitable temperature range are used to locate the pH at which the emf output of the cell is independent of temperature (isopotential pH, pH_i). If the cell is calibrated in the field by immersion in a standard buffer solution at a temperature T_s and is then used to measure the pH *in situ* at a temperature T_t, then the error caused by variation in E' is given by

$$\Delta pH = (T_t - T_s).pH_i/T_t, \tag{20.18}$$

(Whitfield, 1971a; Covington, 1974). If the thermal cell configuration (Fig. 20.9A) is used to determine pH_i then the isopotential point is unaffected by changes in the construction of the reference electrode. With the isothermal cell (Fig. 20.9B), a new isopotential point must be determined for each electrode configuration. These corrections can be made automatically if a fixed cell configuration is used (Taylor, 1961). Negus and Light (1972) have given isopotential data for a wide range of electrodes. This procedure has rarely been used for the calibration of *in situ* probes (see however, Whitfield, 1971b). Ben-Yaakov and Ruth (1974) prefer to use a single point calibration based on the conventional pH values of a dilute phosphate buffer over a suitable temperature range.

The calibration of glass electrodes over a range of pressures has received little attention. It is generally assumed that pressure only affects E' by altering the asymmetry potential (E_{AS}). This potential arises from residual strains and asymmetries in the glass membrane (Whitfield, 1970) and it is measured by immersing the electrode in a solution identical to its filling solution with matched reference electrodes on either side of the membrane. Corrections for this effect may be made by testing the complete cell in a pressure vessel (Distèche, 1959; Ben-Yaakov and Ruth, 1974). The *in situ* pH (pH_x) may be related to the potential measured *in situ* (E_x) by the equation,

$$pH_x = pH_b - kT[E_x - E_{AS}(P) - E_s] \tag{20.19}$$

where pH_b is the pH of the standard buffer at the *in situ* temperature, E_s is the potential produced by the cell in that buffer and $E_{AS}(P)$ is the asymmetry potential at the appropriate pressure. This latter parameter drifts with time.

By suitably modifying the electrode environment, the pH glass electrode can be used to give information about the variation of other pH sensitive parameters in sea water. Weyl (1961) suggested that the glass electrode could be used to measure the degree of saturation of a sea water sample with respect to calcium carbonate by measuring the pH of the sea water before and after the addition of finely powdered (150–300 μm) carbonate material. If the sea water is saturated with calcium carbonate no reaction will occur and the pH will remain constant. If it is *undersaturated* then carbonate will dissolve and the pH will rise. If it is *supersaturated* then precipitation will occur accompanied by a fall in the pH. The degree of saturation can be assessed qualitatively from the direction of the pH shift and quantitatively from the initial and final pH values (Weyl, 1961; Ben-Yaakov and Kaplan, 1969). At normal sea water pH values (8·0 to 8·3) the sensitivity of the saturometer is about 1 mV for each ppm of calcite away from the saturation point. The sensitivity decreases markedly at higher pH values (Weyl, 1961). The response of the saturometer for a particular water sample depends both on the nature of the calcium carbonate material used (e.g. mineral calcite or oolites) and on the pretreatment that the powder is given before it is added to the cell. The technique has been used by Siever (1965) in studies of interstitial water composition and by Ingle *et al.* (1973) to measure the solubility of calcite in sea water at atmospheric pressure. A modified version has been used by Pytkowicz *et al.* (1967) to measure the solubility of calcium carbonate at *in situ* pressures. Ben-Yaakov and Kaplan (1971b) have modified their deep-sea pH assembly to give an *in situ* carbonate saturometer (Fig. 20.10). The glass electrode is permanently surrounded by crushed calcite held in a plexiglass cup. The calcite is thoroughly washed and dried in an oven at 105°C before use. A fairly coarse grain size (approx. 40 mesh) is used to improve the flushing characteristics of the cell. When the assembly has been lowered to the required depth the solenoid valve (Fig. 20.10) is opened and the pump switched on. Sea water then flushes through the cell for five minutes (corresponding to a flow of 2·5 to 3 litres) to remove water previously equilibrated with the calcite. Towards the end of this interval the glass electrode registers the pH of the surrounding sea water to within 0·05 pH units. The pump is then switched off and the solenoid valve is closed for ten minutes while the fresh sea water equilibrates with the calcite. pH readings are taken every minute during this period. The data are recorded on magnetic tape using the set-up described earlier (Ben-Yaakov and Kaplan, 1968b, 1971a). The cells are loath to come to equilibrium (see Weyl, 1961) and most of the data presented so far are qualitative rather than quantitative in character (see discussion by Ben-Yaakov *et al.*, 1974).

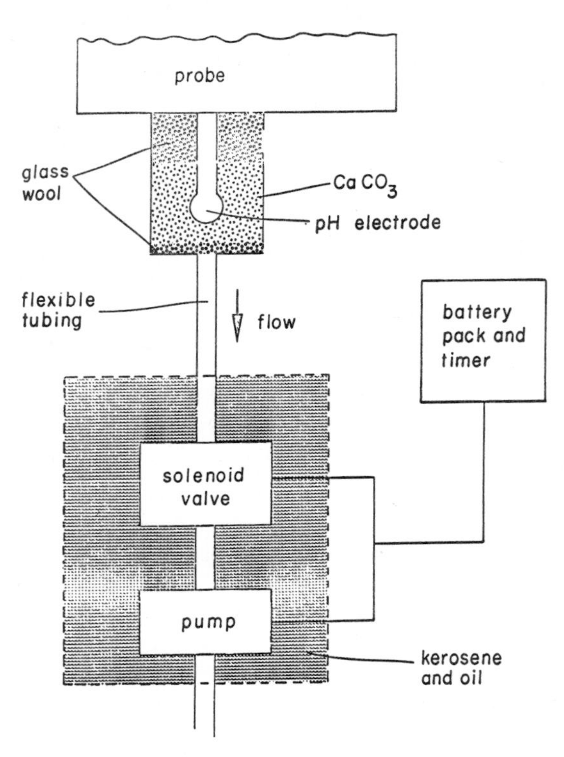

FIG. 20.10. Schematic design of an *in-situ* carbonate saturometer (see text for description). The reference electrode is mounted separately, together with a second pH electrode. Reproduced with permission from Ben-Yaakov and Kaplan (1971b) and the American Geophysical Union.

The pH glass electrode can also be used to sense the partial pressures of gases which hydrolyse in water to give acidic or basic reactions (e.g. carbon dioxide, sulphur dioxide and ammonia). The glass electrode and reference electrode are combined into a single unit which is mounted in an outer body (Fig. 20.11) that contains a buffer solution. A thin layer of the buffer solution is trapped between the glass electrode and a gas permeable membrane which seals off the lower half of the outer body. When the probe is dipped into a solution containing a volatile protolyte, such as carbon dioxide, the gas will pass through the membrane and equilibrate with the thin layer of buffer solution. The resulting pH change registered by the glass electrode is directly related to the partial pressure of carbon dioxide (P_{CO_2}). By suitable manipulation of the buffer solution and the gas permeable membrane a whole range of electrodes can be constructed on this principle (Ross *et al.*. 1973). Only two,

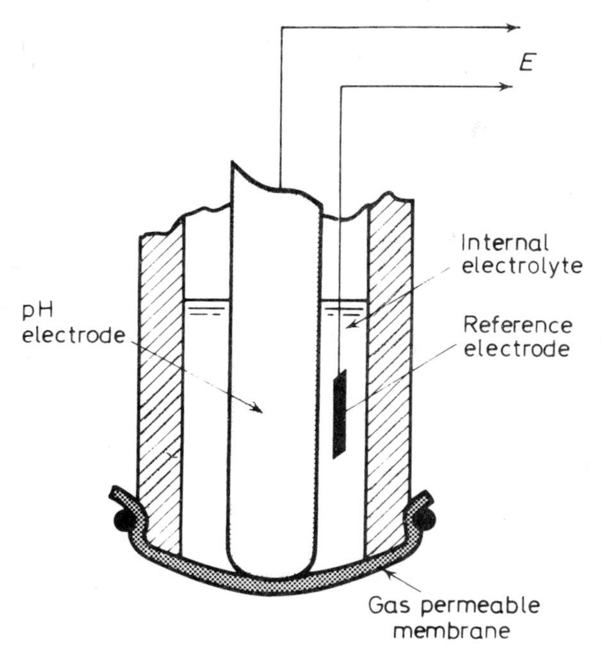

FIG. 20.11. General design for a gas sensing electrode. The pH sensor may be replaced by other ion-selective electrodes for some determinations. Reproduced with permission from Ross *et al.* (1973).

the carbon dioxide probe and the ammonia probe have so far found application in marine chemistry.

The carbon dioxide probe uses a dilute sodium bicarbonate buffer solution and a Teflon or silicone rubber gas permeable membrane (see for example Gertz and Loeschke, 1958; Severinghaus and Bradley, 1958). These membranes are also permeable to other protolytic gases (e.g. SO_2, H_2S). When carbon dioxide is controlling the pH of the internal buffer solution the carbon dioxide partial pressure in the external solution is given by

$$\alpha P_{CO_2} = a_{H^+} \, . \, a_{Na^+}/K_1 \tag{20.20}$$

(Severinghaus and Bradley, 1958). α is the Bunsen coefficient of carbon dioxide in water and K_1 is the first dissociation constant for carbonic acid. For a given buffer solution at a constant temperature and pressure the pH response of the electrode will be directly proportional to log P_{CO_2}. Consequently each ten-fold change in P_{CO_2} will result in a 59 mV signal from the electrode at 25°C. The best *in situ* pH measurements so far (Ben-Yaakov and

Kaplan, 1968c) give reproducibilities of $\pm 0{\cdot}02$ pH units so that the best precision that can be expected for direct P_{CO_2} measurements is 10–20%. This is nearly an order of magnitude less than that currently attainable by gasometric procedures and is inadequate for use together with other parameters in the elucidation of carbon dioxide equilibria (Park, 1969). In addition, the response of the probe is more sluggish than is that of the ordinary glass electrode and is sensitive to stirring because it depends on the establishment of a diffusion equilibrium across the semipermeable membrane. Despite these difficulties, a number of assemblies have been described for measuring P_{CO_2} *in situ* in the water column (Wheeler, 1966; Greene, 1968) and in the interstitial waters of sediments (Jensen *et al.*, 1965; Moore *et al.*, 1962). Such measurements are useful for following the dramatic decreases in P_{CO_2} (300 ppm to 200 ppm) that accompany rapid photosynthesis, or the large increases (300 ppm to 550 ppm) that may accompany the decay of organic matter, particularly in anaerobic environments. Under anoxic conditions precautions must be taken to prevent the migration of hydrogen sulphide across the gas permeable membrane (see Section 20.7.3). The electrode could, in principle, be used to estimate the total concentration of carbon dioxide species in sea water by adjusting the sample to pH 4 or less. However, this measurement would be subject to the same uncertainties as the direct P_{CO_2} determination.

An ammonia (NH_3) sensitive probe can be prepared in a similar way by replacing the sodium bicarbonate buffer solution with 0·01 M ammonium chloride and using a membrane material that is permeable to ammonia molecules. Ammonia interacts with ammonium ions in the buffer according to the equation

$$NH_4^+ \rightleftharpoons NH_3 + H^+$$

for which the equilibrium constant (K_a) may be defined as

$$K_a = a_{H^+}a_{NH_3}/a_{NH_4^+}.$$

The potential of the pH cell within the probe is therefore given by

$$E = E' + kT \log a_{NH_4^+} - kT \log a_{NH_3}. \qquad (20.21)$$

Since K_a is very small, changes in $a_{NH_4^+}$ accompanying ammonia transport across the membrane are negligible (Midgley and Torrance, 1972). The cell potential is then directly related to the ammonia activity in the external solution at constant temperature and pressure,

$$E = \text{const.} - kT \log a_{NH_3}. \qquad (20.22)$$

Because available photometric methods do not always give consistent results with sea water, Gilbert and Clay (1973a) have examined the possibility of using the ammonia probe for the direct determination of total ammonia–nitrogen (i.e. $[NH_4^+] + [NH_3]$) in marine aquaria and in the open sea. Their procedure was simple. They pipetted 100 ml of the sample into a 150 ml beaker that had been previously washed with 6 M HCl and thoroughly rinsed. The sample was stirred rapidly and sufficient sodium hydroxide was added to adjust the pH to a value greater than eleven. This converts the ammonium ions present in solution to ammonia. The ammonia probe was then inserted and the beaker covered with a piece of paraffin film which was wrapped securely about the probe. The potential was read after five minutes and the electrode was then calibrated in the sample by the addition of two separate aliquots of standard ammonium chloride solution. This "standard addition" procedure is to be preferred when *concentration* measurements are required in complex electrolyte mixtures. The probe senses the activity of ammonia in solution and this is sensitive to changes in the temperature, pressure and ionic composition. If the standard aliquots are sufficiently small, the change in ionic composition during the analysis will be negligible. Similar procedures have been described by Barica (1973), Thomas and Booth (1973) and Srna *et al.* (1973). The electrode is rather sluggish in its response to temperature changes, and so it is advisable to thermostat all samples to the same temperature before carrying out a series of analyses. The potentiometric method gives results that are comparable with those obtained using a photometric procedure (Solorzano, 1969) provided that the total ammonia concentration is high ($> 30\,\mu g\,l^{-1}$, Table 20.5). Although lower concentrations may be

TABLE 20.5

Comparison of potentiometric and colorimetric procedures for measuring high levels of total NH_3–nitrogen in sea water (Gilbert and Clay, 1973a).

Sample location	Total NH_3-nitrogen ($\mu g\,l^{-1}$)	
	Potentiometric ammonia probe	Solorzano method
Boston Harbour: treatment outfall	361 ± 3^a	415 ± 71^a
Boston Inner Harbour	180 ± 2	166 ± 19
Off Castle Island	153 ± 8	151 ± 18
New England Aquarium coldwater marine life tank	120 ± 1	115 ± 14
New England Aquarium	54 ± 4	54 ± 18
Massachussets Bay 42° 22′ N, 70° 55′ W	38 ± 5	30 ± 5

[a] Standard deviation for three determinations. Reproduced with permission from Gilbert and Clay (1973a). Copyright by the American Chemical Society.

accessible if a distillation procedure is introduced, the method would no longer have the advantage of simplicity over the more direct photometric methods. None of the components normally found in sea water interferes with the electrode reponse at the high pH values used. The electrode response may become sluggish because of the gradual accumulation of a precipitate containing magnesium hydroxide and calcium carbonate on the membrane surface. This is readily removed by wiping the membrane surface with a tissue soaked in dilute hydrochloric acid. The potentiometric procedure is unfortunately unsuitable for the measurement of the very low ammonia–nitrogen levels normally encountered at sea because the electrode response is sluggish at low concentrations and is non-linear below $10\,\mu g\,l^{-1}$ (Gilbert and Clay, 1973a; Thomas and Booth, 1973). This lower limit may be extended to $1\,\mu g\,l^{-1}$ in fresh waters if great care is taken over the purification of the distilled water used in preparing the standard ammonia solutions (Midgley and Torrance, 1972).

20.3.1.2. *The major electrolyte component*

Since the major ions in any sea water sample are present in fairly constant proportions it is possible to obtain quite an accurate estimate of their concentrations from a simple salinity measurement. The small, but significant changes in the concentration ratios that do occur (Culkin and Cox, 1966; Carpenter and Manella, 1973) cannot be detected by the relatively insensitive direct potentiometric methods. These procedures have been used most effectively in oceanographic work to measure ionic *activities* in attempts to unravel the detailed physical chemistry of sea water. Among the earliest applications of the newer ion-selective electrodes to marine work were the direct measurements of calcium activity (Thompson and Ross, 1966) and magnesium activity (Thompson, 1966) in sea water. Direct measurements have also been made of the activities of sodium (Platford, 1965a), chloride (Platford, 1965b; Lerman and Shatkay, 1968; Ogata, 1972) and sulphate ion (Platford and Dafoe, 1965) activities in sea water and sea water concentrates. These applications have been reviewed in detail elsewhere (Pytkowicz and Kester, 1971; Millero, 1974; Whitfield, 1975).

20.3.1.3. *Electrodes based on insoluble salts—minor components*

The solid state fluoride electrode (Table 20.2) is well suited for analyses in sea water since the only significant interference is from hydroxide ions. Chloride ions have no noticeable effect on the electrode response and it is possible to measure as little as 10^{-7} M fluoride in the presence of 1 M sodium chloride by direct potentiometry (Warner, 1969a). Potentiometric procedures have been described for determining the free *concentration* of fluoride ions in sea water (Warner, 1969b, 1971a; Anfält and Jagner, 1971c), in marine aerosols (Warner and Bressan, 1973) and in estuarine waters (Warner, 1971b).

For the analysis of sea water (Warner, 1969b, 1971a) a 25 ml sample is brought to a uniform pH and interfering species (e.g. aluminium, iron(III)) are complexed by the addition of 5 ml of a citrate buffer solution (TISAB; Frant and Ross, 1968). The potential (E_1) obtained with the fluoride/saturated calomel electrode pair in the treated sample is related to the free fluoride concentration by the equation

$$E_1 = \beta - kT \log C_x. \qquad (20.23)$$

β is determined by standardizing the electrode pair in a 25 ml sample of synthetic sea water (Lyman and Fleming, 1940) with a known fluoride concentration (C_0) after the addition of 5 ml of citrate buffer. The cell potential in this solution (E_3) is related to β by the equation

$$\beta = E_3 + kT \log C_0 \qquad (20.24)$$

To calculate the *total* fluoride concentration (C_S) some allowance must be made for the difference in salinity between the calibrating solution and the sample since both calcium and magnesium ions complex some of the fluoride and thus influence the electrode potential. Warner (1969b, 1971a) assumed that C_S and C_x were related by the equation

$$C_S = Q_a C_x. \qquad (20.25)$$

Q_a can be determined by taking a sample of natural sea water, more concentrated than the standard, and measuring C_x after successive dilutions with distilled water and buffer (Warner 1969b). At the point where both the standard and the sample have the same salinity it is assumed that $C_S = C_x$ so that C_S can be determined from equation 20.25. C_S can then be calculated at any salinity, allowing for the dilution effect, so that Q_a can be determined as a function of salinity. Warner (1971a) gives,

$$Q_a = 0{\cdot}393 + S‰/57{\cdot}6 \qquad (20.26)$$

for the range 32·2‰ to 37·3‰. Q_a could also be calculated directly if the extent of complexing of fluoride with the major cations was accurately known (Warner, 1969a). The main source of error in the procedure arises from shifts in β as the electrode ages. This effect is minimized by frequent (hourly) standardization. The precision of the method (2σ) is $\pm 0{\cdot}7\,\mu mol\,l^{-1}$ at $70\,\mu mol\,l^{-1}$. The relative standard deviation is $\pm 0{\cdot}005$ when used at sea. The method may be biased by an estimated 1 % because it depends on the assumption of constant composition for sea water, and thus the correct value will lie in the range $\pm(0{\cdot}7 + 0{\cdot}01 \times \text{mean}/n^{\frac{1}{2}})\,\mu mol\,l^{-1}$.

A similar procedure for determining the *free* fluoride concentration has been described by Brewer *et al.* (1970). A 50 ml sample of sea water was pipetted into a 100 ml glass beaker and the pH was adjusted to 5·0 to 5·5 by the addition of 1 ml of an acetic acid/sodium acetate buffer. The solution was stirred for ten minutes to allow the electrodes to equilibrate and the reading was taken with the stirrer switched off. The electrode was frequently calibrated by measuring the cell potential in 50 ml samples of sea water with a known salinity and fluoride content (determined colorimetrically) that had been buffered as described above and "spiked" with a 0·1 ml aliquot of a standard fluoride solution containing 50 $\mu g\, F^- ml^{-1}$. No further allowance was made for the effect of solution composition on the relationship between free and total fluoride concentration. The more elaborate procedure outlined by Warner (1969b, 1971a) is likely to yield more accurate results.

The *total* fluoride concentration can be determined directly by the standard addition method. Warner (1971a) has described a procedure in which 5 ml of TISAB citrate buffer was added to 20 ml of the sea water sample in a 100 ml plastic beaker. The solution was thermostatted to 25·0 ± 0·1°C in an air thermostat. The electrodes were inserted and the cell potential was measured while the solution was stirred. Readings stable to within ±0·1 mV were usually obtained within 5 to 15 min. 0·17 ml (±0·001 ml) of 0·01 M sodium fluoride was then added to the solution and the new steady state potential was measured to the nearest 0·1 mV. The *total* fluoride concentration (C_S) was then estimated from the observed potential shift (ΔE) using the equation

$$C_S = \Delta C \,.\, D(e^{\Delta E/S} - 1)^{-1}, \qquad (20.27)$$

where ΔC is the amount of fluoride added in the spike (56·3 $\mu mol\, l^{-1}$ in this case) and D (=1·2) corrects for the dilution with the citrate buffer. If the measurements are made at sea with a digital pH meter a precision (2σ) of ±1·2 $\mu mol\, l^{-1}$ is obtained when C_S is about 70 $\mu mol\, l^{-1}$. The relative standard deviation is ±0·008. As noted earlier the standard addition method is not appreciably influenced by the ionic strength of the solution nor by its detailed composition. Using this procedure Warner (1971a) estimated the F/Cl ratio in normal sea water to be $(6{\cdot}75 \pm 0{\cdot}03) \times 10^{-5}$. This result is in reasonable agreement with values determined earlier using a colorimetric method (Brewer *et al.*, 1970; Greenhalgh and Riley, 1963).

Since the direct potentiometric method measures free fluoride concentrations and the standard addition method measures total concentrations, Warner (1971a) has suggested that the two procedures together might provide the information necessary to explain the abnormally high F/Cl ratios observed by Greenhalgh and Riley (1963) for samples from south of Greenland (see also Riley, 1965b). Brewer *et al.* (1970) postulated that the additional fluoride ions may be associated with colloidal material and would not be sensed by direct potentiometric methods. An investigation on this basis (Bewers *et al.*, 1973) suggested that these high fluoride concentrations are transient in character and the nature of the fluoride injection remains unresolved.

A standard addition procedure has also been described for the analysis of estuarine waters (Warner, 1971b). A rather more elaborate citrate buffer (TISAB IV), containing 1,2-cyclohexane-diamine tetra-acetic acid (CDTA), is used to counteract the interferences that may arise from the presence of high iron and aluminium concentrations in these waters. The procedure is similar to that described earlier for sea water. The buffer must be added to the sample (in a volume ratio of 1:5) twenty hours before the actual analysis to allow for the slowness of the complexation of iron and aluminium with CDTA. The electrode response was very slow for fluoride concentrations around 5 $\mu mol\, l^{-1}$ and potential measurements were arbitrarily taken fifteen minutes after the addition of the spike of standard solution. This procedure has been used to study the application of fluoride as a mixing indicator in estuarine work (Warner, 1972b) and also to investigate the geochemistry of fluoride in coastal waters (Wilkniss *et al.*, 1971).

A further procedure, based on the use of the TISAB IV buffer, has been developed by Warner and Bressan (1973) to determine the very low levels of fluoride found in marine aerosols and marine rains. A direct potentiometric procedure is used which takes advantage of the observation by Baumann (1971) that the fluoride electrode exhibits a Nernstian response down to 10^{-10} M. The electrode response is sluggish at low concentrations and the solutions are analysed by noting the direction of drift when the electrodes are alternately placed in the sample and in a sequence of standards selected to bracket the expected concentration range. Great care must be taken to minimise contamination from the beakers and from the electrodes themselves and the procedure is necessarily rather slow. A note appended to later versions of the method emphasizes the need to use CO_2 equilibrated distilled water, as freshly distilled water exhibits a variable pH which can cause erratic electrode behaviour because of unpredictable hydroxide interference. With an 80 to 90 % probability one can expect measurements between 0·3 and 20 ppb to be within 10–20 % of the true value (except when very near the limit of detection) and those between 20 and 100 ppb to be within 5–20 %.

Anfält and Jagner (1969b, 1970) have observed that carboxylic acid buffers such as TISAB may interfere with the performance of the solid state fluoride electrode. They recommended (Anfält and Jagner, 1971c) a multiple standard addition procedure for the determination of fluoride in sea water that uses the natural carbonate buffer system in sea water to stabilize the pH (see Section 20.3.3.2).

Brewer and Spencer (1970) have described an ingenious device for measuring the F^-/Cl^- ratio *in situ*. The cell is composed of an Orion fluoride working electrode and an Orion solid state chloride reference electrode. Both electrodes are filled with a standard sea water solution which is gelled by adding 2% agar to improve the mechanical stability of the system. Unfortunately the presence of the gel also slows down the thermal response of the electrodes. The effect of temperature and pressure on the activity coefficients of the fluoride and chloride ions in both the internal and external sea water should be closely matched so that the cell response will be directly related to the F^-/Cl^- *concentration* ratio. It is likely however, that the temperature and pressure coefficients in the gel will be somewhat different from those observed in normal sea water. The electrodes themselves are mounted in an O-ring sealed piston which is backed by a silicone oil reservoir which provides both pressure equilibration and electrical insulation. The cell voltage is measured by a digital pH meter and the signal is registered on a galvanometric recorder. These units are mounted, together with their power supplies, in separate glass spheres. It is difficult to assess the reproducibility of this instrument from the data presented (Brewer and Spencer, 1970). The results from a single cast are shown and these indicate a marked asymmetry in the cell response between the readings taken on lowering and raising the probe. The results themselves must be interpreted with care since the proportion of free fluoride ions in solution will depend on the influence of temperature and pressure on the stability of the magnesium and calcium fluoride complexes. This electrode couple might be useful in testing estuarine mixing models as suggested by Warner (1972b) (see also data given by Windom (1971)), and the design philosophy (with the exception of the gelling of the internal solution) may be used with advantage for other *in situ* studies.

The activity of free *sulphide* ions can be measured in the interstitial waters of marine sediments with silver–silver sulphide electrodes (Berner, 1963). The electrode is simply prepared by immersing a polished silver billet, sealed in epoxy resin, into a solution of ammonium sulphide. A black layer of Ag_2S is quickly formed. The electrode is calibrated by immersing it, together with a pH electrode and a reference electrode, into a solution through which hydrogen sulphide gas is constantly bubbled. The pH is adjusted by the addition of acid or alkali and the appropriate sulphide activity is calculated from the equation

$$pS^{2-} = 21{\cdot}9 - 2pH. \tag{20.28}$$

A linear calibration curve with the theoretical slope for divalent ions was obtained down to free sulphide ion concentrations of 10^{-14} M (Berner, 1963). In the field the electrode was used to establish a linear relationship between the potential of the bright platinum electrode (Eh) and the activity of sulphide ions. The good performance of the electrode down to such low concentrations depends on the presence of a fairly high background concentration (millimolar) of complexed sulphide species such as HS^- and H_2S. These components effectively buffer the concentrations of free ions against changes induced by the measuring procedure. A silver–silver sulphide electrode has been incorporated in a probe for *in situ* measurements (Whitfield, 1971b).

Commercial sulphide electrodes, prepared from pressed discs of silver sulphide (Light and Swartz, 1968; Hseu and Rechnitz, 1968) have been used by Allam *et al.* (1972) to determine the *total* sulphide concentration in submerged soils. The electrode couple was calibrated in sodium hydroxide–sodium sulphide solutions. The soil solution was made sufficiently alkaline ($pH > 11$) to convert the HS^- and H_2S components into sulphide ions. The electrode response is restricted to total sulphide concentrations greater than 10^{-7} M. Below this concentration, the electrode response is non-linear (Naumann and Weber, 1971).

Gilbert and Clay (1973b) have described a procedure for determining the *total* acid-soluble sulphide concentration in marine sediments. An aliquot of the sample is acidified and the resulting hydrogen sulphide is stripped from solution in a nitrogen stream and is collected in a strongly alkaline medium. A sulphide electrode and a reference electrode are dipped into this and the total sulphide concentration is estimated by titration with cadmium ions. Analyses below 1 ppm are feasible using this procedure.

The solid state *copper* electrode performs satisfactorily in sea water and it has been used by Van Duin (1972) to investigate the physico-chemical state of copper in sea water. A standard addition procedure for the determination of copper in natural waters has been developed by Smith and Manahan (1973). The samples were treated with an equal volume of a complexing anti-oxidant buffer (CAB) which consists of a mixture of 0·05 M sodium acetate, 0·05 M acetic acid, 2×10^{-2} M sodium fluoride and $2{\cdot}0 \times 10^{-3}$ M formaldehyde. The acetate buffer adjusts the pH to 4·8 which is about optimum for the successful operation of the electrode. The acetate ions also complex copper ions and prevent the removal of copper by precipitation or sorption onto the container walls. The fluoride complexes ferric ions which would otherwise interfere with the electrode response, and formaldehyde acts as an antioxidant which presents oxidising agents present in the solution from interfering with the equilibrium reaction at the electrode surface. Buffer solutions prepared from ultrapure solutions typically contain 1 $\mu g\,l^{-1}$ of copper. Although oceanic copper levels are generally below the lower limit of $3\,\mu g\,l^{-1}$ suggested by Smith and Manahan (1973), the procedure might form the basis of a useful method for determining copper in estuaries and in-shore waters. The concentration of fluoride in the buffer solution would have to be reduced to prevent possible loss of copper by co-precipitation with calcium fluoride. Longer equilibration times are required for the analysis of samples containing small amounts of copper. The times required are typically one minute at $1\,mg\,l^{-1}$, 20 minutes at $30\,\mu g\,l^{-1}$ and 1 hour at $1\,\mu g\,l^{-1}$. Flow-through assemblies have been devised that would enable the analyses to be carried out on a continuous basis (Thompson and Rechnitz, 1972; Fleet and Ho, 1973). Van Duin's observations suggest that sea water can retain a maximum of $50\,\mu g\,l^{-1}$ of dissolved copper (Van Duin, 1972). At high concentrations the copper is present in a colloidal form and will not therefore be directly accessible to electrode measurement. Davey *et al.* (1973) indicate that the copper-selective electrode might be useful for monitoring copper toxicity at these high levels.

20.3.2. POTENTIOMETRIC TITRATIONS

20.3.2.1. *Endpoint location*

The precision of direct potentiometric measurements is restricted by the relatively small Nernstian response to changes in concentration and by the drifts in potential associated with the rather complex structure of the electrodes themselves. These difficulties can be overcome, and selectivity problems can be circumvented to some extent, by using the electrodes to monitor the progress of a titration. Provided that the equivalence point can be accurately located the total concentration of the selected component can be determined with an accuracy of $\pm 0{\cdot}1\,\%$ or better.

To locate the equivalence point some quantitative relationship must be established between the potential measured and the volume of titrant added. This involves a knowledge of the interrelated chemical processes that control the concentration of the electroactive species during the course of the titration and of the response of the electrode in an electrolyte mixture. In the simplest case where the electrode is perfectly selective the cell potential (E) is given by,

$$E = E' + kT \log [\mathrm{M}]$$

or

$$[\mathrm{M}] = 10 \exp (E - E')/kT. \qquad (20.29)$$

The potential associated with the liquid junction and the activity coefficient term are incorporated in E' and are assumed to remain constant throughout the titration. If M is removed quantitatively by titration with a complexing or precipitating agent, X, according to the reaction

$$M + X \rightleftharpoons MX$$

then the concentration of M should decrease in direct proportion to the amount of X added, i.e.

$$[M] \propto (v_{eq} - v)/(v_0 + v). \tag{20.30}$$

v_0 is the volume of the sample, v is the volume of titrant added and v_{eq} is the volume added at the equivalence point. Combining 20.29 and 20.30 and rearranging gives,

$$F_1 = (v_0 + v)\, 10 \exp{(E - E')/kT}$$

or

$$F_1 \propto (v_{eq} - v) \tag{20.31}$$

Plotting F_1 against v gives a straight line which cuts the v-axis where $v = v_{eq}$. This procedure, originally proposed by Gran (1950, 1952), is particularly useful for locating the end point of potentiometric titrations since it makes use of all the data points and does not rely too heavily on readings taken near the equivalence point where the electrode itself will be subject to the greatest interference. It is in this region too that the electrode is likely to show a drifting response as it tries to keep pace with the very rapid changes in the solution environment. In addition, electrodes do not, in general, show a Nernstian response below a total concentration of 10^{-6} M for the selected ion. With the Gran procedure the equivalence point can be located using only data points from solutions with concentrations in excess of this limit where the electrode is exhibiting a significant and reproducible response. When compared with other procedures for end-point location (Anfält and Jagner, 1971b) the Gran method shows a very small systematic error and a high precision. Statistical errors involved in the extrapolation procedure have been considered in great detail (Buffle *et al.*, 1972; Buffle, 1972; Parthasarathy *et al.*, 1972). In sea water analysis the most serious problems arise from purely chemical effects. In most instances the chemistry is not as simple as that assumed above and both M and X will be simultaneously involved in a number of subsidiary reactions. These competing reactions will result in errors in the Gran plot that will cause the F_1 vs v function to deviate from linearity (Ingman and Still, 1966; McCallum and Midgley, 1973) and thus make the location of the endpoint more difficult. The most direct way of overcoming this problem is to take due account of *all* the reactions involved in the titration procedure (Anfält and Jagner, 1969a). The variation in free concentration of the electroactive substance can then be calculated as the titration proceeds using one of the general computer programmes that have been developed for analysing complex ionic equilibria (Ingri *et al.*, 1967; Perrin *et al.*, 1967; Sayce, 1968; Bos and Meersherck, 1972; Van Breemen, 1972; Morel and Morgan, 1972). More appropriate Gran functions can then be defined and the effectiveness of the end point location procedure can be checked by simulated titrations on the computer (Anfält and Jagner. 1971b; Hansson and Jagner, 1973). The programme HALTAFALL (Ingri *et al.*, 1967) has been well documented (Dyrssen *et al.*, 1968a; Wedborg, 1972) and is the only one that has so far seen application in marine work (Hansson and Jagner, 1973). Less generalized curve fitting procedures (see e.g. Isbell *et al.*, 1973) are not so useful in marine chemistry because of the considerable complexity of the sample medium. The use of a general computer programme certainly saves the tedium of preparing complex algebraic equations for particular cases (Anfält and Jagner, 1969a) and also saves the need for introducing various simplifying approximations (Jagner, 1971).

Two further problems associated with equation 20.29 can cause errors in the location of the endpoint of a potentiometric titration. The activity coefficient term in E', and hence the stability constants used in the computer calculations, may vary throughout the titration and other ions may interfere with the electrode's response to the electroactive component (M). McCallum and Midgley (1973) corrected for variations in activity coefficients with ionic strength using an extended Debye–Huckel equation (see Chapter 2).

$$-\log \gamma = Az^2(I^{\frac{1}{2}}/(1 + BaI^{\frac{1}{2}}) - bI) \qquad (20.32)$$

and developed Gran-type functions that should give improved linearity in acid-base and precipitation titrations. These corrections are most significant when the sample solution is relatively dilute. The introduction of activity coefficient corrections into HALTAFALL has been considered by Anfält and Jagner (1973b).

Interference effects are not so easily dealt with because there is, as yet, no agreed convention for expressing the influence of interfering ions on the cell potential (Moody and Thomas, 1972), and very few selectivity studies have been carried out on electrodes over the wide concentration range experienced in analytical titrations (see for example Leyendekkers and Whitfield, 1970, 1971).

The presence of interfering ions results in a shallower potential drop at the end-point and a consequent distortion of the titration curve (Dyrssen *et al.*, 1968b; Whitfield and Leyendekkers, 1969a; Whitfield *et al.*, 1969; Hulanicki and Trojanowicz, 1973). Detailed consideration has been given to the influence of these effects on end-point location in precipitation (Carr, 1971; Schultz, 1971a) and complexation (Carr, 1972; Schultz. 1971b) titrations under conditions in which the contribution of the interfering ions to the cell potential is constant throughout the titration. The more complex and general case in which these parameters vary as the titration proceeds has been analysed by Anfält and Jagner (1973a) using HALTAFALL.

It would appear that the simplest and most general approach to the elucidation of titration procedures involving ion-selective electrodes is to define the chemical basis of the procedure as fully as possible using a programme like HALTAFALL. The influence of side reactions, the finite solubility of precipitates and the influence of shifts in activity coefficients and variations in electrode selectivity can then be dealt with on a uniform basis without recourse to complex algebraic expressions setting out the detailed conditions for each particular case (Jagner, 1971). The appropriate Gran functions can then be defined and their effectiveness tested.

A number of automatic digital titrators have been described (Jagner, 1970; Anfält and Jagner, 1971a; Johansson and Pehrsson, 1970) that remove the tedium from the titration procedures and provide the large numbers of precise data points over the whole titration range that are required to take full advantage of these accurate analytical procedures. A titrator of this kind has been used successfully on board ship (Almgren *et al.*, 1974).

20.3.2.2. *Alkalinity titrations*

Titrimetric procedures can be used to estimate the total alkalinity (A_t) and the total carbon dioxide content (C_T) of a sea water sample. By subtracting the contribution of boric acid to A_t the carbonate alkalinity (A_C) can be estimated. A_C and C_T together can be used to calculate the total concentrations of the various components of the carbon dioxide system (CO_2, HCO_3^- and CO_3^{2-}) and the pH of the sample. If the pH is measured separately the three parameters A_C, C_T and pH can be combined in five different ways (Park, 1969) to calculate the concentrations of these components, thus providing a rigorous test of the internal consistency of the measurements.

Following the original suggestion of Dyrssen (Dyrssen, 1965; Dyrssen and Sillén, 1967) Edmond (1970) has developed a precise titrimetric method for the determination of A_t and C_T using the Gran linearization procedure for the detection of the carbonate and bicarbonate end-points. The titration is carried out in a 150 ml closed flask (Fig. 20.12). A piston (D) can be displaced to accommodate the acid added from the burette and a vertical capillary (*E*) is used to equalize the pressure when the vessel is closed initially. The flask is completely filled with the sea water sample ($\sim$170 ml at 25°C) which is titrated with 0·3 M hydrochloric acid that has been made 0·53 M in chloride by the addition of sodium chloride. The solution is stirred throughout the titration using a Teflon coated bar magnet. The end-points are located by plotting the Gran functions (Dyrssen and Sillén, 1967).

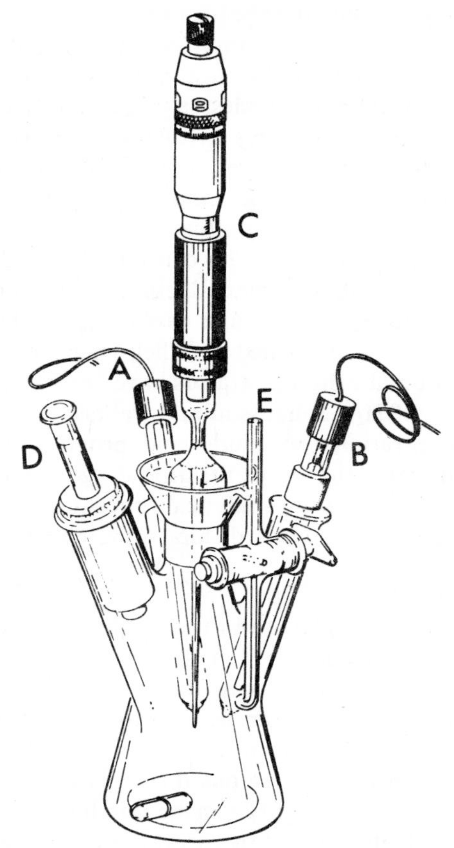

FIG. 20.12. Titration flask for the determination of C_T and A_t (see text for details). A. Glass electrode. B. Calomel reference electrode. C. 2·5 ml micrometer burette. D. Displacement piston. E. Pressure equalizing capillary. Reproduced with permission from Edmond (1970).

$$F_1 = (v_2 - v)\,10\exp(E - E')/kT, \tag{20.33}$$

$$F_2 = (v_0 + v)\,10\exp(E - E')/kT \tag{20.34}$$

versus v (Fig. 20.13). Equation 20.34 is used to locate the second end-point (v_2) which corresponds to the proton condition

$$[H^+] + [HSO_4^-] \equiv [HCO_3^-]$$

The value of v_2 obtained is then used in equation (20.33) to locate the first end-point (v_1) for which the proton condition is

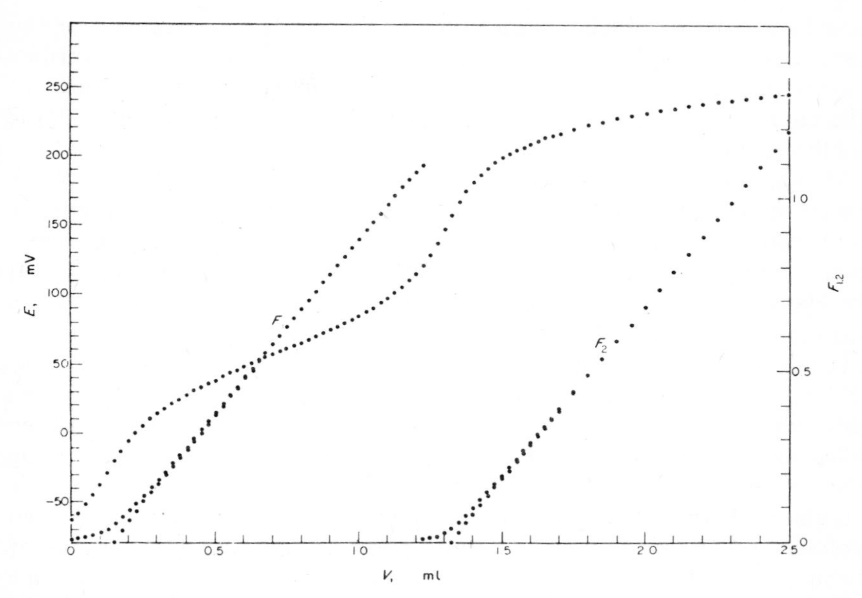

FIG. 20.13. Complete curve for the titration of a sea water sample from Scripps Pier with 0·3 N HCl. The Gran functions F_1 and F_2 are also plotted. Reproduced with permission from Edmond (1970).

$$[H_2CO_3] \equiv [CO_3^{2-}] + [B(OH)_4^-]$$

If T is the concentration of acid in the titrant then

$$A_t = v_2 T/v_0 \tag{20.35}$$

and

$$C_T = (v_2 - v_1)\, T/v_0 \tag{20.36}$$

Edmond (1970) suggested that accuracies of $\pm 0{\cdot}17\,\%$ for A_t and $\pm 0{\cdot}68\,\%$ for C_T at the 95% confidence level should be possible at sea. Measurements made by Edmond during the Geosecs Intercalibration study (Takahashi *et al.*, 1970) indicated a working precision and accuracy of $\pm 0{\cdot}3\,\%$ for A_t and a precision of $\pm 0{\cdot}3\,\%$ and an accuracy of $\pm 0{\cdot}6\,\%$ for C_T.

A detailed consideration of the Gran extrapolation procedure by Hansson and Jagner (1973) indicated that better precision could be attained by paying careful attention to small deviations from linearity in the exponential plots. Improved Gran plots can be obtained by allowing for the contribution of other protolytic species (e.g. $B(OH_4)^-$, F^- and HSO_4^-) to the total alkalinity, and by the use of stoichiometric equilibrium constants to account for the influence of the major cations in sea water on the distribution of the various carbon dioxide species. The improved functions are necessarily complex, but the computation procedure as outlined by Hansson and Jagner (1973) is quite straightforward.

Automatic titration procedures (Anfält and Jagner, 1971a) improve the precision of the titrations and yield data that can be processed automatically on board ship. Almgren *et al.* (1974) have taken advantage of these developments, and of the clearly defined system of stoichiometric constants provided by Hansson (1973b,c,d), to carry out a series of precise automatic alkalinity titrations on board the R/V *Dmitry Mendeleev.* The titration vessel consisted of a 250 ml flanged beaker with a Teflon lid sealed in place with an O-ring. The electrodes, the burette tip, a thermistor and a Teflon plunger were inserted through the lid and were in turn sealed by O-rings. After the vessel had been filled, any excess air was expelled through a small hole in the lid by depressing the Teflon plunger. This hole was then sealed off with a screw closure. The volume enclosed was reproducible to within 0·01 %. The relationship between volume and temperature was determined ashore and was fed to the on-line computer which controlled the titration. A thermistor was also placed in the hydrochloric acid stock flask and the mean temperature throughout the titration was used to correct the acid concentration. At the beginning of the titration the temperature in the reaction vessel was measured and the computer calculated the sample weight from the salinity and volume data. The titrant was then added in increments and at each step the temperature was measured and was used to correct the volume data and the Nernst factor of the electrodes. The titration end-points were located using the refined Gran functions of Hansson and Jagner (1973). At the end of the titration values of A_t, C_T, pH, $[CO_2]$, $[HCO_3^-]$ and $[CO_3^{2-}]$ were printed out. Using this procedure Almgren *et al.* (1974) were able to achieve a precision of $\pm 0{\cdot}1\,\%$ in their estimations of C_T on board ship. It is likely that this represents the limit of precision for this particular technique. At this level it becomes necessary to consider the contributions to A_t of protolytes other than carbonic and boric acids (Edmond, 1970; Almgren *et al.*, 1974). For example 85 % of the phosphate (totalling 1–2 μmol kg^{-1} sea water) is present as HPO_4^{2-} and this will be titrated to $H_2PO_4^-$.

A simpler, though less precise procedure for estimating A_t may be mentioned here. The technique, developed by Culberson *et al.* (1970) from the original method of Anderson and Robinson (1946), is based on the measurement of the pH of a sea water sample after the addition of a known aliquot of strong acid. 6 ml of 0·01 N HCl are added to 20 ml of sea water at 25°C and the carbon dioxide released is driven off by purging with water-saturated air for five minutes. A_t can be related to the final pH (assuming pH $= -\log a_{H^+}$) by the equation

$$A_t = (1000/dv_s)(vN - (v_s + v)\,a_{H^+}/f_{H^+}). \qquad (20.37)$$

where v_s is the volume of the sea water sample, v is the volume of HCl added and N is its normality. d is the density of the sea water at 25°C and the activity coefficient f_{H^+} is given the value 0·741 (Culberson *et al.*, 1970). With care this procedure will give a precision of ±0·6% for A_t (Takahashi *et al.*, 1970). The accuracy of the method is strongly dependent on the validity of the value assumed for f_{H^+}.

20.3.2.3. *Major cations*

Jagner (1967, 1971) has reported a potentiometric procedure for the determination of magnesium in sea water. The magnesium was precipitated as magnesium by titration with sodium hydroxide and the progress of the titration was followed with a pH electrode. The solution was flushed with nitrogen to prevent interference from atmospheric carbon dioxide, and ethanol was added to the sample (60% sea water by volume and 40% ethanol) to decrease the solubility of the precipitate. The equivalence point was located using a modified Gran procedure. No evidence was found for the incorporation of chloride ions in the precipitate. Fluoride is co-precipitated with the magnesium hydroxide although the amounts normally present in sea water would cause the recorded magnesium concentration to be only 0·07% too low, even if it were all incorporated. The original report (Jagner, 1967) gave 0·05269 mol kg^{-1} of magnesium in standard sea water compared to the range 0·05292 to 0·05331 mol kg^{-1} (corrected to salinity 35‰) found by Riley and Tongudai (1967) for eighty samples of sea water from the major oceans. Later work (Jagner, 1971) suggested that a value of 0·05280 mol kg^{-1} might be more accurate for standard sea water.

Attempts have also been made to determine magnesium and calcium potentiometrically in sea water using a calcium liquid ion-exchange electrode (Orion 92–20) to detect the end-point (Dyrssen *et al.*, 1968b; Whitfield *et al.*, 1969). The sea water sample was divided into two aliquots and one was titrated with DCTA (1,2-diaminocyclohexane-*N,N,N′,N′*-tetraacetic acid) to determine calcium plus magnesium. The other aliquot was titrated with EGTA (Ethyleneglycol bis-(β-amino ethyl ether)-*N,N′*-tetraacetic acid) to determine the calcium selectively. Unfortunately the calcium electrode did not have a very good selectivity for calcium over sodium and magnesium (Table 20.2) so that the resulting titration curves were rather distorted. Analysis of the chemical equilibria involved in the titration using the conditional constant approach of Ringbom (Whitfield and Leyendekkers, 1969a;

Ringbom, 1963) and the general programme HALTAFALL (Dyrssen *et al.*, 1968b) enabled titration curves to be calculated for a variety of solution compositions taking due account of the imperfect response of the electrode to calcium. Theoretical predictions were borne out by the experimental titrations (Dyrssen *et al.*, 1968b; Whitfield, *et al.*, 1969), although the agreement was not quantitative since the selectivity coefficients themselves (equation (20.14)) varied during the course of the titration (Leyendekkers and Whitfield, 1971; Whitfield and Leyendekkers, 1970). The end-points could be located by the Gran method with an overall accuracy of 0·5 to 1 %. Similar performance was obtained with the Beckman 39608 calcium electrode, (Table 20.1; Jagner, 1971).

Ruzicka *et al.* (1973) have since developed a calcium electrode with a much better selectivity for calcium in the presence of sodium and magnesium. The ion-exchanger is incorporated in a PVC matrix (Table 20.1). The electrode selectivity can be predicted from quite simple equations and the electrode stability is superior to that of the Orion electrode in the presence of an excess of interfering ions (e.g. near the end-point of the titration). Undoubtedly compleximetric titration procedures for calcium and magnesium will shortly be developed to take advantage of this significant development.

20.3.2.4. *Major anions*

Considerable attention has been paid to the determination of the chlorinity of sea water by potentiometric titration with silver nitrate. The general procedure is to titrate the bulk (99 %) of the halides by the addition of a strong silver nitrate solution and then to approach the equivalence point slowly by the addition of a more dilute solution. Generally, the end-point of the titration is assumed to coincide with the point of maximum slope in the potentiometric titration curve. Vigorous stirring is required throughout the titration to prevent the coagulation of the silver chloride curds that are produced.

The procedure used for the preparation of standard sea water samples (Hermann, 1951) involves the use of a silver indicator electrode and a silver counter electrode that is mounted in the burette tip. The titration is continued until the cell potential reaches the value that has been shown, by a previous series of titrations, to correspond to the point of maximum slope on the titration curve (known as a dead-stop titration). With care, the procedure gives a reproducibility of $\pm 0{\cdot}008$ %. The indicator electrodes have to be carefully rinsed after each titration and wiped free of adherent silver chloride at the end of each day.

Bather and Riley (1953) use a cell containing a platinum indicator electrode and a mercury–mercurous sulphate reference electrode. This indicator electrode apparently gives more stable readings in the vicinity of the end-point than does the silver electrode and can be used for 100 to 200 titrations before it starts to show a sluggish response. Again they used a dead-stop procedure and reported an accuracy of ±0·006% for their determinations. Silver–silver chloride electrodes have also been used to indicate the end-point (Hindman *et al.*, 1949; Harwell, 1954; Reeburgh and Carpenter, 1964). They are not well suited for dead-stop titrations as the end-point potential gradually shifts as bromide ions are assimilated into the silver chloride lattice (Table 20.2; Proctor, 1956; Ives and Janz 1961). The differential procedure of Reeburgh and Carpenter (1964, see also Brand and Rechnitz, 1970) overcomes this difficulty since it uses a matched pair of silver–silver chloride electrodes to indicate the end-point. One electrode is attached to the piston of a syringe. When the piston is withdrawn the electrode is in contact with a small volume of sea water in the syringe barrel and is effectively isolated from the bulk of the sample. As the titration proceeds a potential difference arises between this electrode and its partner which is immersed in the bulk sample. Near the end-point, the syringe is flushed out after each addition of titrant. The potential difference between the isolated electrode and the "bulk" electrode reaches a maximum at the end-point. Since both electrodes are affected in the same way by the presence of bromide ions and relative, rather than absolute, potential measurements are made this procedure makes better use than does the dead-stop procedure of the properties of the silver–silver chloride electrode.

Dyrssen and Jagner (1966) suggested a potentiometric procedure for determining chlorinity based on the Gran method for locating the end-point. This has been refined by Jagner and Årén (1970) and adapted for use with a semi-automatic titrator (Jagner, 1970). The electrode couple consisted of a polished silver billet and a saturated calomel electrode that was connected to the sample *via* an acidic potassium nitrate salt bridge. The sample (approx. 10 ml) was weighed into a polythene beaker and was diluted first with 50 ml of acidic potassium nitrate solution, and then with 50 to 100 ml of ethanol. The acid nitrate solution helps to ensure that free silver and chloride ions are the predominant species in the vicinity of the end-point and the ethanol decreases the solubility of the silver chloride so that a sharper end-point is observed. The Gran procedure used in this titration is of particular interest since it only uses the portion of the titration curve that is recorded *after* the end-point has been passed. The experimental procedure is consequently quite rapid requiring only fifteen minutes for each titration, including computation. The Gran function plotted is given by equation 20.31.

The authors presented a very thorough discussion of the chemical processes involved in the titration and of the errors associated with the various manipulations and calculations. They concluded that a precision of ±0·02% is obtainable only if extreme care is exercised during sampling and titration. If the samples are pipetted, rather than weighed, into the titration vessel this error rises to $\pm 0{\cdot}04\%$. It is surprising that the errors involved in this rather sophisticated procedure should be so much higher than those reported for conventional dead-stop titrations.

Mascini (1973) has reported a titrimetric procedure for the determination of sulphate ions in sea water using a solid-state lead selective electrode as an endpoint indicator. The sea water was first passed through a cation exchange column in the silver form to remove halides and then through a second cation exchanger in the hydrogen form to remove silver ions. The final eluate was adjusted to pH 5–6 to remove bicarbonate ions and was then diluted with an equal volume of 1,4-dioxan and titrated with lead nitrate. The end-point was was assumed to coincide with the point of maximum slope on the titration curve. Errors as large as $\pm 4\%$ were noted when the method was applied to sea water. A similar procedure, using a solid lead amalgam electrode has been reported for use in saline soil waters (Robbins *et al.*, 1973).

A potentially more useful procedure has been described by Jasinski and Trachterberg (1973) who employed a chalcogenide glass electrode to respond to uncomplexed Fe(III) in solution. Fe(III) was added to acidified sea water in which it formed stable complexes with sulphate. Barium chloride was then titrated into the solution and decomposed the complex. After the end-point the concentration of uncomplexed Fe(III) was constant and the potential levelled off. There was some interference from calcium ions but this could be removed by 10:1 dilution of the sea water.

20.3.3. MULTIPLE STANDARD ADDITION METHODS

20.3.3.4. *Principle*

The standard addition method is the complement of the conventional titration procedures discussed in Section 20.3.2 in that the component to be analysed is added to the sample in a series of known increments rather than being removed by a chemical reaction with a complexing or a precipitating agent. It is particularly useful for the determination of ions in samples with unknown ionic strengths and/or with undetermined levels of interfering ions.

If the electrode is responding specifically to the electroactive ion M the cell potential is given by equation (20.29). If a small increment of $M(\Delta C)$ is added to the solution, then the potential will shift by an amount ΔE so that

$$E + \Delta E = E' + kT \log (C + \Delta C), \qquad (20.38)$$

where C is the original concentration of M in the solution.

Assuming that neither the activity coefficient of M nor the liquid junction potential are affected by the increment of M, equation (20.29) can be subtracted from equation 20.38 to give

$$C = -\Delta C(1 - 10 \exp(\Delta E/kT))^{-1} \quad (20.39)$$

C can therefore be calculated directly from ΔE provided that the accompanying volume change is too small to produce significant dilution effects and that the extent of complexing of M has not been altered by the additions. The response slope must be accurately known, and the potential change must be monitored to within a few hundredths of a millivolt. This simple procedure has been used by Garrels (1967) to measure the concentration of sodium ions in sea water to within $\pm 1\%$.

If the electrode is sensitive to more than one ion in the sample then an equation analogous to equation (20.39) can be derived (Whitfield, 1971a) from equation (20.14), viz

$$C = -\Delta C(1 - 10 \exp(\Delta E/kT))^{-1} - \Sigma_j K_{ij}(a_j)^{n/y}/\gamma_M \quad (20.40)$$

Interfering ions will therefore alter the slope and intercept of the electrode response. If two known additions are made then two equations corresponding to (20.39) are obtained and it is possible to estimate the slope of the electrode response directly in the sample solution. Procedures for determining fluoride (Warner, 1971a) and ammonia–nitrogen (Gilbert and Clay, 1973a) in sea water using this method have been described in earlier sections. An alternative procedure in which known volumes of the sample are added to a known volume of the standard solutions has also been described by Durst (1969b).

The standard addition method can be made more precise by carrying out a series of known additions so that a number of independent determinations of C and kT can be compared. The resulting titration curve can be analysed by a non-linear least squares method (Brand and Rechnitz, 1969), or the data can be treated as a normal titration curve and analysed by the Gran linearization procedure (Section 20.3.2.1).

20.3.3.2. *Application*

Anfält and Jagner (1971c) have described a standard addition procedure for estimating the concentration of fluoride in sea water. The maximum buffering region of the carbonate system is attained for sea waters around 35‰ by adding about 50 mmol of hydrochloric acid for each 100 ml of sea water giving a final pH in the region of 6·6. To a sample of approximately 150 g of sea water, accurately weighed into a polyethylene beaker, 0·5 to 0·75 mls of 0·1 M hydrochloric acid was added and the potential reading registered after 1 minute was recorded. Titrant increments of 0·05 ml of 0·01 M sodium fluoride were then added until the emf reading was approximately −75 mV (with an Orion Model 94–09 fluoride electrode and an Orion Model 90–01 single junction reference electrode). The total volume of titrant added was about 2 ml and at least 15 seconds was allowed between consecutive additions.

The total concentration of fluoride in the solution was determined using the Gran function given in equation (20.31). The total fluoride concentration (x) in sea water was calculated from the equation,

$$x = -v_{eq}t/v_0, \tag{20.41}$$

where t is the molality of the sodium fluoride standard. The influence of dilution on the formation of complexes between calcium and magnesium ions and fluoride (Anfält and Jagner, 1971c) can also be accommodated using a more elaborate function

$$F_2 = F_1(1 + 1{\cdot}07v_0S/35(v_0 + v)), \tag{20.42}$$

where

$$1{\cdot}07 = \beta_{MgF^+}[Mg^{2+}] + \beta_{CaF^+}[Ca^{2+}]. \tag{20.43}$$

The β-terms are the formation constants of the complexes indicated and S is the salinity of sample. With standard sea water a precision of 0·6 % was possible for the determination of fluoride by this procedure. The precision is lower for less saline waters and falls to 7 % for a water with a salinity of around 2‰.

The data from a single titration can be used to calibrate the electrode for direct potentiometric analysis since,

$$E_s^0 = E - kT\log t(v - v_{eq})/(v + v_0) \tag{20.44}$$

for each titration point. The mean value of E_s^0 for the particular temperature and salinity ($E_{s,m}^0$) can then be substituted in the cell equation to give,

$$[F^-]_{tot} = 10\exp((E_1 - E_{s,m}^0)/kT) \tag{20.45}$$

where E_1 is the cell potential before the addition of titrant. Fluoride concentrations can then be calculated from a single reading using equation (20.45), provided that the value of $E_{s,m}^0$ is appropriate to the temperature and salinity of the sample. The procedure is much simpler than the standard addition method and yields a precision of 0·88 % for standard sea water, falling to 8·2 % for water with a salinity of 2·3‰. Anfält and Jagner (1971c) have also presented computer calculations to verify their procedure and indicate that interference from the formation of AlF^{2+} is negligible in normal sea water at pH 6.

Anfält and Jagner (1973b) used the standard addition method for determining potassium in sea water using the Philips IS 500 K potassium electrode. This procedure was preferred to a conventional titration (say with tetraphenyl borate ion) because the sodium response of the electrode would tend to distort the titration curve in the vicinity of the end-point in a manner analogous to that observed with the calcium electrode (Section 20.3.2.2). The titration was carried out using an automatic titration system (Anfält and Jagner, 1971a). The titrant volume and concentration were adjusted to give increments of one or two millivolts after each addition of standard. The potential was measured after each addition until the average of thirty consecutive potential readings did not differ from the average of the next thirty by more than 0·002 mV (corresponding to an approximately forty second waiting time). The potassium concentration was estimated from a series of ten incremental readings using a modified Gran plot.

Some difficulty was experienced with the potassium electrode because both the electrode slope and the sodium selectivity coefficient tended to shift with age even when the electrode was stored in a sea water solution of 12‰ salinity. For calibration in 0·6 M NaCl, the net effect was to shift the electrode slope from 58·7 to 19·5 mV per logarithmic unit over a period of four days. An average of 97 titrations (with the electrode being prepared afresh every tenth titration) gave the concentration of potassium in standard sea water as $399{\cdot}6 \pm 1{\cdot}1$ mg kg^{-1} at a salinity of 35‰.

Jasinski *et al.* (1974) have reported a standard addition procedure for the determination of copper in sea water at the 1 ppb level using the copper(II) ion selective electrode. However the electrode response is unusual at these low levels and slopes of 50 to 120 mV/decade are observed (cf. Nernstian slope of 30 mV/decade!) that appear to vary with the nature of the sample in an unpredictable way. It is difficult to assess the precision and accuracy of the method from the data presented.

20.3.4. MISCELLANEOUS POTENTIOMETRIC APPLICATIONS

A number of potentiometric procedures have been described that provide useful information for the interpretation of chemical processes in the sea although they are not designed to analyse for a specific chemical component.

Koske (1964) and Gieskes (1967) developed an electrode system for the direct measurement of salinity. The cell was formed by two membrane electrodes with standard sea water filling solutions and silver–silver chloride internal reference electrodes. One of the membranes is a cation exchanger and the other an anion exchanger. These membranes have a relatively low resistance (approx. $10^4 \Omega$) so that impedance matching amplifiers are not needed. When the cell is immersed in an electrolyte solution the potential generated is related to the total salt content in a manner that is dependent on the permselectivity of the ion-exchange membranes and on the charge and the activity coefficients of the ionic components. Empirically, the cell response may be related to the salinity of a sea water sample by

$$E = (CkT)\log S_1/S_2, \tag{20.46}$$

where S_1/S_2 is the ratio of the salinity of the external sea water to that of the filling solution, and C is an empirical correction factor for the Nernstian slope. For a typical sensor (CkT) has a value of about 90 mV for each tenfold change in salinity at 25°C. A portable probe based on this design has been described by Wilson (1971) (see Chapter 6). Because of its logarithmic response the membrane cell is useful for measuring salinity under estuarine conditions where large variations are encountered. It is particularly useful for measurements in waters of low salinity for which the conductometric procedures are not so reliable. Conductometric methods have a higher precision and sensitivity than do membrane salinometers for *in situ* measurements in the open sea where only small variations are encountered about a relatively high value. Under field conditions without strict temperature control the accuracy is ~1% of the salinity reading and the precision is ~0·1% for replicates run as a batch.

Leyendekkers (1973) has taken the concept of a membrane salinometer a stage further by suggesting that an electrode array could be used with one electrode selectively responding to each of the major ionic components. She used an array consisting of sodium, potassium, calcium + magnesium and chloride selective electrodes versus a saturated calomel reference electrode. The potentials of the individual cells can be summed to eliminate the liquid junction potential at the reference electrode and to give an overall cell potential related to the mean ion activities ($a_{\pm\text{MCl}}$) of the chloride salts in sea water, i.e.

$$\begin{aligned} E_{\text{tot}} &= (E_{\text{Na}^+} + E_{\text{K}^+} + E_{(\text{Ca}^{2+}+\text{Mg}^{2+})}) - 3(E_{\text{Cl}^-}) \\ &= E^\circ_{\text{tot}} + kT\log[a^2_{\pm\text{NaCl}}\cdot a^2_{\pm\text{KCl}}(a^3_{\pm\text{MgCl}_2} + a^3_{\pm\text{CaCl}_2})] \end{aligned} \tag{20.47}$$

where E_X refers to the potential of the electrode couple sensing the ion X. E°_{tot}, the total standard electrode potential of the array, is defined in an analogous manner to E_{tot}. The overall response of this array is equivalent to 325 mV for each tenfold change in salinity so that it is three or four times more sensitive than the membrane salinometer. Because of its complexity this electrode system cannot compete with the more conventional conductometric procedures for the *in situ* measurement of salinity. Conductivity is poorly understood but easily measured whereas chemical potential can be accurately defined but only measured with difficulty. The real value of Leyendekker's approach is that it provides a means of determining a thermodynamically well defined parameter that can be used for assessing the influence of the "sea salt" component on the progress of chemical reactions in the sea. This is likely to prove valuable as marine chemists try to piece together quantitative models for chemical processes and as the chemical limitations of the arbitrarily defined salinity parameter become more obvious. Mangelsdorf and Wilson (1971) have described a beautifully sensitive device for sensing very small changes in the relative concentrations of the major cations in sea water (see Section 19.8.25).

Considerable interest has been shown in the possibility of measuring the redox potential of sea water and interstitial waters of marine sediments since Cooper's elegant statement of the problem (Cooper, 1937). Initial attempts were centred around the use of the platinum electrode as an inert sensor although ZoBell (1946) had clearly indicated that the potential generated at this electrode can, at best, be seen as an empirical parameter and should not be related to specific redox equilibria in solution (Whitfield, 1969, 1971a). Later criticisms (Stumm, 1967; Morris and Stumm, 1967) have thrown doubt on the very concept of a single redox potential that is representative of a particular natural system since it is unlikely that the numerous redox equilibria active in the solution will be so intimately interlinked. Breck (1972) brought a fresh approach to this problem by suggesting that a small sample of a well characterized redox couple (O^*/R^*) brought into electrical contact with a large excess of the natural sample would eventually adopt the redox potential of the dominant redox couple (O/R) in the environment. For simplicity Breck's redox electrode may be visualized as a U-tube (Fig. 20.14) with a small sample of O^*/R^* in one limb and a large excess of the sea water in the other. The two solutions are connected at the base by a potassium chloride salt bridge and they are short circuited by a plug of graphite connected between the limbs. According to Breck (1972) the end of the graphite plug immersed in the sea water will adopt a potential which is determined by the redox potential of the dominant couple, O/R, in that solution. Since the graphite is a good conductor, electrons will flow round the circuit until the end of the graphite plug immersed in the reference redox solution causes the O^*/R^* ratio in this solution to shift so that it registers this same potential. This potential can then be sensed by immersing a platinum electrode into the reference compartment and, say, a calomel electrode in the sea water compartment. This is a very neat idea, but Ben-Yaakov and Kaplan (1973a) have argued that the circuit would only work in the way that Breck had assumed if the couples O^*/R^* and O/R were reversible at the graphite electrode. Breck (1973) has countered that as the graphite was used only as an

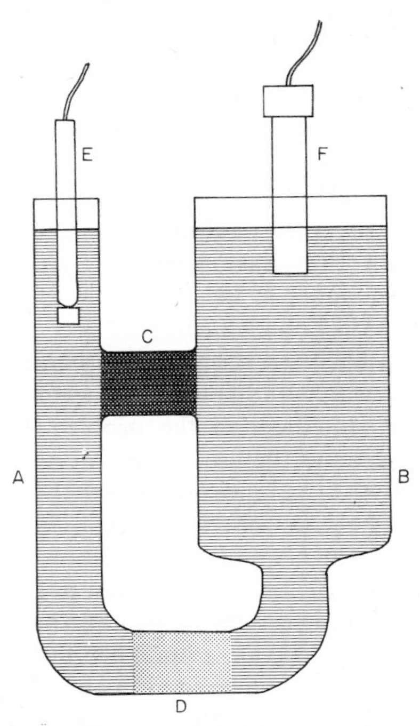

FIG. 20.14. Schematic diagram of Breck's equilibrating redox electrode (Breck, 1972). A. Limb containing standard redox couple (O^*/R^*). B. Limb containing a large sea water sample with a dominant redox couple (O/R). C. Graphite plug. D. Liquid junction. E. Platinum electrode. F. Reference electrode.

electronic conductor the problem of reversibility does not arise since the graphite is not used at any point as a measuring electrode. The validity of the objection raised by Ben-Yaakov and Kaplan (1973a) can be seen more clearly if we consider briefly the processes that occur as soon as the sea water is poured into the right-hand limb (Fig. 20.14). The potential established at the right hand end of the graphite plug will depend on the reversibility of the various competing redox couples at the graphite surface. The potential will be dominated by the most reversible couple, and the actual value of the potential will depend on the exchange current density that the particular couple can exert at the interface (see the discussions by Stumm, 1967; Morris and Stumm, 1967). This in turn depends on the properties of the electrode itself. In sea water it is likely that the final potential will represent the resultant of a number of irreversible processes, and will consequently depend largely on the characteristics of the graphite surface. Once this potential has been established the same potential will be established at the graphite surface in contact with O*/R*. However only if this reference couple behaves reversibly at the graphite–solution interface will the O*/R* ratio in the bulk solution accurately reflect this potential. If the couple is irreversible the O*/R* ratio, and hence the potential of the platinum electrode, will depend on the kinetics of the charge–discharge process at the graphite–solution interface. The electrode behaviour in the reference compartment is, in effect, the mirror image of that in the sea water compartment. Seen in this light the new procedure has few advantages over the direct use of a platinum (or graphite) electrode. In fact, a number of disadvantages arise because the electrode takes several hours to equilibrate and shows a marked insensitivity to changes in the environment. Direct attempts to measure the redox potential by determining the concentrations or activities of the oxidized and the reduced forms of particular redox couples (e.g. Sb(III)–Sb(V), I^-–IO_3^-) are likely to yield more useful results.

20.4. Net Current Flow in Quiescent Solutions—Polarography

20.4.1. Analysis of Raw Sea Water

In principle, any component in sea water that can be oxidized or reduced at the mercury electrode can be determined by polarography. Sea water is a good supporting electrolyte for polarographic work since it has a high ionic strength and contains surface active agents which should help to suppress polarographic maxima (Kečkeš and Pucar, 1966). Wherever possible it is preferable to determine the electroactive component directly in sea water without prior extraction. In this way, the risk of contamination during sample manipulation is avoided and the analyses can be carried out on small samples which can be processed rapidly. Such direct procedures also open up the possibility of *in situ* analyses for trace components.

One of the more troublesome aspects of polarography is the need to de-oxygenate the samples since oxygen is reduced at the dropping mercury electrode in two irreversible stages (Bond, 1973a). In the first stage ($E_{\frac{1}{2}} = -0{\cdot}25$ V versus the saturated calomel electrode) oxygen is reduced to hydrogen peroxide in acid media,

$$O_2 + 2H^+ + 2e^- \rightleftharpoons H_2O_2 \qquad (20.48)$$

and to hydrogen peroxide and hydroxide ions in neutral or alkaline media,

$$O_2 + 2H_2O + 2e^- \rightleftharpoons H_2O_2 + 2OH^-. \qquad (20.49)$$

In the second stage ($E_{\frac{1}{2}} = -0{\cdot}5$ V to $-1{\cdot}3$ V versus the saturated calomel electrode) oxygen is reduced to water in acid media,

$$O_2 + 4H^+ + 4e^- \rightleftharpoons 2H_2O \tag{20.50}$$

and to hydroxyl ions in neutral or alkaline media

$$O_2 + 2H_2O + 4e^- \rightleftharpoons 4OH^-. \tag{20.51}$$

The sensitivity of the mercury electrode to oxygen in solution can be used to good effect to provide a relatively simple means of monitoring oxygen partial pressure (Section 20.7). When other constituents are being measured in solution the oxygen reduction peaks are a nuisance. If the concentrations of the other electroactive components are high and their half wave potentials do not coincide with those of oxygen, then it is often feasible to carry out measurements in the presence of oxygen with little loss in sensitivity (Bond 1973a). In direct sea water analysis the components of interest are usually present in trace amounts, and the oxygen reduction peaks will give rise to irregular base lines and to a loss of sensitivity. In addition, the release of hydrogen peroxide at the mercury surface and the rise in pH associated with the oxygen reduction processes can seriously affect the electrochemical behaviour of heavy metals so that their reduction waves may be severely distorted or even suppressed. In unbuffered systems in the presence of chloride, hydrogen peroxide and hydroxide ions are produced even in the absence of an applied potential because of the reaction,

$$O_2 + 2Hg + 2Cl^- + 2H_2O \rightleftharpoons H_2O_2 + Hg_2Cl_2 + 2OH^- \tag{20.52}$$

In trace analysis, it is therefore advisable to de-oxygenate the solution before introducing mercury into the cell.

The normal de-oxygenation procedure is to purge the solution for fifteen minutes with a rapid stream of oxygen-free nitrogen that has passed through a presaturator containing the supporting electrolyte to prevent concentration changes in the cell caused by the evaporation of water. This is the most time consuming step in the analysis. With sea water this procedure removes carbon dioxide as well as oxygen and the resulting rise in pH (approximately 0·5 units after twenty minutes vigorous bubbling) can, on occasion, significantly alter the characteristics of the other electroactive components (see e.g. Odier and Plichon, 1971). Procedures for overcoming this difficulty include deoxygenation with specially prepared nitrogen–carbon dioxide mixtures, adding auxiliary pH buffers to the sample or using chemical reagents such as ascorbic acid to remove oxygen from the mixture (Florence and Farrar, 1973).

Since the concentrations of electroactive components are normally too low for conventional dc polarography (see Fig. 20.3) a number of instrumental techniques have been developed to enhance the sensitivity of the method (see the discussions by Meites, 1965; Schmidt and von Stackelberg, 1963; Maienthal and Taylor, 1968; and Kambara, 1966). The aim of these procedures is to increase the ratio of the analytically useful diffusion current (i_D) to the capacitance current (i_c) which is associated with the growth of the mercury drop. This ratio is at a maximum towards the end of the drop life. All cells have to be thermostatted during the analysis (usually at $25 \pm 0{\cdot}1$ °C) since the temperature coefficient of the diffusion current is around 2%/°C.

Whitnack (1961, 1966) suggested the use of *cathode ray polarography* for the direct analysis of a number of trace constituents (copper, lead, zinc, cadmium, nickel, cobalt, manganese, iodate and chromate) in untreated sea water samples. This form of polarography (Rooney, 1963) is closely related to voltammetry at a stationary electrode (Section 20.2.3) since the whole of the voltage sweep is applied during the life of a single drop. Each electroactive component is characterized by a peaked wave at the appropriate potential (Meites, 1965; Schmidt and von Stackelberg, 1963; Rooney, 1966). When used in the *derivative* mode the rate of change of current with potential is plotted so that the individual peaks stand out more clearly against a uniform background and much better resolution is possible. In a typical example, the mercury drop has a life of 7 s and the voltage sweep is applied during the last 2 s of the drop life when the drop size is approximately constant and the background currents are stable. Because of the transient nature of the current–voltage signal the whole trace is recorded on a storage oscilloscope and a complete polarogram is recorded every 7 seconds. The procedure is rapid and has a higher sensitivity and resolution than dc polarography. It is possible to look at small changes in solution composition by running the polarograph in the *subtractive* mode (Davis and Rooney. 1962). Two cells are mounted side by side with synchronized mercury drops. One cell contains a reference solution (e.g. standard sea water) and the second a working solution of slightly different composition. Signals from the two cells are recorded simultaneously from synchronized voltage sweeps and the *difference* between the two signals is displayed on the oscilloscope. In this mode the instrument can be tuned to a high sensitivity and it is possible to discern small changes in composition that would not be distinguishable in the trace from a single cell.

Whitnack (1961) used a Davis Differential Cathode-Ray "Polarotrace" (Model A1660, Shandon-Southern Analytical Instruments Ltd., England). The cell contained about 2 ml of solution which was flushed with nitrogen for three minutes before recording the polarogram. Using a dropping mercury working electrode and the mercury pool both as counter electrode and reference electrode he was able to show that at least seven metals (Cu, Pb, Cd, Zn, Ni, Co, Mn) give useful peaks in sea water as a supporting electrolyte so long as their free concentrations are greater than 5×10^{-8} M. The half wave potentials reported are shown in Table 20.6. This report was exploratory and no attempt was made to assess the precision or the accuracy of the method. Although the sensitivity is not high enough for a direct determination of most of the metals in pelagic sea water the technique looks promising

TABLE 20.6

Polarographic half wave potentials in untreated sea water[a].

Component	$E_{\frac{1}{2}}$ (V)					
	b	*c*	*d*	*e*	*f*	*g*
Cd(II)	−0·35	−0·67	—	−0·67	—	—
Co(II)	−1·44	−1·46	—	—	—	—
Cu(II)	+0·17	−0·25	−0·26	−0·23	—	—
Mn(II)	−1·56	−1·58	—	—	—	—
Ni(II)	−1·06	−1·17	—	—	—	—
Pb(II)	−0·14	−0·50	−0·51	−0·57	—	—
Zn(II)	−1·01	−1·09	—	−1·05	−1·03	−1·00
IO_3^-	—	—	—	−1·09	−1·09	−0·30

[a] Potentials versus the saturated calomel electrode (SCE) recorded at the dropping mercury electrode:
SCE versus mercury pool electrode = −0·058 V at 35‰ salinity (Whitnack, 1966)
SCE versus silver–silver chloride electrode = −0·005 V at 35‰ salinity (Gilbert and Hume, 1973).
[b] Whitnack (1961).
[c] Whitnack (1966).
[d] Odier and Plichon (1971).
[e] Whitfield, unpublished measurements with Shandon–Southern A3100 pulse polarograph.
[f] Liss *et al.* (1972); Herring and Liss (1974).
[g] Petek and Branica (1969a, b) at pH 3.

for inshore or estuarine waters in which higher concentrations might be expected. If the mercury drop/mercury pool cell is replaced by a silver–silver chloride pair then two peaks are observed for sea water at +0·27 V and +0·38 V that are proportional to the chlorinity of the sample. The peak heights are insensitive to stirring and the procedure is rapid but the resolution is coarse (Whitnack, 1961).

Using the cathode-ray polarograph in the subtractive mode, together with a standard addition technique, Whitnack (1966) was able to extend the analyses for the same array of metals down to about 10^{-9} M (Table 20.6). Results were reported for samples of Atlantic, Pacific and Arctic waters but no assessment was made of the accuracy and precision of the procedures. Iodate ions were shown to give a clear peak in untreated sea water at a potential very close to that of zinc. Whitnack (1973) also indicated that it is possible to determine separately the concentrations of Cr(III) and Cr(VI) and of As(III) and As(V) in sea water if these elements are present in sufficiently high concentrations. Such analyses might provide interesting sidelights on the ever present problem of assessing the effective oxidation–reduction potential of sea water.

The polarographic traces presented (Whitnack, 1966, 1973) indicate that some skill is required in locating and measuring the peaks of many metals and suggest that this technique is stretched to its limits in untreated sea water. Whitnack's exploratory studies, however, have indicated the wide range of potential applications of polarographic methods in oceanography. These various applications are now being considered in some detail with different instrumental techniques.

Odier and Plichon (1971) used ac polarography to analyse copper directly in sea water and to determine its chemical form. In ac polarography (Breyer and Bauer, 1963), an alternating voltage of small amplitude is superimposed on the normal linear voltage ramp. The resulting alternating current is measured as a function of the applied voltage. The trace is built up from the readings taken on successive drops and the discharge of the various electroactive components results in peaked waves against a fairly uniform background (Schmidt and von Stackelberg, 1963; Meites, 1965). The ac current is readily amplified and noise can be effectively filtered from the signal thus enhancing the sensitivity. The technique is some 50 times more sensitive than conventional dc polarography. Because of the rapid changes in the direction of current flow at the mercury surface this procedure is not particularly sensitive to irreversible electrode reactions which exhibit a low net current flow per unit concentration (Berge and Brügman, 1972). ac polarography is therefore less sensitive than are other polarographic techniques to the presence of dissolved oxygen (Bond and Canterford, 1971; Bond, 1973a).

Odier and Plichon (1971) used a Tacussel (Solea, Lyon, France) PRT 30-01 low noise potentiostat with a Tacussel UAP2 sinusoidal signal generator. A 27Hz ac voltage was applied with an amplitude of 10 mV and the baseline voltage was swept at the rate of 0·25 to 0·5 $mV\,s^{-1}$. They were able to estimate the stability constants of the $CuCl^+$ and $[Cu(HCO_3)_2OH]^-$ complexes by observing the shift in half wave potential (Fig. 20.2, equation (20.5)) resulting from the addition of the complexing agent (see Crow, 1969). Although the sensitivity of the method can be extended somewhat by de-oxygenating the samples with CO_2 to reduce the pH to 5 they set a lower limit of 5×10^{-8} M for the direct analysis of copper in sea water. Standard addition procedures are feasible for the determination of copper in estuarine or inshore waters in which the concentrations are likely to be above this threshold. No information is given from which the accuracy or precision of the procedure may be assessed.

Berge and Brügman (1972) have described an indirect procedure for the determination of fluoride in sea water using ac polarography. An acidic solution of the zirconium–Alizarin S complex (reagent A) is added to the sea water sample and the resulting hydrofluoric acid reacts with a portion of the complex to release some free alizarin into solution according to the reaction

$$4HF + \text{(zirconium–Alizarin S complex, } SO_3Na\text{)} \longrightarrow \text{(Alizarin S, } OH, OH, SO_3Na\text{)} + H_2(ZrOF_4) \qquad (20.53)$$

The ease with which the quinonoid form of the Alizarin S molecule is reduced to the hydroquinoid form at the mercury electrode enables the Alizarin S concentration and hence the fluoride concentration to be determined polarographically.

After addition of the reagent solution the sea water sample is allowed to stand for 2 hr in a water bath at $25 \pm 0{\cdot}1°C$ to allow reaction (20.53) to go to completion. The high acidity of the mixture should ensure that most of the complexed fluoride is taken up in this reaction. The mixture is then placed in the polarographic cell which has a dropping mercury working electrode, a mercury pool counter electrode and a saturated calomel reference electrode. The applied potential is swept from $-0{\cdot}05$ to $-0{\cdot}4$ V (versus the saturated calomel electrode) at a rate of $1{\cdot}7\,mV\,s^{-1}$. The superimposed ac potential has an amplitude of 20 mV and a frequency of 78 Hz. Degassing is not necessary since the oxygen waves are irreversible and produce very small ac signals. The peak current of the Alizarin S reduction is linearly related to the fluoride concentration over the range 5×10^{-6} M to 1×10^{-4} M with a slope of $207\,\mu A\,mmol^{-1}$. The maximum fluoride concentration normally found in sea water (8×10^{-5} M) falls well within this range. The procedure is quite sensitive to changes in fluoride concentration even though only one Alizarin S molecule is released for every four fluoride ions reacting (equation 20.53). The only significant interference in sea water is provided by sulphate ions which are also capable of displacing zirconium from the complex at this acidity. The sulphate ions in sea water will in effect add to the reagent blank which is composed partly of the sulphate ions introduced with reagent A and partly of the slight excess of Alizarin S introduced with this mixture to ensure a linear calibration curve down to low fluoride concentrations. Since there is a linear relationship between the sulphate concentration in sea water and the Alizarin S peak height a correction can be applied by estimating the sulphate concentration from the salinity of the sample. Replicate analyses gave precisions of 0·6% for an artificial sea water (42 analyses) and 0·7% for North Atlantic sea water (40 analyses). Since the initial reaction between hydrofluoric acid and the zirconium complex (equation 20.53) takes two hours to complete, the standard addition procedure cannot be used and so the concentrations must be estimated from a prepared calibration chart. Although the waiting time between sample preparation and analysis is long, the polarographic procedure itself is simple and rapid and the authors claim that a single operator can complete 50 analyses within four hours. This procedure appears to be more precise, although more cumbersome than alternative potentiometric procedures (Section 20.3.1.3 and 20.3.3.2). Only extensive use will decide which is the more convenient in practice. Whether or not it becomes a routine procedure the principle of indirect analysis is a useful addition to the polarographic armoury.

Branica *et al.* (1969) have used derivative pulse polarography to study the chemical speciation of a number of metals in sea water. In this procedure (Brinkman, 1968) square wave voltage pulses of a uniform magnitude (usually 20 to 50 mV) are superimposed on the linear voltage ramp. A single pulse of about 40 ms duration is applied towards the end of the life of each drop at a point where the background currents associated with the growth of the drop are almost constant and the capacitance current has decayed to a negligible value. The increase in current resulting from the application of the pulse is integrated over the last 20 ms of the pulse duration. Since successive drops are always sampled at the same instant in their life the resulting train of pulses acts as a stroboscopic filter making the dropping mercury electrode appear as a solid electrode of fixed geometry but with a renewable and highly reproducible surface. As the voltage sweep proceeds, the trace is revealed as a series of current peaks against a uniform background. The limit of detection of this procedure is about 10^{-9} M and, under favourable circumstances, peaks as little as 25 mV apart can be resolved. More rapid analyses with a high reproducibility may be obtained by knocking the mercury drop off the electrode after a fixed time interval rather than allowing it to break off naturally. One disadvantage of the procedure is that it is particularly sensitive to the presence of dissolved oxygen.

Branica and his co-workers have used a Shandon-Southern A3100 pulse polarograph and have generally examined sea water samples enriched with the appropriate trace components (to approx. 10^{-4} M) to determine the dominant chemical forms under near natural conditions. The chemistry of indium (Barić and Branica, 1969), zinc and cadmium (Barić and Branica, 1967; Piro *et al.*, 1973) have been studied in this way. Kečkeš and Pucar (1966) suggested, following the observations of Whitnack (1961, 1966), that zinc and iodate ions might be present in natural sea water at sufficiently high concentrations for direct analysis by pulse polarography. At the normal sea water pH the peaks attributed to zinc and iodate lie very close together so that some procedure must be adopted to separate them. Iodate was determined in the presence of zinc by taking advantage of the pH sensitivity of the half wave potential of iodate under acid conditions (Petek and Branica, 1969; Branica and Petek, 1969a,b). According to the method outlined by Branica (1970) the sea water sample is acidified with hydrochloric acid to pH 3 and placed in the polarographic cell with a dropping mercury electrode and a mercury pool counter and reference electrode. After oxygen has been removed by nitrogen bubbling, the iodate peak at $-0{\cdot}3$ V and the zinc peak at $-1{\cdot}0$ V (versus mercury pool) are scanned at the rate of $4{\cdot}8\,\mathrm{mV\,s^{-1}}$ with a forced mercury drop time of 2 s and a pulse height of 35 mV (Petek and Branica, 1969b). The concentrations of iodate and zinc are then estimated by the standard addition procedure. This technique has been used for measurements on a range of samples from the Northern Adriatic (Petek and Branica, 1969b) and for detailed studies of the speciation of zinc in the marine environment (Piro *et al.* 1973). The Adriatic samples gave values ranging from 28 to 141 $\mu\mathrm{g\,l^{-1}}$ for zinc and from 36–172 $\mu\mathrm{g\,l^{-1}}$ for iodate.

Liss *et al.* (1973) have used pulse polarography to determine iodate and iodide in sea water in an attempt to assess the redox potential characteristic of sea water (Section 20.3.4). They eliminated the zinc peak by the addition of Na–EDTA to a final concentration of $2{\cdot}5 \times 10^{-4}\,\mathrm{M}$ (Herring and Liss, 1974). A Princeton Applied Research polarograph (PAR Model 174) was used in conjunction with a three electrode cell containing a dropping mercury working electrode, a platinum wire counter electrode and a calomel reference electrode. Two filtered subsamples were taken. One was deaerated for 15 minutes with oxygen-free nitrogen after the addition of the EDTA. This long equilibration time is essential since zinc, like cadmium, reacts only slowly with EDTA in natural sea water (Branica *et al.*, 1969; Malijković and Branica, 1971). The polarogram was recorded at a scan rate of $2\,\mathrm{mV\,s^{-1}}$ over the range $-0{\cdot}5$ V to $-2{\cdot}0$ V (versus the saturated calomel electrode) with a forced drop time of 1 sec and a pulse amplitude of 25 mV. The iodate concentration was estimated by standard addition using the same instrument settings throughout. The procedure was shown to have a precision of $\pm 1\,\mu\mathrm{g\,l^{-1}}$ at the $40\,\mu\mathrm{g\,l^{-1}}$ level. Herring and Liss (1974) observed no detectable change in the half-wave potential of the iodate peak over the range pH 5 to 8 but they found that the peak height decreased by 25% per pH unit as the solution became more acid in this range. They suggested that this decrease in sensitivity is likely to become critical below pH 5. The pH shifts resulting from deaeration (increase in pH) and EDTA addition (decrease in pH) should therefore be kept as small as possible and only a five minute deaeration period is used between addition of standard aliquots. A more detailed study of the polarographic behaviour of iodate in sea water is required to reconcile the apparent contradiction between this behaviour and that described by Petek and Branica (1969b). Sea water off the coast of Southern California gave a range of iodate concentration from 38 to 60 $\mu g\,l^{-1}$ depending on the depth.

The second subsample was exposed to UV radiation for three hours (Armstrong *et al.*, 1966) to oxidize the iodide in the sea water to iodate. If the content of organic matter in the sample was high a few drops of 30% hydrogen peroxide were required to prevent loss of iodate. Iodide spikes in the range 5–20 $\mu g\,l^{-1}$ were recovered with a 99% efficiency using this procedure. The iodate content of the irradiated sea water (equivalent to iodate plus iodide in the original sample) was then determined using the procedure outlined above. The concentration of iodide in the original sample is equivalent to the difference in iodate concentration between these two subsamples. The standard deviation of the iodide determination was 1·4 $\mu g\,l^{-1}$ and the range observed off Southern California was from 1 to 14 $\mu g\,l^{-1}$. Iodate dominates over the low and erratic iodide levels at all depths and the general levels observed are in agreement with the earlier measurements of Tsunogai (1971).

Whitnack (1973) has extended his work using the cathode ray polarograph to include shipboard analyses of a number of ionic components in sea water including iodate and iodide. The iodate is measured directly in a 2 ml sample of sea water using the standard addition method without pretreatment to remove zinc interference. The iodide in the sample is then oxidized to iodate using chlorine water. The iodide plus iodate concentration is then determined polarographically and the iodide concentration calculated from the difference between the two results. The iodide values observed ranged from 1 to 25 $\mu g\,l^{-1}$ and those for iodate from 2 to 35 $\mu g\,l^{-1}$. These values are much lower than those observed by Petek and Branica (1969b) or by Liss *et al.* (1973). This might be attributed in part to the irreversibility of the iodate peak; the discharge process might be very inefficient at the high sweep rates used in cathode ray polarography. This irreversibility could probably be put to good effect since it should be possible to analyse for zinc in the presence of iodate using ac polarography or derivative pulse polarography with reverse sweep. These procedures will be relatively insensitive to the irreversible discharge of iodate but will respond in full to the reversible discharge of zinc ions.

Investigations so far have indicated that ten metals (As, Cd, Co, Cr, Cu, In, Mn, Ni, Pb, Zn) and two anions (IO_3^- and CrO_4^-) can give usable reduction waves at the dropping mercury electrode using untreated sea water as a supporting electrolyte. Only for iodate, and in some instances zinc, are the concentrations in the open sea high enough for direct analysis. Considerable attention is now being directed to coastal and estuarine waters and it is likely that situations will arise where the concentrations of these components are above the threshold for direct analysis by ac or pulse polarography while the background salinity is still high enough for direct determination. If this is so it is tempting to consider *in situ* analysis. The dropping mercury electrode works well over a range of pressures (Hills, 1965; Hills and Ovenden, 1966). Whitnack (1973) has successfully used polarographic techniques on board ship and Coleman *et al.* (1972) have described a dc polarographic method for the field analysis of lead, cadmium, zinc and iron in fresh water samples with a precision of $\pm 0{\cdot}1$ ppm. They used a standard addition technique and measured the current flowing at a predetermined potential on the diffusion limited current plateau (Fig. 20.2). A residual current compensation circuit was included so that the waves for each component could be measured using a two point reading cycle. Solutions were degassed using volatile fluorocarbons dispensed from a pressure pack.

Much of the technology is already available for the development of *in situ* methods. A wide variety of flow cells have been devised for use with the dropping mercury electrode (see Janata and Mark, 1967; Scarano *et al.*, 1970 and references quoted therein) and techniques have been described for the accurate and direct digital recording of current potential curves under field conditions (Goldsworthy and Clem, 1971, 1972; Clem and Goldsworthy, 1971) so that the data can be manipulated digitally without intermediate signal conditioning stages. Computer techniques are available for enhancing the sensitivity of the measurements by numerical filtering of the data (Keller and Osteryoung, 1971; Barker and Gardener, 1973; Hayes *et al.*, 1973), for the deconvolution of overlapping peaks (Gutknecht and Perone, 1970) and for the use of learning machines for the analysis of mixtures (Sybrandt and Perone, 1971). Since the signals can be recorded digitally a wide range of signal conditioning procedures can be applied including baseline compensation and digital subtractive polarography using a single electrode assembly (Clem and Goldsworthy, 1971). Multimode polarographs are now being designed with integral data processing units (Barker *et al.*, 1973). Many of these advances have been reviewed by Matson *et al.* (1972).

Rapid scan ac polarography (Bond, 1973a, b; Bond and Canterford, 1972b) would appear to be the most appropriate technique for *in situ* work since it is relatively insensitive to the presence of dissolved oxygen and the analyses are completed in a few seconds. The solution must, however, be strongly acidic or well buffered if hydrolysis effects (equations (20.49), (20.51) and (20.52)) are likely to be significant. Problems of temperature sensitivity, cell calibration and the mutual interference of electroactive components will have to be resolved as the individual analytical methods are developed.

20.4.2. ANALYSIS AFTER PRECONCENTRATION

For constituents with concentrations too low for direct analysis it is often convenient to preconcentrate the material by selective extraction from a large volume of water. Electrolytic extraction procedures will be considered in Section 20.5 and the present section will consider the polarographic analysis of selected components after a simple chemical extraction procedure. Although such procedures are less convenient than direct analysis they do give the analyst the opportunity to choose his own supporting electrolyte from the detailed compilations available (e.g. Meites, 1963) so as to maximize the sensitivity of the analysis and minimize the possible interferences. Relatively few analysts have taken full advantage of this opportunity.

Korkisch *et al.* (1956) extracted uranium from sea water as its negatively charged acetate and chloride complexes by passing sea water, buffered to pH 5 with an acetate buffer, through an Amberlite IRA-400 anion exchange column. The uranium was then eluted with strong acid (0·8 N HCl) and determined directly using the catalytic effect of uranyl ions on the reduction of nitrate at the dropping mercury electrode (Hecht *et al.*, 1956). A number of other components in the sea water also influence the reduction of nitrate ions so that the method is susceptible to many interferences and has not been widely used. Ishibashi *et al.* (1961) developed a rather more complicated procedure involving a number of extraction and purification steps with the uranium being determined by conventional dc polarography using the combined diffusion currents obtained from free uranyl ions and the uranyl complex of Mordant blue 2R. No details were given of the polarographic procedure.

Milner *et al.* (1961) removed the uranium from acidified sea water by solvent extraction, using di-(2-ethylhexyl)-phosphoric acid in carbon tetrachloride, followed by back extraction into 11 M HCl. After the wet oxidation of any remaining organic matter this extract was further purified by the extraction of uranium as uranyl nitrate into ethyl acetate from a solution saturated with aluminium nitrate. The remaining organic matter was then destroyed by wet oxidation and the uranium was taken up in a sodium tartrate supporting electrolyte. The efficiency of the overall extraction produced was checked by radiotracer techniques. The purified extract in the tartrate buffer was de-oxygenated with nitrogen for 15 min and then analysed by derivative pulse polarography. The potential was swept at a rate of

0·7 mV sec^{-1} from −0·1 to −0·7 V versus the mercury pool electrode and the uranyl ion peak was recorded at −0·39 V. The uranium concentration was determined by standard addition.

Copper and lead show peaks quite close to that of the uranyl ion (−0·21 V and −0·59 V respectively) but they do not lead to an error of more than 1% in the determination of 1 μg l^{-1} of uranium provided that the copper and lead concentrations do not exceed 1 μg l^{-1}. The separation of the peaks is dependent on the chloride ion concentration and the copper and uranyl peaks become unresolveable in solutions 1 M in chloride ions. The most serious interference arises from the presence of molybdate ions which give peaks at −0·18 V and −0·44 V linked by a saddle which tends to swamp the uranyl peak. The extraction procedure is designed to remove molybdate ions and to reduce the lead and copper concentrations to manageable proportions. The method is time consuming and cannot claim any advantages over the more direct isotope dilution procedures (see Section 19.10.5.29).

Tikhonov and Shalimov (1965) have described a method for determining nickel and manganese in sea water. The metals were extracted into chloroform from filtered sea water as their diethyldithiocarbamates. The extract was evaporated to dryness redissolved in hydrochloric acid and buffered with ammonia–ammonium chloride. The elements were determined by conventional polarography with the manganese step appearing at −1·66 V and the nickel step at −1·10 V versus the saturated calomel electrode.

Harvey and Dutton (1973) used radio tracer techniques to show that more than 90% of the radioactive cobalt added to sea water could be scavenged by co-precipitation with manganese dioxide under the influence of ultraviolet radiation. In the presence of 50 to 100 μg l^{-1} of manganese quantitative co-deposition of ^{60}Co occurred within eight hours using a 1l-Hanovia reactor fitted with a 2 W low pressure mercury vapour discharge lamp. The irradiation period had to be extended to 24 h to ensure complete decomposition of the stable cyanocobalamin complex. The manganese dioxide precipitate adheres to the quartz thimble surrounding the lamp and was washed off with 0·15 M HCl containing a trace of sulphur dioxide. The solution was evaporated to dryness and the residue dissolved in HCl and ammonia solution was added to give a supporting electrolyte 0·5 M in ammonium chloride and 0·5 M in ammonia. 0·1 ml of 1% ethanolic dimethylglyoxime was added to this medium to complex the cobalt. This solution was then analysed on the Shandon-Southern model A3100 pulse polarograph after five minutes deoxygenation with nitrogen. The polarograph was used in the derivative mode with a 5·0 mV pulse height and a 1 s drop life. The voltage sweep was begun at −1·0 V versus the mercury pool electrode and the cobalt peak appeared at −1·10 V. The concentration of cobalt in the sample was determined by standard addition. Some difficulty was experienced in obtaining a reproducible blank reading. High blank readings result from traces of cobalt in the reagents (in particular the hydrochloric acid) and a negative bias is introduced at low concentrations apparently because of trace organic components in the ammonia that form stronger complexes with cobalt than does dimethylglyoxime. The detection limit of the method is around 0·6 ng Co l^{-1} and the recovery of added cobalt was effectively constant (about 95·7 %) up to 0·5 μg l^{-1}. Above this level the recovery rate decreased to 80% at 2 μg l^{-1}. If high levels of cobalt are experienced it might therefore be necessary to repeat the scavenging process to remove the cobalt quantitatively. This procedure could be extended to other elements co-precipitated with manganese under the influence of ultraviolet light.

The unique advantages of the polarographic method are more clearly demonstrated by the ingenious procedure developed by Abdullah and Royle for the determination of copper, lead, cadmium, nickel, zinc and cobalt in a single aliquot of sea water (Abdullah and Royle, 1972; Abdullah *et al.*, 1972). They extracted these elements from ten litres of filtered sea water (0·45 μm membrane filter) on to a chelating ion-exchange resin (Chelex-100 in 5 cm × 1 cm columns) in the calcium form. The columns were then washed with distilled water and eluted with 70 ml of 2 M nitric acid. The eluate was irradiated for one hour under a 1 kW mercury vapour lamp to destroy any organic matter present. The solution was then evaporated to dryness several times with hydrochloric acid to remove the nitrate and the final residue was acidified with hydrochloric acid and made up to 25 ml. The resulting solution was approximately 0·2 M in calcium chloride with only minor amounts of sodium, potassium and magnesium and acted as a supporting electrolyte for the first stage of the analysis. A 2·5 ml aliquot of this solution was placed in the polarographic cell and after de-oxygenation a derivative polarogram was recorded using a Shandon-Southern model A 3100 pulse polarograph. The potential was scanned over the range −0·2 V to −1·1 V (versus the mercury pool anode) with a 50 mV pulse and a 1 sec drop life for the determination of copper (−0·45 V), lead (−0·70 V) and cadmium (−0·92 V). 0·2 ml of a standard containing 10 μg l^{-1} each of copper lead and cadmium was then added and a second polarogram was recorded. 0·6 ml of a 2 M ammonia solution was added to the same sample in the cell making the solution 0·2 M with respect to the ammonia–ammonium chloride buffer. A further polarogram was recorded between −0·8 V and −1·7 V for the determination of nickel (−1·15 V) and of zinc (−1·43 V). The cobalt wave in this medium (−1·40 V) coincides with the zinc wave. However the cobalt reduction is irreversible whereas that for zinc is reversible so that the zinc concentration can be accurately determined in the presence of cobalt by employing a reverse potential sweep. A second polarogram was recorded in this medium after the addition of 0·2 ml of a standard solution containing 10 $\mu g\,l^{-1}$ nickel and 100 $\mu g\,l^{-1}$ zinc.

Finally, to the same sample in the cell, 0·2 ml of a 1% ethanolic solution of dimethylglyoxime was added and a polarogram was recorded between −1·1 V and −1·5 V for the determination of cobalt (−1·44 V). A final polarogram was recorded after the addition of 0·2 ml of a standard solution containing 1 $\mu g\,l^{-1}$ of cobalt. The concentration of the various components could be calculated from the recorded peak heights making due allowance for the dilution of the sample by the addition of the reagents and the standard solutions. The results presented suggest that the method should give a precision of about 6% for the determination of copper, lead and cobalt and about 3% for cadmium, nickel and zinc.

The fact that six metals can be determined in a single aliquot indicates the flexibility of the polarographic method and the ease with which it is possible to alter the polarographic characteristics of a sample by simple chemical manipulation. This is both the strength and the weakness of the polarographic method. In well characterized samples in the laboratory it enables mutual interferences to be eliminated by alterations in the pH of the medium or by selective complexing of the various components (see for example Nakagawa and Tanaka, 1972). In field measurements, however, it implies that marked changes in the supporting electrolyte, or in the trace metal pattern, might give rise to large variations in the polarographic characteristics of the system. These changes may not always be immediately apparent unless a wide variety of parameters is being monitored simultaneously.

20.5. Net Current Flow in Quiescent Solutions Following Electrolytic Preconcentration—Stripping Analysis

20.5.1. Introduction

In stripping analysis (Shain, 1963; Barendrecht, 1967; Ellis, 1973; Zirino *et al*., 1973) trace constituents are concentrated onto a stationary electrode from a rapidly stirred, de-oxygenated solution by controlled potential electrolysis (voltammetric or *pre-electrolysis phase*). After a suitable period of electrolysis the stirring is stopped and the solution is allowed to become quiescent with the potential still applied (*rest phase*). The sign of the potential is then reversed and the material deposited during the pre-electrolysis phase is stripped off using some suitable amperometric technique (*stripping phase*). The various constituents in the sample will give peaks at characteristic potentials in the resulting current versus potential trace. The heights of these peaks can be used after appropriate calibration to determine the concentrations of the constituents in the original solution. Anodic stripping techniques are more sensitive than are polarographic methods to the presence of dissolved oxygen because the metals, once plated, are susceptible to spontaneous reoxidation and removal from the amalgam (Lang, 1973). It is normal for nitrogen bubbling to precede and to accompany the pre-electrolysis phase and for nitrogen to be flushing over the surface of the solution during the rest phase and the stripping phase. The current produced in the stripping phase is also sensitive to changes in solution temperature and the cells are normally thermostated at $25 \pm 0{\cdot}1°C$.

Although a wide range of electrode substrates can be used in stripping analysis (Adams, 1969; Brainina, 1971) oceanographic applications have been restricted almost exclusively to the use of mercury electrodes either in the form of a stationary drop or as a thin film on a carbon substrate. Such electrodes are relatively simple to prepare, are stable and reproducible in their behaviour and enable a range of metallic ions to be studied. Stripping analysis with mercury electrodes is only applicable to those metals which are capable of forming amalgams. Since mercury is readily oxidized in the anodic region the applications have been largely confined to cathodic deposition followed by anodic stripping. In sea water, the useful range of the mercury electrode is from $-1{\cdot}3$ V to $-0{\cdot}05$ V versus the saturated calomel electrode. A useful summary of the problems involved in the application of stripping analysis to sea water has been given by Whitnack and Sasselli (1969).

The pre-electrolysis phase serves to concentrate the electroactive constituents on to the mercury electrode and thus provides a marked increase in the sensitivity of the subsequent analysis. For a simple spherical electrode (e.g. a hanging mercury drop electrode, HMDE) the solution of Fick's second law of diffusion gives the concentration ratio (r_c) as (Shain, 1963),

$$r = C_a/C_s = 3i_c t/4\pi nFr_0^3, \tag{20.54}$$

where C_a is the concentration of the metal in the amalgam after pre-electrolysis, C_s is the concentration in the solution, i_c is the pre-electrolysis current, r_0 is the radius of the electrode and t is the pre-electrolysis time. In practice, a concentration ratio of about one hundred is commonly achieved at this stage. By careful selection of the pre-electrolysis potential it is possible to obtain some degree of separation between the electroactive components in the sample solution. However, the greatest control over the selectivity of the procedure is exercised in the stripping phase. The potential at which a particular metal plates out will depend not only on the ionic medium, but also on the concentration of the amalgam produced at the working electrode. Polarographic half-wave potentials (Meites, 1963; 1965) refer to the equilibrium condition where the activity of the metal ion in the solution is equal to the activity of the metal in the amalgam. These potentials therefore only provide a rough guide to the appropriate plating potentials, and Shain (1963) has recommended that the actual plating potential should not be less than $0{\cdot}12/n$ volts cathodic of the polarographic half wave potential observed in the plating medium. In conventional electrochemical cells, stoichiometric removal of the selected electroactive constituent is a protracted business and the problem becomes more acute the more dilute the solution. For trace analysis it has become the usual practice to plate out only a fraction of the electroactive species present. The pre-electrolysis is carried out for a fixed period of time at a carefully monitored potential in a cell with a rigidly reproducible geometry.

The sensitivity of the anodic stripping procedure can be enhanced by making the amalgam concentration as high as possible by employing very small mercury electrodes, long electrolysis times and a rapid stirring rate (equation 20.6). For a typical hanging mercury drop electrode (volume $10^{-3}\,cm^3$) this can involve pre-electrolysis times of 15 minutes for 10^{-6} M solutions and 60 min for 10^{-9} M solutions. Throughout this period accurately reproducible rates of stirring must be maintained while applying a precisely controlled potential. To reduce the pre-electrolysis time the mercury is sometimes deposited as a thin film on a carbon electrode. When the conditions for the pre-electrolysis phase are being considered a balance must be struck between the speed and the sensitivity of the analysis (equation 20.6). The procedure must be calibrated using a series of spiked samples to ensure that a fixed proportion of the electroactive component is being accurately recovered. The concentration of the metal in the sample (C_x) can then be calculated from the concentration of the standard (C_s) using the equation

$$C_x = i_1 v C_s/(i_2 v + (i_2 - i_1)\,V), \tag{20.55}$$

where i_1 is the original peak height, i_2 is the "spiked" peak height, v is the volume of the standard added and V is the original sample volume.

The relationship between the efficiency of recovery of the spike and that of the original electroactive species will depend to some degree on the relationship between the chemical forms of the original and spiked material. If metal in the ionic form is used in the spike its recovery characteristics might be quite different from those of metal in the original sample which is present either in a colloidal form or as a strongly bound organo-metallic complex (Piro *et al.*, 1973; Zirino *et al.*, 1973). This discrepancy if unrecognized can lead to erroneous results. However, if used carefully such differences in behaviour can be used to gain a more detailed understanding of the distribution of trace metals between various chemical forms (Piro *et al.*, 1973; Zirino *et al.*, 1973). The use of thin layer cells with very small sample volumes (Reilley, 1968; Hubbard and Anson, 1970; Hubbard, 1973) enables the stoichiometric recovery of electroactive components to be achieved in a very short time. The use of such cells with thin film mercury electrodes might repay further investigation in the marine context. They would certainly provide a natural starting point for the investigation of cathodic stripping analysis from alternative electrode substrates (Brainina, 1971).

In the rest state, the amalgam concentration tends to become more homogeneous throughout the mercury drop or film. Intermetallic compounds (e.g. Ni–Zn, Zn–Cu, Cd–Cu: Shain, 1963; Neiman and Brainina, 1973; Kozlovskii and Zebreva, 1972) may form if several metals have been plated out simultaneously. The resulting stripping curve will consequently contain additional peaks representative of the intermetallic compounds and the peak heights for the pure metals will be correspondingly depressed. These problems become most severe if the sensitivity of the method is enhanced by the production of very high amalgam concentrations by prolonged electrolysis or by the use of thin film mercury electrodes. The formation of intermetallic compounds effectively sets the upper concentration limit for anodic stripping analyses at 10^{-5}M.

The stripping process itself provides a further enhancement of the overall sensitivity of the method which is comparable to that obtained in the pre-electrolysis phase. Normally the metals in the amalgam are re-oxidized by applying a linearly varying voltage ramp to the mercury electrode at a rate of about $30 \mathrm{mV s^{-1}}$ and monitoring the resulting current. This simple procedure has a number of drawbacks when used with the stationary mercury drop electrode. The dc signal obtained during the stripping process cannot be amplified very greatly since one would simultaneously amplify the accompanying background noise. However, for mercury drop electrodes the peak current is proportional to the square root of the rate of the voltage scan (Reinmuth, 1961; de Vries, 1965) so that some increase in sensitivity is possible by scanning the potential span very rapidly. However the background current itself is *directly* proportional to the scan rate so that it gradually swamps the Faradaic current at high sweep rates. The current peaks also become distorted and tend to overlap at high scan rates because the metal atoms in the amalgam take a finite time to migrate to the mercury surface where they can be discharged (Roe and Toni, 1965). Considerable enhancement in the sensitivity of anodic stripping can be obtained by modulating the linear stripping ramp with an ac voltage or a series of square wave pulses and amplifying the resulting modulated current signals, leaving the background noise unaffected.

If the resulting signals are recorded digitally a wide range of signal processing procedures can be applied (see Section 20.4.1).

The method of *medium exchange* removes a number of restrictions that may result when the pre-electrolysis and stripping steps are carried out in the same medium. Metals that might plate out at closely similar potentials in the sample medium may be resolved if the electrode is transferred to a different medium for the stripping process. High residual currents that may be associated with high concentrations of readily oxidizable species (e.g. organic compounds) in the sample medium can also be avoided. The procedure also offers the possibility of using anodic stripping voltammetry to determine the stability constants of metal complexes at very low metal concentrations. Care must be taken to avoid oxidation of the amalgam during the transfer process and the mercury electrode must be held at an appropriate potential to avoid spontaneous stripping of the trace constituents when it is first immersed in the new medium. Zieglerová *et al.* (1971) have described a flow cell to simplify this procedure.

20.5.2. HANGING MERCURY DROP ELECTRODES (HMDE)

Ariel and Eisner (1963) used a hanging mercury drop suspended from a mercury coated platinum wire (HMDE(Pt): Shain, 1961) together with a saturated calomel counter electrode to determine the concentrations of zinc ($-1{\cdot}00$ V) and cadmium ($-0{\cdot}7$ V) in Dead Sea brines. The presence of lead ($-0{\cdot}51$ V) and copper ($-0{\cdot}32$ V) was also confirmed, but no quantitative estimates were made. The HMDE(Pt) electrode contained approximately 20 mg of mercury and had a surface area of about 3 mm^2. No erratic effects were noted due to the dissolution of platinum in the mercury, and the electrodes were mechanically stable provided that the platinum surface was first given a coherent coating of mercury. Pre-electrolysis periods of 3–8 min were employed and the scan rate in the stripping phase was between 200 $mV\,s^{-1}$ and 800 $mV\,s^{-1}$. The stripping curves were recorded on a dual beam oscilloscope. If slower scan rates compatible with a strip chart recorder were used (approximately 3 $mV\,s^{-1}$), then a pre-electrolysis time of 30 min to 90 min was required to attain the same sensitivity. Because of the large volume of solution employed (50 ml) a de-oxygenation time of 20 min was required before the pre-electrolysis phase. The concentrations of zinc and cadmium found in the Dea Sea brine were $5{\cdot}7 \times 10^{-7}$ M ($\pm 2{\cdot}5$ %, 6 determinations) and $4{\cdot}4 \times 10^{-8}$ M ($\pm 0{\cdot}8$ %, 4 determinations) respectively. The cadmium peak was measured at a scan rate of 380 $mV\,s^{-1}$ and the zinc peak at 220 $mV\,s^{-1}$.

Copper could not be determined by this method because the copper peak merges with the mercury oxidation peak at high scan rates and prolonged electrolysis (80 min) is required to obtain adequate sensitivity at scan rates at which the mercury and copper peaks could be resolved (Fig. 20.15). Ariel *et al.* (1964) therefore developed a separate procedure for the determination of copper. The pre-electrolysis was carried out at −0·48 V so that only copper and a very small amount of lead were deposited (Fig. 20.15). The technique of

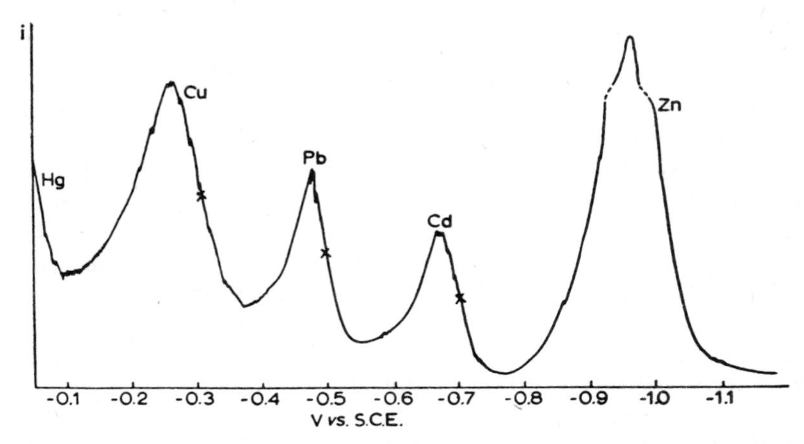

FIG. 20.15. Anodic stripping trace showing the presence of copper, lead, cadmium and zinc in Dead Sea brine. Linear scan rate 220 mV s^{-1}, pre-electrolysis time 80 minutes. Reproduced with permission from Ariel *et al.* (1964).

medium transfer (Section 20.5.1) was then used to determine the copper selectively. After pre-electrolysis the HMDE(Pt) electrode was transferred into a separate stripping medium (0·5 M ammonia, 0·5 M ammonium chloride) in which the copper reduction peak is sufficiently well separated from the mercury oxidation peak. The copper peak height was measured at −0·48 V *versus* the saturated calomel electrode after stripping at a scan rate of 220 mV s^{-1}. The concentration of copper in the Dead Sea brine was estimated to be $4{\cdot}5 \times 10^{-8}$ M with a standard deivation of $\pm 15\,\%$.

Whitnack and Sasselli (1969) used the HMDE(Pt) electrode to study procedures for the analysis of other trace metals in sea water. Their cell was constructed entirely of Teflon and fused quartz and they used a Teflon spoon mounted on a quartz rod to transfer the mercury drop from the dropping mercury electrode to the HMDE(Pt) electrode mounted in the same cell. A mercury pool at the base of the cell acted as both a counter and reference electrode. The various components were inserted through the Teflon lid *via* O-ring seals. For the determination of zinc a 50 ml de-oxygenated sample was electrolysed at −1·20 V versus the mercury pool electrode (0·2 V more negative than the peak observed using cathode ray polarography (Whitnack, 1966) for ten minutes with vigorous stirring (175 rev min^{-1} with a magnetic stirrer). After a 10 sec rest phase the potential was scanned from $-1{\cdot}20$ V to more anodic potentials at 16 mV s^{-1} and the resulting trace was recorded on a Sargent Model FS Polarograph. The analysis was then repeated with a fresh mercury drop, after the addition of a spike of standard zinc solution to the cell. Since nickel forms intermetallic compounds with zinc and shows a closely similar peak potential in natural sea water it is likely that the reported zinc concentration also included nickel. Copper ($-0{\cdot}3$ V), cadmium ($-0{\cdot}65$ V) and lead ($-0{\cdot}48$ V) peaks were also observed in analyses of raw sea water. The peak heights for the standard addition of these metals were proportional to their concentration over the range 10^{-7} to 10^{-9} M in sea water. For the simultaneous determination of copper, cadmium and lead the pre-electrolysis potential was fixed for 30 minutes at $-0{\cdot}80$ V to avoid the deposition of zinc, and the anodic scan was again made at 16 mV s^{-1}. At this scan rate a pattern of well formed and well separated peaks was obtained. Whitnack and Sasselli (1969) indicated that more rapid scan rates resulted in distorted and overlapping peaks. An average of five determinations on a sample of Pacific surface water gave $3{\cdot}0 \times 10^{-5}$ g l^{-1} for zinc (± 5 %) and $9{\cdot}0 \times 10^{-6}$ g l^{-1}, $1{\cdot}0 \times 10^{-6}$ g l^{-1} and $4{\cdot}0 \times 10^{-6}$ g l^{-1} for copper, cadmium and lead respectively (± 10 %). The results for copper and zinc were in reasonable agreement with simultaneous measurements made using cathode ray polarography (see Whitnack, 1966; Section 20.4.1). The value for zinc is substantially higher than that reported by other investigators using anodic stripping techniques, and that for lead is two orders of magnitude greater than currently accepted values.

Zirino and Healy (1972) attempted to improve the sensitivity of the HMDE(Pt) cell by using a subtractive technique for the anodic stripping phase (see Section 20.4.1.) with a cell containing two HMDE(Pt) electrodes The sizes of the mercury drops on the two electrodes had to be carefully matched. During the pre-electrolysis phase only one HMDE(Pt) electrode was used for amalgam accumulation. The stripping potential, however, was applied between both electrodes and the platinum counter electrode and the *difference* between the two stripping currents was monitored by the polarograph (a Heath Co., Controlled Potential Polagraph). With matched electrodes the residual current was therefore subtracted from the stripping signal and the oxidation currents of the electroactive species could be amplified without a concomitant increase in the background noise. As mentioned earlier (Section 20.4.1) modern digital techniques enable subtractive analysis to be carried out using a single working electrode with consecutive scans to measure the background current and the background plus sample current (Clem and Goldsworthy, 1971). The sensitivity limit of the procedure was enhanced to 0·005 μM for zinc and 0·001 μM for lead, cadmium and copper with a 5 min pre-electrolysis phase, a 60 s rest phase and a stripping scan of 8 $mV\,s^{-1}$. Such slow stripping rates enhance the resolution of the current peaks and tend to give a much flatter baseline. Carbon dioxide/nitrogen mixtures were used to de-oxygenate the samples and to reduce the pH to 5·8 at which interferences from Ni(II) and Fe(II) were negligible. The reported current sensitivities were surprisingly low compared to those of other anodic stripping procedures (Table 20.7).

The working cell was rigidly aligned in a metal framework to ensure geometrical stability during the plating and stripping cycles. The cell was only half-filled (25 ml) to prevent loss of sample and the scavenging of dust from the cell lid due to frothing during the purging of the sample.

The sample was de-oxygenated for 5 min and both HMDE(Pt) electrodes were run through the stripping scan to remove any contaminants that might have accumulated at the mercury surface. The stripping peak heights were calibrated by the addition of 0·1 ml aliquots of a solution containing 5×10^{-6} M zinc and copper and 1×10^{-6} M cadmium and lead. The solution was allowed to re-equilibrate for 15 mins after the addition of the standard, and the analysis was repeated using the *same* pair of HMDE(Pt) electrodes. Ariel and Eisner (1963) used fresh mercury drops after each standard addition since the conditions which they used in the stripping stage did not ensure the stoichiometric removal of the plated metals from the mercury drop. Since repeated plating and stripping analyses using the same electrodes produced reproducible zinc peak currents ($\pm 3\%$ at the 2·6 $\mu g\,l^{-1}$ level: Zirino and Healy, 1972) it would appear that their stripping process was nearly stoichiometric although this was not explicitly stated. The standard deviation for zinc analyses in sea water was 0·7 $\mu g\,l^{-1}$ for samples containing 2 to 6 $\mu g\,l^{-1}$. At the 1 $\mu g\,l^{-1}$ the coefficient of variation was 50% (Zirino and Healy, 1972). For lead a value of 0·9 $\mu g\,l^{-1}$ was recorded ($\sigma = 0{\cdot}3\,\mu g\,l^{-1}$) and for cadmium 0·4 $\mu g\,l^{-1}$ ($\sigma = 0{\cdot}1\,\mu g\,l^{-1}$). The peak potentials recorded are listed in Table 20.8. At concentrations up to 0·1 $\mu mol\,l^{-1}$ there appears to be negligible interaction between zinc, cadmium, copper and lead caused by intermetallic compound formation. The large analytical errors were attributed to problems of sample manipulation rather than to errors in the electrochemical analyses. A number of precautions were taken to minimize these errors. Before analysis all containers were rinsed with dilute nitric acid and deionized water and all surfaces were then equilibrated for several hours with subsamples of the sea water sample. Rapid pH changes must be avoided as they give rise to unpredictable desorption errors. The adsorption and desorption of zinc by a range of materials has been investigated by Bernhard *et al.* (1972).

TABLE 20.7

Summary of representative anodic stripping procedures based on the parameters used in the analysis of zinc in sea water.

Working[a] Electrode	Mercury deposit (mol cm^{-2})	Pre-electrolysis phase		Stripping phase			Reference
		duration (s)	stirring rate (rev min^{-1})	rate (mV s^{-1})	detection mode	current sensitivity (μA/μg l^{-1})	
HMDE(Pt)	—	300	200	8	dc	$1{\cdot}5 \times 10^{-4}$	Zirino and Healy (1972)
HMDE(Hg)	—	50	146	250	dc	$5{\cdot}1 \times 10^{-3}$	Macchi (1965)
HMDE(Hg)	—	600	—	7	dc	0·042	Smith and Redmond (1971)
HMDE(Hg)	2·22[c]	120	2400	2·5	ac	0·022[b]	Rojahn (1973)
MCGE	$<2{\cdot}5 \times 10^{-7}$	300	120[d]	16·5	dc	1·11[e]	Bradford (1972)
MCGE	8×10^{-6}	600	1000	3	dc	0·83	Lang (1973)
Flowing MCGE	$7{\cdot}5 \times 10^{-9}$	600	160[f]	17	dc	0·71	Liebermann and Zirino (1974)
GCE	10^{-7}	1800	2000	50	dc	0·32[b]	Florence (1972)
GCE with collector	—	480	2900	133	dc	2·70[b]	Laser and Ariel (1974)

[a] The different electrode types are described in detail in the text.
[b] Acetate added to sea water.
[c] Surface area (mm^2).
[d] Nitrogen gas flow rate (ml min^{-1}) used to stir solution.
[e] Tartrate added to sea water.
[f] Solution flow rate ml min^{-1}.

TABLE 20.8

TABLE 20.8. *Peak potentials*[a] *recorded by anodic stripping voltammetry at mercury electrodes in untreated sea water.*

	Hanging mercury drop							Thin film		
Component	*b*	*c*	*d*	*e*	*f*	*g*	*h*	*i*	*j*	*k*
Bi(III)	—	—	—	—	—	—		−0·09	—	—
Cd(II)	−0·60	−0·65	−0·70	−0·63	−0·61	−0·59	−0·70	−0·69	−0·62	−0·76
Cu(II)	−0·13	−0·31	−0·25	−0·15	−0·16	−0·16	−0·32	−0·24	−0·21	−0·34 to −0·42
In(III)	—	—	—	—	—	—		−0·58	—	—
Ni(II)	{−0·97, −0·20}	—	—	—	—	—	—	—	—	—
Pb(II)	−0·42	−0·48	−0·52	−0·43	−0·39	−0·39	−0·51	−0·48	−0·43	−0·60
Sb(III)	{−0·28, −0·97}	—	—	—	—	—	—	−0·16	—	—
Sn(II)	−0·61	—	—	—	—	—	—	—	—	—
Tl(III)	—	—	—	—	—	—	—	−0·62	—	—
Zn(II)	−0·97	−1·01	−1·00	−0·98	−0·99	−0·95	−1·00	−1·05	−1·05	−1·15

[a] Versus SCE, see footnote *a*, Table 20.6.
[b] Smith and Redmond (1971). They also reported the following peak potentials: Co(II) −0·22; Cr(VI) −0·58; V −0·14.
[c] Whitnack and Sasselli (1969).
[d] Macchi (1965).
[e] Zirino and Healy (1972).
[f] Rojahn (1972).
[g] Donadey (1969).
[h] Ariel and Eisner (1963) for Dead Sea brine.
[i] Florence (1972).
[j] Lang (1973) versus Ag/AgCl electrode in saturated KCl.
[k] Liebermann and Zirino (1973).

The procedure is almost insensitive to ship's motion and has been used by Zirino and Healy (1971) to study the distribution of zinc in the Northeastern Tropical Pacific. The high standard error in the procedure makes detailed interpretation of the results difficult. Zirino and Healy (1970) have used the shifts in the peak potential observed with changes in the stripping medium to investigate the stability of inorganic complexes of zinc in sea water.

A microcell based on a HMDE(Pt) electrode (Huderová and Štulík, 1972) and capable of analysing sample volumes as small as 0·01 ml might be useful for studying the interstitial waters extruded from sediments and cellular fluids taken from live animals. The mercury drop (approximately 5 μl) rests at the bottom of the cell and contact to the auxilliary electrodes is made via capillary salt bridges. De-oxygenation is extremely rapid and a pre-electrolysis time of 5 min is sufficient for stoichiometric removal of electro-active components from such a small volume. Consequently, variables such as stirring rate, pre-electrolysis time and temperature are not critical. The analysis can be performed rapidly with the minimum of auxiliary equipment.

Macchi (1964) employed an ordinary dropping mercury electrode with a capillary (380 mm × 0·01 mm) designed to give a very long drop life (68 s) to determine zinc directly in sea water (see also Branica, 1970). The cell was mounted in a specially made casting set in a massive steel frame and great care was taken to ensure the reproducible positioning of the capillary in successive experiments. The apparatus automatically stepped through a cycle of operations that was initiated by the fall of the previous mercury drop. The pre-electrolysis potential (−1·3 V versus the saturated calomel electrode) was applied for the first 50 s of the drop life and the stirring was then stopped for 10 s. The plated metal was then stripped off with a rapid, linear voltage scan (250 mV s^{-1}) and the resulting trace was recorded on a cathode ray polarograph (Model 451, AMEL, Milan). The recording circuit compensated for the background current associated with the growth of the drop and the charging of the double layer capacitance. The zinc peak was observed at −1·00 V versus the saturated calomel electrode. The procedure was rapid, simple and direct. The concentration of zinc in the sample was estimated by a standard addition procedure using spikes of zinc sulphate solution. Cadmium, copper and lead gave peak potentials of −0·7 V, −0·25 V and −0·52 V respectively (cf. Table 20.8) and did not interfere with the determination of zinc. Excessive concentrations of Co^{2+} and Ni^{2+} (more than twenty times the concentration of zinc) tended to distort the baseline. A range of other ions (Mn^{2+}, Cr^{3+}, Fe^{3+} and Al^{3+}) apparently did not interfere. The maximum sensitivity obtained for zinc was 3×10^{-9} M. A total of 15 min was required for each analysis including standardization and degassing. A concentration of 3 $\mu g\,l^{-1}$ was found for a Mediterranean sea water sample with a coefficient of variation of 4·5 %.

The procedure for analysis at the slowly dropping mercury electrode has been refined by Velghe and Claeys (1972). They increased the drop life to 18 min by constricting the capillary and therefore enabled much longer pre-electrolysis times to be employed. The sensitivity of the stripping phase was enhanced by the use of a Tacussel type PRG 3 three electrode alternating current polarograph with a saturated calomel reference electrode and a platinum wire counter electrode. This unit applied the ac potential with phase-sensitive modulation. When an alternating potential was applied, the phase of the Faradaic (or diffusion) current due to a reversible reaction was shifted 45° relative to the applied ac modulation, whereas that associated with the capacitance current was shifted 90°. The phase-sensitive detection system allowed the Faradaic current to be recorded selectively thus giving the stripping trace much higher sensitivity. The resolution of the stripping phase was much better than that of the corresponding linear dc procedures and the sensitivity towards cadmium, for example, was enhanced eighteen times.

Branica and his co-workers have repeated the work of Macchi (1964) and have also investigated the use of a modified Kemula hanging mercury drop electrode (the HMDE (Hg) electrode) for the determination of zinc in sea water (Bernhard *et al.*, 1972). The HMDE (Hg) electrode (Kemula and Kublik, 1963) was prepared by extruding a mercury drop from a capillary using a micrometer head. Electrical contact was made to the mercury drop via the mercury thread in the capillary. This electrode is simpler to prepare than the HMDE (Pt) electrode but its reproducibility is rather critically dependent on the ability of the micrometer head to deliver the same volume of mercury each time. To prevent erratic behaviour because of the intrusion of the electrolyte, the capillary should be given a hydrophobic coating. Zieglerová *et al.* (1971) recommended drawing a 6 % solution of dimethyl dichlorosilane in benzene through the capillary followed by air drying for 1 h and baking to 240–260°C for 4 hours. Bernhard *et al.* (1972) chose to strip out the zinc *quantitatively* by prolonged electrolysis at a potential of −0·2 V after the normal stripping cycle had been completed. The zinc amalgam diffuses into the body of the mercury drop and on occasion into the thread of the mercury capillary, so that the time required to remove the last traces of zinc depended on the amount taken up during the pre-electrolysis phase. In sea water 10 to 15 min are normally required for drops 0·8 mm in diameter. Pre-electrolysis is carried out at −1·3 V (versus the saturated calomel electrode) for 90 s followed by a 30 s rest phase. The original pH of the sample is restored at this stage by the addition of a few drops of Zn-free hydrochloric acid. The stripping phase was carried out at a linear rate of 250 $mV\,s^{-1}$ over a 1 V scan. The zinc peak appeared at $-1{\cdot}0$ V versus the saturated calomel electrode. After calibration by standard addition, the precision of the method was $\pm 4{\cdot}6\%$. This procedure has been used to study the physico-chemical state of zinc in sea water (Piro *et al.*, 1973; Branica *et al.*, 1972).

Smith and Redmond (1971) used a HMDE (Hg) electrode to study interference problems that might be encountered in the analysis of sea water. They used a Radiometer PO4g Polariter dc polarograph together with a Radiometer model E69b HMDE (Hg) electrode and a saturated calomel counter electrode. The sea water samples were untreated and had a pH of $8{\cdot}10 \pm 0{\cdot}05$ after de-oxygenation. Pre-electrolysis at $-1{\cdot}20$ V (*versus* saturated calomel electrode) was carried out for 10 min followed by a 60 s rest phase. The stripping phase used a linear voltage scan of 7 mV s^{-1}. Standard solutions were nearly neutral and were prepared by dissolving suitable salts in distilled water immediately before use. Of the four elements most commonly studied in sea water (zinc, cadmium, lead and copper) Smith and Redmond found potentially serious interference effects for zinc (with nickel), cadmium (with tin) and copper (with vanadium). The tin concentrations normally encountered (less than $0{\cdot}02$ μg l^{-1}) are not likely to produce serious effects (Table 20.9). The current sensitivity for zinc is fifteen times that for nickel (Table 20.9). The error introduced in the zinc determination by the presence of an equal quantity of nickel will only be about 7%. This is in general, less than the reproducibility of the method between replicate samples. The current sensitivity for vanadium is about one seventh of that for copper (Table 20.9). It must be stressed that these interference factors refer to the response of the anodic stripping procedure to the addition of the

TABLE 20.9

Interference effects in the anodic stripping voltammetry of untreated sea water using a hanging mercury drop electrode[a]

Peak potential (V *vs* SCE)	Peak current (μA)	Element	Concentration[b] (μg l^{-1})
−0·97	0·081	Zn(II)	1·93
		Ni(II)	27·9
−0·60	0·0018	Cd(II)	0·18
		Sn(II)	22·5
−0·42	0·00015	Pb(II)	0·21
−0·13	0·0075	Cu(II)	1·10
		V(V)	8·5

[a] Smith and Redmond (1971). Reproduced with permission.
[b] Concentration of element required to yield peak current shown in second column.

various elements in the ionic form. Analysis followed immediately after the addition of the standard aliquot so that no time was allowed for the ionic form to equilibrate with the various fractions of the element that might be present in natural sea water. It is likely that most of the elements in natural samples will not be present predominantly in the ionic form so that interference effects may not be so acute.

The sensitivity of anodic stripping analyses with the HMDE (Hg) electrode can be enhanced by suitable modulation of the potential ramp used in the stripping phase. Rojahn (1973) used a Tacussel type PRG3 three electrode ac polarograph in a cell with a platinum coil counter electrode and a saturated calomel reference electrode. The working electrode was a Metrohm E410 HMDE (Hg) electrode with a surface area of 2·22 mm^2. For the determination of copper, lead and cadmium the filtered sea water samples were acidified with nitric acid (15 mmol l^{-1}). 25 ml of the sample was placed in the cell and de-oxygenated by nitrogen purging for 15 minutes. Pre-electrolysis was carried out at −0·8 V for 2 min with a stirring rate of 40 rev s^{-1}. The stripping phase was carried out with a scan rate of 2·5 mV s^{-1} with a 15 Hz ac signal superimposed with a voltage amplitude of 20 mV. The copper, lead and cadmium peaks were recorded (Table 20.8). The metal concentrations were determined by a standard addition procedure. The reproducibility of replicate measurements for a single sea water sample was ±2% for copper and lead and ±3% for cadmium. However, the standard errors for measurements on 5 aliquots of a single sea water sample taken from Oslofjord were much higher (7% for lead, 8% for copper and 21% for cadmium) thus emphasizing the problems inherent in sample manipulation (see Zirino and Healy, 1972). This method apparently shows a distinct enhancement in the sensitivity of copper analyses over the conventional linear sweep methods. For the determination of zinc, 2·5 ml of a 1 M acetate buffer (pH 4·6) are added to 25 ml of acidified sea water and the pre-electrolysis is carried out at −1·2 V for 2 min to minimize contamination from other electroactive species. Once again the reproducibility of measurements on a single sample (±2%) was much better than the reproducibility observed between replicate samples (±16%). Rojahn (1972) attributes this large error in part to the contamination of the acetate buffer with copper. The results of the analyses were in reasonable agreement with values determined by atomic absorption after extraction of the trace metals (Brooks *et al*., 1967).

Donadey (1969) and Donadey and Rosset (1972) used differential pulse anodic stripping (see Section 20.4.1., also Flato, 1972; Christian, 1969; Siegerman and O'Dom, 1972; Copeland *et al*., 1973) to enhance the sensitivity of analyses with the HMDE (Hg) electrode. As with ac polarography, the metals taken up during the pre-electrolysis step are reoxidized with a wave form that minimizes the contribution of the charging current and enables the output signal to be filtered electronically to remove excess noise. The theory of pulsed anodic stripping has been considered in detail by Osteryoung and Christie (1974). Donadey (1969) used a Metrohm type E410 hanging mercury drop electrode and a silver–silver chloride working and counter electrode. A Shandon-Southern Model A3100 pulse polarograph was used to apply the pulsed signals (see Section 20.4.1.) during the stripping phase. The solutions were degassed for 20 min prior to electrolysis. With a 2·2 mm^3 drop size a pre-electrolysis time of 3 min at −1·4 V (versus the silver–silver chloride electrode) was used for the analysis of zinc, copper, cadmium and lead in a 50 ml sample of sea water. After a rest phase of 60 s the trace metals were stripped from the mercury drop at 2·2 mV s^{-1}. 35 mV pulses with a duration of 40 ms were applied to the linear voltage ramp at intervals of one second. The resulting current was recorded for the last 20 ms of the pulse life. The resulting trace for sea water (Fig. 20.16) shows a high sensitivity and a clear

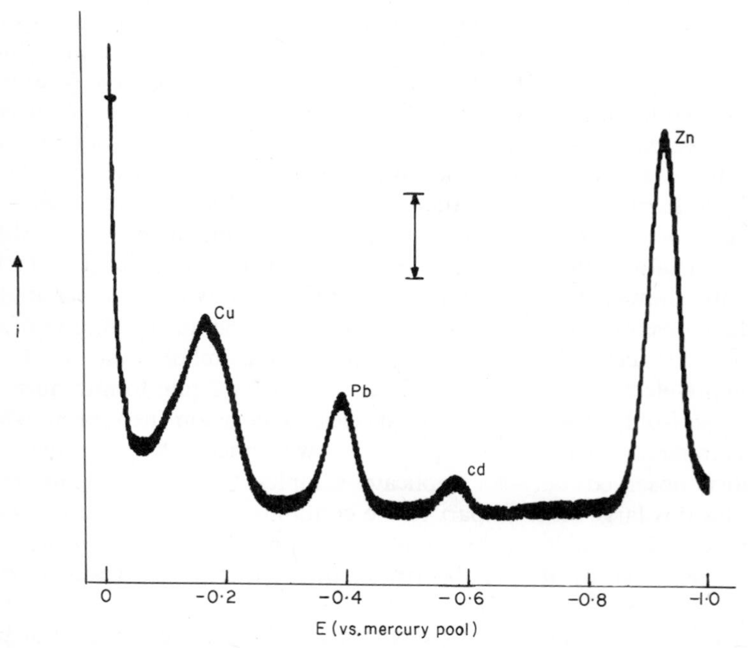

FIG. 20.16. Differential pulse anodic stripping trace showing the presence of copper (4×10^{-8} M, −0·14 V), lead (8×10^{-9} M, −0·39 V), cadmium (3×10^{-9} M, −0·59 V) and zinc (9×10^{-8} M, −0·95 V) in untreated sea water. The arrows indicate one major scale division on the recorder chart. The instrument was at 1/120th maximum sensitivity (cf. Fig. 20.15). Pre-electrolysis time 3 minutes. Linear scan rate 2·2 mV s^{-1}. Reproduced with permission from Donadey (1969).

resolution of the peaks against a uniform background. The limit of detection for cadmium was 4×10^{-11} mol l^{-1} and the reproducibility of analysis was $\pm 3{\cdot}5\%$ at the $3{\cdot}5 \times 10^{-8}$ M level. No details are given from which the current sensitivity of the procedure may be calculated. Fresh mercury drops are used for every analysis, including the analysis of the sample after the addition of standard aliquots, and the precaution is taken of flushing out several drops of mercury each time to prevent problems that may arise because of the diffusion of deposited metals into the capillary thread.

20.5.3. THIN FILM MERCURY ELECTRODES

The relatively large volume of mercury used in the HMDE restricts the sensitivity of the anodic stripping procedure. Fairly long pre-electrolysis periods are required to produce amalgams with a useful concentration factor. In addition, metals dissolved in the amalgam tend to diffuse into the bulk of the drop. Consequently, attempts to enhance the sensitivity of the method by increasing the scan rate of the stripping step run into trouble because the peaks are distorted by the relatively slow diffusion of metal from the bulk of the drop to the surface. These problems can be reduced by preparing the mercury as a thin film on a conducting substrate. The resulting working electrode has a large surface area, which enhances the efficiency of the electrolysis step, and a small volume, which increases the concentration factor achieved in a given pre-electrolysis time (equation (20.6)). A further advantage arises because the peak current on the stripping trace is directly proportional to the scan rate for thin film film electrodes (Roe and Toni, 1965), whereas it is proportional to the square root of the rate for mercury drop electrodes (Reinmuth, 1961). The sensitivity of the analysis can, therefore, be enhanced more effectively for thin film electrodes by stripping at high scan rates.

Matson developed the mercury coated graphite electrode (MCGE) for environmental work (see Matson *et al.*, 1965; Matson, 1968; Perone and Bromfield, 1967; Carter and Hume, 1972). This electrode is prepared by vacuum impregnating a graphite rod with a soft, non-crystalline wax with a high boiling point. Ceresine wax (Gilbert and Hume, 1973) or Sonneborn wax (Clem *et al.*, 1973) have been used in later versions. The wax apparently fills in cracks and irregularities in the graphite and divides the electrode surface into a large number of discrete plating sites. After the electrode has been polished these sites can be uniformly coated with mercury (at a density of 10^{-7} mol cm^{-2}) by electrolysis at a potential above the point where copper will plate out in the particular medium. The graphite substrate must be uniformly coated with mercury otherwise the electrode behaviour of the graphite is superimposed on that of the mercury and broad, unresolved stripping peaks result. If, on the other hand, the mercury coating is too thick the mercury droplets tend to coalesce and run down the electrode surface and again expose unplated graphite. The electrodes only have a limited life. Matson (1968) suggested that this may be as long as two or three weeks under quiescent conditions, but would fall to 4 or 5 days if subjected to stirring rates of 350 rev min^{-1}. Clem *et al.* (1973) found that the useful life of their electrodes under experimental conditions was only three or four days. The end of the active life is heralded by the development of a low overvoltage for hydrogen evolution, an increasingly high residual current and a somewhat reduced short term stability. This deterioration is accelerated if the electrodes are exposed to the air after coating with mercury or if the electrode is used at potentials where hydrogen evolution is appreciable. This type of electrode is, therefore, less convenient to use than the HMDE for the determination of zinc in acid solutions. Matson (1968) plated his electrode in a separate acidic mercury(II) nitrate solution and then transferred it, after rinsing to the sample solution. Later workers have preferred to deposit the mercury coating in the sample solution during the pre-electrolysis phase of the analysis. Clem *et al.* (1973) give a detailed discussion of this procedure.

Matson (1968) used a quartz cell with a Teflon lid. The MCGE had a geometric surface area of 2·2 cm^2. The silver–silver chloride reference electrode and platinum counter electrode were mounted in reservoirs containing 0·6 M NaCl which were connected to the cell solutions *via* 4 mm diameter porous glass liquid junctions. The solution was stirred by vigorous bubbling of nitrogen gas (0·21 l min^{-1}). The MCGE was held at −0·2 V versus the silver–silver choride electrode while the solution was purged of oxygen to prevent the spontaneous oxidation of the mercury coating (equation (20.52)). The formation of intermetallic compounds is likely to create problems in the use of the MCGE because of the high amalgam concentrations that are produced. For example, Matson (1968) found that in a medium containing 1 M sodium chloride and a 0·26 M acetate buffer (pH 5–6) nickel and cobalt both interfere with the determination of zinc, copper and bismuth when plating is carried out below −1·1 V versus the saturated calomel electrode. Further details of the preparation and use of the MCGE in the analysis of fresh waters will be found in the papers by Matson and Roe (1967) and by Allen *et al.* (1970). A number of applications of the technique for sea water analysis have been reported in recent years.

Fitzgerald (1970) studied the speciation and total concentration of lead and copper in sea water using the MCGE. His work on the speciation of copper has been discussed in some detail by Siegel (1971). Seitz (1970) has shown that the procedure can be used for the determination of lead and cadmium in sea water at the 10^{-8} to 10^{-10} M level. The use of an ac modulated stripping phase enhanced the sensitivity of these measurements although interfering currents due to the absorption and desorption of Cl^- and Br^- were observed. Using the ac stripping technique it was also possible to resolve the zinc stripping peak from that associated with irreversible hydrogen evolution. Intermetallic compound formation with nickel, copper and cobalt, however, restricts the reliability of zinc analyses using the thin film electrode. It is sometimes possible to minimize intermetallic compound formation by carefully selecting the pre-electrolysis potential. For example, copper can be determined without interference from nickel and zinc if the pre-electrolysis is carried out at potentials more positive than $-0{\cdot}1$ V versus the silver–silver chloride electrode in sea water (Seitz, 1970). Thallium gives a useable stripping peak in sea water if its concentration is greater than 5×10^{-11} M. However, the half wave potential of this peak is very close to that of cadmium. If a large MCGE is used, and the cadmium is first masked by the addition of EDTA then thallium can be determined directly in sea water (Gilbert, 1971). Some electronic compensation of the background current is required to achieve the necessary instrumental sensitivity. No thallium was observed in natural sea water above the detection limit of the method. Silver can also be determined directly in sea water if the copper present is first complexed with EDTA (Gilbert, 1971). Natural concentrations between $0{\cdot}9 \times 10^{-9}$ M and 3×10^{-9} M were reported. The determination of indium in sea water (acidified to pH 3·5 with acetic acid) is subject to interferences from cadmium and lead.

Gilbert and Hume (1973) have described procedures for the determination of antimony and bismuth in sea water using the MCGE. The cell design (Fig. 20.17) was the same as that described by Matson (1968) except that the silver–silver chloride reference electrode was immersed in sea water giving a potential of $+5$ mV versus the saturated calomel electrode at a salinity of 35‰ The potentials quoted are therefore effectively the same as potentials versus the saturated calomel electrode. The MCGE was plated with mercury at a potential more negative than $-0{\cdot}5$ V from 10 ml of acidified (pH 2·5) sea water to which sufficient mercury(II) nitrate had been added to give a mercury coverage of 5×10^{-7} mol cm^{-2}. This mercury coverage is heavier than that recommended by Matson (1968) since traces of oxygen may convert some of the mercury to calomel (equation (20.52)) in the acid conditions used in this analysis. When not in use the electrode is held at $-0{\cdot}2$ V in a cell filled with sea water at pH 2·5. Bismuth is determined in a sea water sample made 1 M with hydrochloric acid by bubbling HCl gas through the solution until the calculated weight increase is observed. At this acidity antimony in the solution is apparently involved in the formation of polymeric hydroxy chlorides which are electrolytically inactive. 20 ml of the acidified sample is transferred

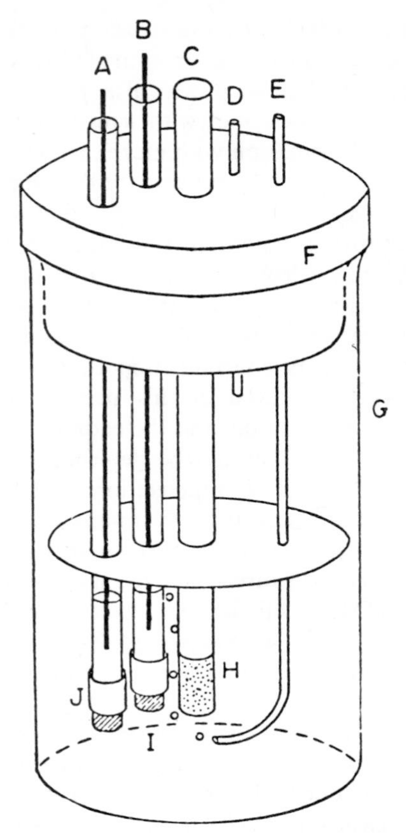

FIG. 20.17. Cell used for the determination of antimony and bismuth in sea water by anodic stripping voltammetry (Gilbert and Hume, 1973). A. Platinum counter electrode. B. Silver–silver chloride reference electrode. C. Wax impregnated graphite rod. D. Nitrogen vent. E. Nitrogen inlet. F. Teflon stopper. G. Quartz cell. H. Mercury coating on graphite electrode. I. Unfused Vycor plugs. J. Teflon tubing. Reproduced with permission from Gilbert and Hume (1973).

to the quartz cell and purged with nitrogen for ten minutes. The cell is then assembled and the nitrogen purging is continued for a further ten minutes with the MCGE at a potential of $-0{\cdot}1$ V. The pre-electrolysis is carried out at $-0{\cdot}4$ V for 15 minutes with the solution stirred by steady nitrogen bubbling (1 ml s^{-1}). After a 20 s rest period the MCGE is scanned to $-0{\cdot}10$ V at 17 mV s^{-1}. The bismuth peak appears at $-0{\cdot}2$ V. A compensation circuit is used to subtract a linearly varying background current from the stripping trace. The slope of this compensating ramp is adjusted just before the anodic scan. The bismuth concentration is determined by the standard addition procedure using 10 μl aliquots of a 10^{-6}–10^{-7} M bismuth solution in hydrochloric acid. This solution is prepared daily from a $1{\cdot}5 \times 10^{-4}$ M stock solution of Bi_2O_3 in 1 M HCl. At the 10^{-8} M level there is no interference from a nine-fold excess of antimony, and the acidified sample is stable for at least ten days.

If the sea water is made 4 M with respect to hydrochloric acid the inactive antimony complexes are apparently broken down and the stripping peak at $-0{\cdot}3$ V includes contributions from both the antimony and the bismuth in solution. Since the bismuth current peak is displaced slightly from that of antimony, the peak profile is distorted somewhat and peak area rather than peak height is proportional to the sum of the bismuth and antimony concentrations. The pre-electrolysis step is carried out at $-0{\cdot}5$ V for ten minutes and after a rest phase of 20 s the stripping potential is swept from $-0{\cdot}2$ V at at 17 mV s^{-1}. Both Sb(III) and Sb(V) are reduced to metallic antimony under these conditions. The concentration of antimony was determined by standard additions of a 10^{-5} to 10^{-6} M Sb(III) solution in 2 M HCl prepared daily from a millimolar stock solution of Sb_2O_3 in 2 M HCl. The antimony concentration is calculated by subtracting the previously determined bismuth concentration from the antimony + bismuth total determined here on the assumption that both metals have identical current sensitivities in 4 M HCl. The limit of detection of the method is around 0·1 μg kg^{-1}. The 95% confidence limits for triplicate samples were within ±30% for bismuth and ±20% for antimony. Significant amounts of antimony and bismuth may be lost to polythene containers if the sample is not acidified immediately. However, much of this is recoverable if the sample is allowed to remain in the same container for a few days after acidification.

Some interference might be expected from copper and silver which can plate out at the potentials used in the pre-electrolysis phase. However a thousandfold excess of copper and a twohundredfold excess of silver can be tolerated without a noticeable effect on the stripping currents of either bismuth or antimony + bismuth. The values obtained for the concentrations of these elements in sea water agree in general with the concentrations measured by Portmann and Riley (1966a, b) using spectrophotometric procedures following preconcentration. Inter-laboratory calibration studies also show that the procedure gives results for antimony that are in good agreement with estimates made using neutron activation analysis on freeze-dried samples (Gilbert and Hume, 1973).

Conditions for the analysis of zinc, cadmium, copper and lead in sea water using the MCGE have been considered in some detail by Lang (1973). He determined the optimum conditions for the analysis of these metals by carefully investigating the influence of each of the important parameters on the sensitivity and precision of the procedure.

The cell used accommodated a 50 ml sample and was almost identical with that described by Matson (Fig. 20.17). The platinum counter electrode was immersed in a 0·1 M potassium chloride solution and the silver–silver chloride reference electrode in a saturated potassium chloride solution. A somewhat heavier mercury coverage is required in sea water than that recommended by Matson (1968) because of the spontaneous oxidation of mercury (equation 20.52, Gilbert and Hume, 1973). The concentration of 8×10^{-6} mol cm^{-2} recommended by Lang (1973) is an order of magnitude greater than that used by Gilbert and Hume (1973). A pre-electrolysis phase of 10 minutes duration was used at a potential of $-1{\cdot}25$ V with nitrogen bubbling at a rate of 40 cm^3 min^{-1}. The solution was stirred at 1000 rev min^{-1} during this phase. Following a rest phase of 50 s the metals were stripped off the mercury film with a linear voltage ramp at 3 mV s^{-1}. All materials coming in contact with the samples were acid washed, rinsed and then conditioned with a subsample of the sea water to be analysed. The current sensitivities (μA/μg l^{-1}) observed at the stripping stage under these conditions were: zinc, 0·83; cadmium, 1·33; lead, 1·02; copper, 1·53. The limit of detection for zinc (0·2 μg l^{-1}) is actually greater than that obtained by Macchi (1965, 0·05 μg l^{-1}) using a slowly dropping mercury electrode (Section 20.5.2). Lang (1973) attributes this to the greater sensitivity of the system used by Macchi to record the stripping current. The precision is between 10% and 15% in the range 0·2 to 10 μg l^{-1} for all four elements. Lang (1973) indicated that serious interference effects might be expected because of intermetallic compound formation if the ratios of the electroactive metal concentrations exceeded the following values:

for zinc $[Cu^{2+}]/[Zn^{2+}] > 10$ (48% error when ratio = 12);
for copper $[Zn^{2+}]/[Cu^{2+}] > 10$ (15% error when ratio = 12);
for cadmium $[Zn^{2+}]/[Cd^{2+}] > 10$ (8% error when ratio = 12).

Since in non-polluted sea waters the ratio $[Zn^{2+}]/[Cu^{2+}]$ can be as high as 33, the direct analysis of copper is already running into difficulties. By pre-electrolysis at $-0{\cdot}4$ V it is possible to deposit copper in the absence of zinc and thus remove this source of interference.

Bradford (1972) used the MCGE to determine zinc in Chesapeake Bay water. The electrode was prepared in the manner described by Matson with mercury plated from a solution 5×10^{-5} M in Hg (II) and 0·01 M in $HClO_4$. A fresh electrode was prepared each day. The cell was set up in a 150 ml Teflon beaker with a Teflon lid and the electrode system was identical to that used by Matson (1968). With the MCGE potential held at $-0{\cdot}15$ V (versus a silver–silver chloride electrode immersed in 0·1 M sodium chloride) a 130 ml sample was degassed with nitrogen flowing at 180 ml min^{-1} until the oxygen reduction current fell nearly to zero. The potential was then increased to −0·6 V and 1 ml of 1·0 M sodium tartrate was added to the solution. When the oxygen reduction current was again close to zero (5 min) the potential was adjusted to −1·27 V for the pre-electrolysis phase with the nitrogen gas flowing at 120 ml min^{-1} to stir the solution. After a brief rest phase the zinc was stripped off at a scan rate of 16 m V s^{-1}. The pre-electrolysis and stripping phases were repeated two or three times until a reproducible stripping current was observed. The procedure was repeated after the addition of standard aliquots of ionic zinc. Interference effects were noted for zinc with copper, cobalt and nickel, and for copper with cobalt and nickel. Interference factors, showing the apparent change in the concentration of the metal being analysed for each unit concentration of the interferant added, were calculated by Bradford (1972; Table 20.10). These factors are lower for the tartrate medium than for the natural sea water. If only a small correction needs to be made for the presence of copper in the sample, then the precision of the analysis for zinc is about ±4%. However, when the copper concentration is high the reproducibility is generally no better than 20 to 50%.

The MCGE electrode has been used by Muzzarelli and Sipos (1971) to study the effectiveness of chitosan for the collection of ionic zinc, cadmium copper and lead from sea water. A pre-electrolysis time of 5 min was used followed by a rest phase of 30 s. Pre-electrolysis at −1·2 V was used for the determination of zinc and at −0·85 V for the determination of the other

TABLE 20.10

Current sensitivity and interference factor data for the determination of zinc and copper in untreated sea water using anodic stripping voltammetry at a thin film electrode[a].

		Mean	σ	No. of measurements
Current sensitivity ($\mu A(\mu g\, l^{-1})^{-1}$)	Zn	1·115	0·190	29
	Cu	0·540	0·193	26
Interference factor[b]	ΔZn/ΔCu	0·430	0·076	26
	ΔZn/ΔCo	0·38	—[c]	6
	ΔZn/ΔNi	0·252	0·118	21
	ΔCu/ΔCo	0·602[d]	—[c]	10
	ΔCu/ΔNi	0·084[d]	—[c]	6

[a] Data from Bradford (1972).
[b] Defined in the text.
[c] Insufficient data for variance calculation.
[d] Estimated in the presence of tartrate. This treatment reduces the interference factor to approximately 50% of that observed in untreated sea water.

metals to avoid contamination from the large excess of zinc present. A scan rate of 13 mV s^{-1} was used in the stripping phase. The chitosan extraction is apparently quantitative for copper and cadmium, but not for zinc and lead.

Liebermann and Zirino (1974) have developed a *tubular* MCGE that can be used for the analysis of continuously flowing sea water samples. Using fast flow rates (up to 950 ml min^{-1}) it is possible to analyse large volumes in a relatively short time and thereby increase the sensitivity of the technique. Since this high flow rate induces a fairly rapid deterioration of the mercury film it is necessary to interpolate a recoating step between each measurement cycle. The electrodes were prepared from 6 mm o.d. spectroscopic grade graphite with a 3·2 mm diam. hole bored axially. Electrical contact was made via silver wires sealed around the tubes with epoxy cement. A similar tube containing a coil of chloridized silver wire was used both as the reference and counter electrode. A 1 litre polyethylene bottle contained the mercury plating solution (2×10^{-5} M mercury(II) nitrate in nitric acid) and a 100 ml stoppered measuring cylinder contained the sample. Both solutions were deoxygenated by a continuous flow of purified nitrogen. The flow of sample or plating solution through the cell was controlled by manipulating the three way stop cocks (1 to 3, Fig. 20.18). The mercury plating solution was allowed to flow through the cell for 5 min at 160 ml min^{-1} while a potential of −1·4 V was applied at the working electrode. Less stable results were obtained if the mercury film and the electroactive metals were plated simultaneously onto the graphite electrode from the sea water sample (Florence, 1970, 1972). The system was then switched to the sample and after a 15 s flushing time the potential was re-applied for two to five minutes. While the sample was still flowing the trace metals and the mercury were anodically stripped from the electrode at a scan rate of 17 mV s^{-1}. The scan was stopped at +0·5 V versus the silver–silver chloride electrode and the procedure was repeated.

The mercury film employed ($7{\cdot}5 \times 10^{-9}$ mol cm^{-2}) was much thinner than that recommended by other workers (Matson, 1968; Lang, 1973; Gilbert and Hume, 1973). In the flowing system thicker films apparently increase the hydrogen current on stripping and produce a cathodic peak of 10^{-5} to 10^{-4} coulomb at −0·5 V. Thinner films give peak heights that vary with the degree of mercury coverage. The peak current for zinc was shown to be linear with concentration by a standard addition procedure using fresh aliquots of sea water and fresh mercury electrodes for each addition. The procedure gave a relative standard deviation of ±9·5% for 17 aliquots of a sample of San Diego Bay water (concentration 2 μg l^{-1}). This precision is much greater than that reported previously with the HMDE (see Zirino and Healy, 1972; Rojahn, 1972). The sensitivity of the method can be increased fifty fold if the

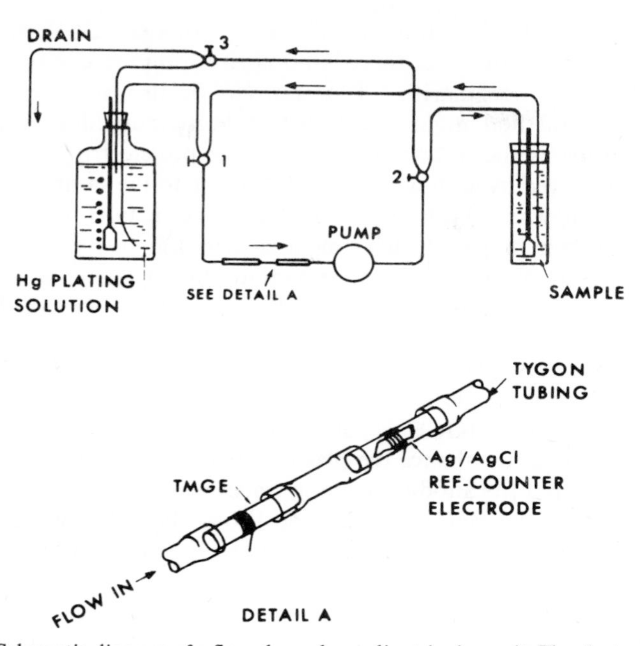

FIG. 20.18. Schematic diagram of a flow-through anodic stripping unit. The electrode assembly, including the tubular mercury coated graphite electrode (TMGE), is shown in detail in the lower half of the figure. Reproduced with permission from Liebermann and Zirino (1974). Copyright of the American Chemical Society.

differential pulsed anodic stripping is used to reoxidize the trace metals (Zirino *et al*., 1973). The stripping potentials observed versus the silver–silver chloride electrode were, zinc −1·15 V, cadmium −0·76 V. copper −0·34 to −0·42 V and lead −0·060 V. The plating efficiency is much greater for the tubular MCGE than for the conventional MCGE, and under comparable conditions the flow system gives a peak current which is nine times greater than that observed with the normal Matson cell. The continuous flow system has a number of other advantages. The analysis time is short since the plating efficiency is enhanced, and the samples can be changed with the minimum of disturbance to the electrodes and with the minimum risk of contamination. Since a fresh electrode is prepared for each determination there is no risk of the erratic behaviour often observed when these film electrodes are exposed to air after the mercury plating (Perone and Davenport, 1966). The system is readily automated and is ideally suited for continual shipboard analysis. A closely similar system (Seitz *et al*., 1973) has been used for the determination of thallium in sea water.

Florence (1970) suggested that glassy carbon might provide a more easily prepared substrate for thin film electrodes. This material is dense and can be given a high polish so that there is no need to impregnate the electrode to remove surface irregularities. Florence (1970) also suggested that the sensitivity of the method might be enhanced if the mercury film was deposited at the same time as the trace metals during the pre-electrolysis procedure. This can be achieved by adding mercury(II) nitrate to the sample and plating at a potential at which both Hg^{2+} and the trace metal ions are reduced. A fresh electrode is prepared each time and there are no problems over the long term stability of the rather fragile mercury film. Clem *et al.* (1973) have since shown that *in situ* plating works equally well with the MCGE. The glassy carbon rod is set in an epoxy resin and one end is polished to give a flat electrode surface. A platinum spiral counter electrode and a saturated calomel reference electrode were used. During the pre-electrolysis phase the glassy carbon electrode (GCE) is rotated at 2000 rev min^{-1} to break down the diffusion layer and enhance the plating efficiency. The resulting mercury film is very thin and the anodic stripping traces produced show much better resolution and greater sensitivity than do those observed for other anodic stripping techniques under comparable conditions (Fig. 20.19).

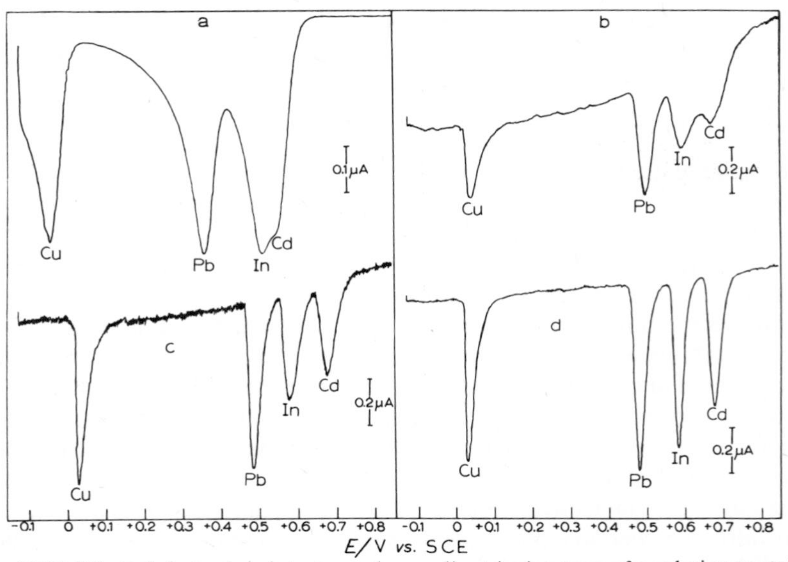

FIG. 20.19. Effect of electrode substrate on the anodic stripping trace of a solution containing 2×10^{-7} M Cd(II), In(III), Pb(II), and Cu(II) in 0·1 M KNO_3. (a) HMDE (Pt), (b) Pyrolytic graphite, (c) Unpolished GCE, (d) Polished GCE. 30 min pre-electrolysis was used for (a) and 5 min for (b)–(d). Solutions (b)–(d) contained 2×10^{-5} M Hg(II). Scan rate 5 mV s^{-1}. Reproduced with permission from Florence (1970).

Florence (1972) described procedures for the determination of bismuth, copper, lead, cadmium and zinc in sea water samples and in the ashed remains of marine organisms. He also indicated the possibility of determining antimony(III), indium and thallium in these samples. For the determination of bismuth, copper, lead and cadmium a 50 ml sea water sample was made 4×10^{-5} M in mercuric nitrate. 10 ml of this sample was transferred to the cell and degassed with helium for 10 min with the GCE rotating at 2000 rev min^{-1}. The gas was then allowed to flush over the surface of the solution and a deposition potential of −0·9 V was applied for 5 min with the electrode still rotating at 2000 rev min^{-1}. After a brief rest period the anodic sweep was applied from −0·9 V to zero volts at a rate of 50 mV s^{-1}. The first sweep preconditions the electrode and the trace is not recorded. The procedure is then repeated with a pre-electrolysis period of 30 min and the cadmium, lead, copper and bismuth stripping peaks are recorded. The concentrations of these elements are then determined after the analysis has been repeated on a separate sea water sample which has been made 1×10^{-5} M in all four elements by a standard addition. After each analysis the mercury film, together with remaining trace metals, was removed by simply wiping the electrode with a moist tissue, rinsing in distilled water and drying on a second tissue. No carry over of mercury or trace metals is observed if this procedure is used. Clem *et al.* (1973) preferred to remove the excess mercury electrolytically by continuing the stripping process in the manner described by Liebermann and Zirino (1974).

For the determination of zinc, 5 ml of acidified filtered sea water are pipetted into a 50 ml volumetric flask. Reagents are added to the flask so that the final solution is 4×10^{-3} M in hydrochloric acid, 4×10^{-5} M in mercury(II) nitrate and 0·02 M in sodium acetate. The analytical procedure is the same as that described for the other metals except that a deposition potential of −1·3 V is used and the anodic scan is stopped at −0·85 V. A reagent blank must be determined. Since the mercury coating is codeposited with the trace metals, the problems observed by Lang (1973) with the long term stability of thin mercury films at low pH are not encountered. Film stabilities over 30 minutes or so are all that are required. No data are provided for assessing the reproducibility of the analyses in sea water.

Florence (1972) observed that the zinc peak height in acetate buffer (pH 5·7) was depressed by nickel and copper, probably by the formation of intermetallic compounds. Since the percentage depression was apparently proportional to the nickel and copper concentrations but independent of the zinc concentration, Florence suggested that these artefacts could be removed simply by diluting the sample. A tenfold dilution was therefore adopted as a standard procedure.

Florence (1974) later extended his work on the GCE to the determination of bismuth in sea water using the same experimental set-up. Filtered sea water samples were stored in acid washed 1 litre polythene bottles containing 5 ml of 10 M hydrochloric acid. The resulting acid concentration in the sea water is about one twentieth of that used by Gilbert and Hume (1973). 0·1 ml of 1×10^{-2} M mercury(II) nitrate is added to 50 ml of acidified sea water. A 10 ml sample is then placed in the cell and de-oxygenated for 10 minutes with nitrogen before pre-electrolysis at −0·25 V for 5 minutes. The deposited metals are stripped off at a scan rate of 83 mV s^{-1}. This step pre-conditions the glassy carbon surface. The procedure is repeated with a pre-electrolysis time of thirty min and of recording the peak height of the bismuth wave at −0·10 V. The concentration is determined by repeating the procedure on a second 50 ml aliquot of acidified sea water that has been "spiked" with 0·1 ml of a 1×10^{-5} M standard bismuth solution.

Florence (1974) used the GCE procedure to ascertain the efficiency of ion-exchange procedures for the removal of bismuth from sea water (Portmann and Riley, 1966a). Measurements with a sea water sample containing 0·1 mol l^{-1} of hydrochloric acid and 0·11 μg Bi l^{-1} indicated that a 6 × 0·5 cm col. of BioRad AGl-X8 (100–200 mesh) chelating exchange resin retained 90% of the natural bismuth from a 50 ml sample passed through at 0·7 ml min^{-1}. The residual 10% of bismuth in the effluent could not be removed by passage through a second column of the same dimensions. Florence concluded that tracers in the ionic form that are normally used to check the efficiency of extraction procedures may not equilibrate adequately with all of the chemical forms of a particular element in the time allotted prior to the analysis. This contention is given circumstantial support by the work of Piro *et al.* (1973) and Branica (1970) on the speciation of zinc in sea water.

An interesting instrumental improvement on the GCE has been described by Laser and Ariel (1974). A glassy carbon rod was mounted coaxially inside a hollow cylinder also made of glassy carbon and the two electrodes were insulated from one another by a layer of silicone rubber. This composite probe was then pressed into a Teflon cylinder which acted as an outer insulator. When the end of the probe was polished the end of the inner rod appeared as a disc electrode analogous to the GCE. The end of the outer cylinder appeared as a concentric ring surrounding the disc (a ring disc electrode, Alberg and Hitchman, 1971). The disc is used as the conventional working electrode during the pre-electrolysis phase using the procedures described by Florence (1970, 1972). When the components are stripped from the disc, the ring is maintained at a constant cathodic potential corresponding approximately to the pre-electrolysis potential. A proportion of the metal stripped from the disc is therefore collected at the ring. By measuring the *collection* current at the ring during the stripping process a trace is obtained that is practically free of the charging current since the ring potential is essentially constant throughout. Consequently, it is possible to enhance the collection current markedly by using very rapid scan rates during the stripping process. Scan rates up to 333 mV s^{-1} were used by Laser and Ariel (1974) but higher rates are possible. The collection current was shown to be directly proportional to the stripping current for zinc, copper and lead.

20.5.4. ANODIC STRIPPING AND CHEMICAL SPECIATION (Tables 20.7 and 20.8)

Anodic stripping techniques in untreated sea water are able to measure the concentration of a range of metals that are capable of forming amalgams and can be deposited within the potential range −1·3 V to −0·05 V (Tables 20.7 and 20.8). If the metal is present in solution in a form that cannot be reduced at the mercury surface then it will not be analysed. Studies with anodic stripping voltammetry indicate that only a fraction of each of the electro-active metals zinc, copper, cadmium and lead is normally present in an electroactive form. Zinc has been studied in some detail since it is present in relatively high concentrations, and it has a radioactive isotope with a useful half-life so that electrochemical studies can be supported by tracer techniques (Branica *et al.*, 1972; Piro *et al.*, 1973). From this work, and that of Zirino and Healy (1970) and Bradford (1972), it is clear that zinc is present in sea water in a number of fractions that are all intricately interrelated (Fig. 20.20).

The electroactive portion of the total zinc concentration comprises three fractions—the free ionic metal and the labile organic and inorganic complexes. The peak potential and the peak current on stripping will depend on the relative proportions of these fractions in sea water and, indeed on the proportions of the individual complexes that make up each fraction (Matson, 1968; Lang, 1973; Bradford, 1972). If the reaction of the complexing species with the free metal ion is rapid compared to the rate at which metal is reoxidized from the mercury electrode then the anodic stripping technique can be used to estimate the stability constants of these individual complexes (Zirino and Healy, 1970; Bradford, 1972, 1973) and to assess the nature and the concentration of the individual ligands (Bradford, 1972; Matson, 1968; Clem, 1973).

The calibration of anodic stripping analyses of zinc by the addition of standard aliquots of ionic zinc is based on the assumption that the ionic zinc will distribute itself between the various electroactive fractions in the same way as the "natural" zinc. The procedure of standard addition can,

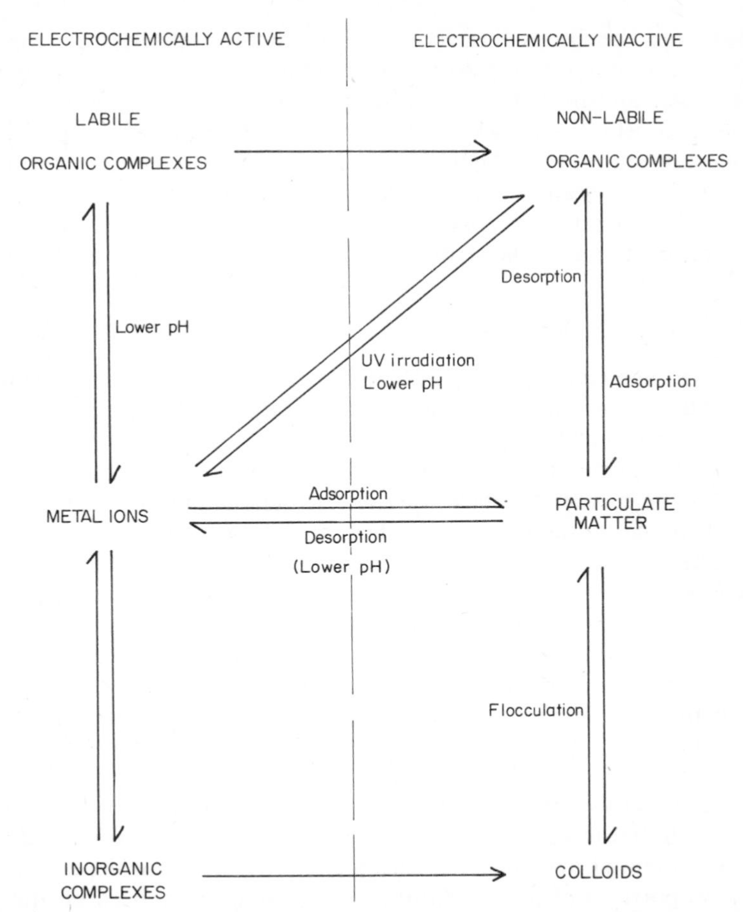

FIG. 20.20. A tentative picture showing interrelationships between the various forms of a trace metal in sea water and their relationship to electroanalytical methods. The definitions of some of the categories (e.g. labile, non-labile complexes) have not yet been discussed quantitatively while the definitions of others (e.g. particulate matter) depend on experimental convenience rather than physical reality.

in fact, be used to titrate excess ligands in sea water so that some estimate of the natural complexing capacity of the system can be made (Matson, 1968; Shuman and Woodward, 1973). For example, Clem (1973) titrated the lead complexing ligands in a natural sea water sample by adding consecutive aliquots of ionic lead (Fig. 20.21). The plot of stripping current peak height *versus* the aliquots of standard solution added shows a distinct break indicating that the ligand responsible for reducing the lead stripping current has been

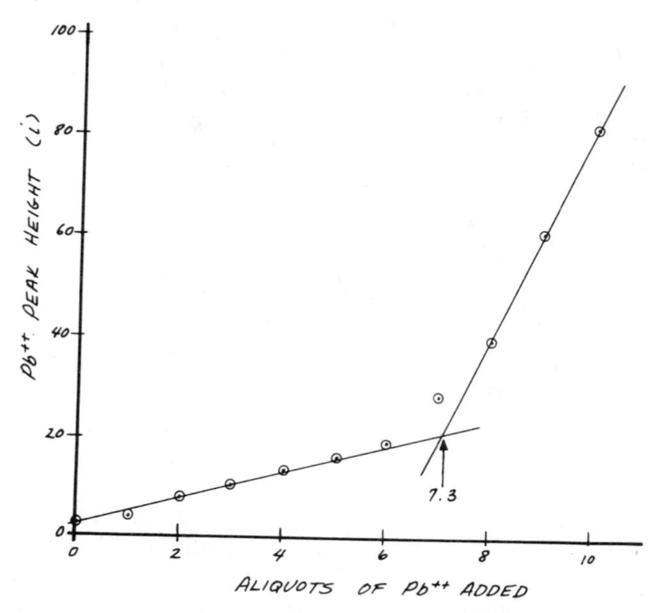

FIG. 20.21. Amperometric titration (Section 20.6) of natural sea water, using anodic stripping voltammetry, for ligands that effectively complex lead. 15 ml of sea water containing 27 μg Hg(II) were titrated with successive 25 μl aliquots containing 130 ng Pb(II) and an anodic stripping voltammogram was recorded after each addition. The ligand concentration was equivalent to 7·3 aliquots ($\sim 3{\cdot}2 \times 10^{-7}$ M). Reproduced with permission from Clem (1973).

effectively complexed. The ligand concentration in sea water was equivalent to $3{\cdot}2 \times 10^{-7}$ M of lead.

Additional information can be obtained about the nature of the electroactive components by comparing the influences of plating potential, plating time and stripping sweep rate on the final current–potential curve with that predicted by some theoretical analyses of the stripping procedure (e.g. Roe and Toni, 1965). Such procedures (Matson, 1968; Bradford, 1972) enable the rate of formation of the electroactive complexes to be estimated and provide diagnostic criteria for identifying the presence of different ligand species in the environment (Bradford, 1972).

If the sea water sample is acidified to pH 2 or 3 or irradiated with ultraviolet light then most of the organic complexes will break down to release the metal in an ionic form that is readily detected by anodic stripping analysis. By analysing an aliquot of untreated sample and then an aliquot of an acidified or irradiated sample it is, therefore, possible to estimate respectively the "free" and "bound" metal in solution. Since trace metals are readily lost to the container walls in neutral or alkaline samples such analyses are preferably carried out *in situ* or on board ship as soon as the samples are received. It is not possible at present to distinguish between colloidal material and material bound in strong organic complexes by this procedure. Coagulation of colloidal material by the addition of very long chain polymers followed by centrifugation might prove a useful tool for investigating this important fraction.

Branica *et al.* (1972) used the acidification procedure to study zinc fractionation, and they found that only 10–15% of the zinc in Mediterranean sea water was present in the labile form. The remaining 85–90% is released at pH2. In earlier experiments they were able to identify an intermediate fraction (35–40% of the total zinc) which was released at pH 6 from particulate matter in the sample. This fraction was not observed by Piro *et al.* (1973).

Fitzgerald (1970) studied the fractionation of copper in a series of sea water samples using acidification and ultraviolet oxidation techniques (Table 20.11 see Siegel, 1971). The unusual behaviour of the Sargasso Sea sample (Table 20.11) is possibly due to the desorption of copper from the surface of particulate matter in the unfiltered sample under acidic conditions. Holm-Hansen *et al.* (1970) used the MCGE to study the distribution of copper and lead in association with the development of a ciliate red tide at Barrow, Alaska. They determined ionic or weakly bound copper directly and then determined the organically bound material after oxidation with persulphuric acid. The total concentration of copper was significantly higher inside the red tide area, and 70 to 90% of this copper was present in an organically bound form.

Lang (1973) examined the fractionation of zinc, cadmium, lead and copper in Mediterranean water off the coast of Monaco, Hyère and Sète during June and July. He found great variability in the fractionation between the electroactive and the acid labile forms. Bradford (1972) has shown that in Chesapeake Bay there is a gradual build up of the acid labile component over the summer months and that large variations in the electroactive/acid labile concentration ratio are to be expected. These fractionation studies have been reviewed in some detail by Zirino *et al.* (1973). Table 20.12 briefly summarizes the ways in which anodic stripping voltammetry can be used to give some information about the chemical speciation of the amalgam forming metals.

20.5.5. CATHODIC STRIPPING

Under certain circumstances it is possible to deposit anions electrochemically as insoluble compounds on the surface of solid working electrodes (e.g. platinum or graphite electrodes). During the deposition step the working electrode acts as the anode. The insoluble film formed in this way may

TABLE 20.11

Fractionation of trace metals in filtered sea water observed by anodic stripping voltammetry.

Metal	Sample location	% electroactive	% released at pH2[a]	% released by UV oxidation[a]	Reference
Cd	Sargasso Sea (unfiltered)	50	—	50	Siegel (1971) Fitzgerald (1970)
	Mediterranean Coast (Monaco)	49(±34)	51(±34)	—	Lang (1973)
Cu	Sargasso Sea (unfiltered)	7	93	44	Siegel (1971) Fitzgerald (1970)
	Buzzard's Bay (near shore)	30	70	70	Siegel (1971) Fitzgerald (1970)
	Wood Hole ("polluted")	36	36	64	Siegel (1971) Fitzgerald (1970)
	Mediterranean Coast (Monaco)	28	72	—	Lang (1973)
Pb	Sargasso Sea (unfiltered)	83	17	−29(?)	Siegel (1971) Fitzgerald (1970)
	Monaco	51(±18)	49(±18)	—	Lang (1973)
	Hyère	44(±21)	56(±21)	—	Lang (1973)
	Sète	51(±25)	49(±25)	—	Lang (1973)
Zn	Sargasso Sea (unfiltered)	52	—	48	Siegel (1971) Fitzgerald (1970)
	N. Adriatic sea[b]	10–15	40–50	—	Branica *et al.* (1973)
	Monaco	48(±21)	52(±21)	—	Lang (1973)
	Hyère	43(±10)	57(±10)	—	Lang (1973)
	Sète	59(±23)	41(±24)	—	Lang (1973)

[a] From separate experiments on different aliquots. The two categories are not mutually exclusive.
[b] An intermediate fraction (35–40%) is sometimes released from particulate matter at pH 6.

subsequently be stripped off the working electrode by reversing the potential so that the working electrode is now the cathode in the cell. This cathodic stripping technique has been discussed in detail by Brainina (1971) and it is analogous to the anodic stripping of metals apart from the need to precipitate the anion as a compound rather than as the free element.

Brainina and Sapozhnikova (1966) have used this technique for the determination of iodide in sea water. The following sequence of reactions can take place at a graphite anode in the presence of chloride ions and rhodamine C,

TABLE 20.12

Some procedures for investigating the chemical speciation of electroactive metals in sea water using anodic stripping voltammetry (ASV).

Vary	Plot	Information obtained[a]	References
I. *ASV characteristics*[b]			
(a) Pre-electrolysis potential (E_d)	ip_c vs E_d^c	Strength of organic sequestering	Bradford (1972) Matson (1968)
(b) Pre-electrolysis time (t_d)	ip vs t_d	Kinetics of dissociation of non-labile complexes	Matson (1968) Lang (1973)
(c) Stripping scan rate (v)	ip vs v	Rate of formation of labile complexes	Matson (1968) Lang (1973)
II. *Chemistry of sample*			
(a) Concentration of electroactive metal (C_m)	Ep_d vs $C_m{}^d$	Extent of complex formation	Bradford (1972)
	ip vs C_m	Total ligand concentration capable of suppressing electroactive metal concentration	Matson (1968) Clem (1973) Shuman and Woodward (1973)
(b) Concentration of ligands (C_L)	Ep vs C_L	Stability constants of complexes	Bradford (1972) Zirino and Healy (1970)
	ip vs C_L	Effect of ligand on availability of electroactive metal	Zirino *et al.* (1973)
(c) pH[e]	ip vs pH	Release of metal from organic complexes and colloids	Piro *et al.* (1973) Branica *et al.* (1972) Zirino *et al.* (1973) Lang (1973) Fitzgerald (1970)
(d) duration(t) of UV irradiation[e]	ip vs t	As above	Fitzgerald (1970) Bradford (1972, 1973)

[a] Entries in this column only give a brief indication of the kind of information that can be obtained. The references given should be consulted for further details.

[b] Much useful information can be obtained by comparing the measured parameters with those predicted from some theoretical model of the ASV procedure (e.g. Roe and Toni, 1965; Zakharov and Trushina, 1966; Osteryoung and Christie, 1974).

[c] Peak current in resulting stripping scan.

[d] Peak potential in resulting stripping scan.

[e] The other procedures listed can be used before and after this manipulation to give more detailed information about its effects.

$$2I^- \leftrightharpoons I_2 + 2e^-$$

$$I_2 + Cl^- \leftrightharpoons I_2Cl^-$$

$$I_2Cl^- + R^+ \leftrightharpoons R[I_2Cl]\downarrow$$

where R^+ is the rhodamine C cation. 10 µg of hydrazine hydrochloride was added to 30 ml of sea water to reduce any free iodine present to iodide. The solution was then stirred while 0·5 ml of a 0·2% rhodamine C solution were added and the volume was made up to 40 ml with 0·2 N hydrochloric acid. The insoluble compound was then deposited on to the graphite anode by electrolysis for ten minutes at +0·8 V versus the saturated calomel electrode while the solution was stirred. The scan rate for the subsequent cathodic dissolution process was not given. Analysis of a sea water sample by this procedure gave a value of 62·5 $\mu g\,l^{-1}$ for iodide plus iodine. This is much greater than the values reported by other workers (Tsunogai. 1971; Whitnack, 1973; Liss *et al.*, 1973; see Section 20.4.1) suggesting that iodate ions are also determined by this procedure.

20.5.6. SUMMARY

The high sensitivity of the stripping technique provides a unique opportunity for the investigation of trace metals at natural concentrations. The analyses however tend to be rather time consuming since the solution must be effectively degassed (5–15 min) and a sufficiently long plating time must be employed (5–10 min) for adequate preconcentration of very dilute samples. The stripping process itself is normally quite rapid. These procedures must be repeated for the sample and for the one or two standard additions used in the calibration procedure. Unless the metals are quantitatively removed from the working electrode during the stripping phase the electrode itself must be renewed between each analysis. On this basis the complete analysis of a single sample can take 30 min to 1 hour. Often two or three metals can be determined simultaneously in this time. The total time for the analysis of a sequence of samples can be reduced by running a series of three or four cells in parallel (Lang, 1973) and the pre-electrolysis time can be reduced considerably by decreasing the volume and increasing the surface area of the mercury electrode by the use of thin films (Matson, 1968; Florence, 1972). The pre-electrolysis time can also be reduced if the sensitivity of the stripping phase is enhanced by the use of rapid linear scan (Macchi, 1965), ac modulated scan (Rojahn, 1973), differential pulsed scan (Donadey, 1969) or a 'step and hold' scan (Clem *et al.*, 1973). A range of other stripping procedures (Barendrecht, 1967) is also available. Clem and his co-workers have described a cell configuration with extremely efficient stirring that enables the de-oxygenating time to be drastically reduced (Clem *et al.*, 1971, 1973; Clem, 1971). The stirring unit is designed to throw the solution by centrifugal force against the walls of the cell where it forms a thin layer which can be sparged of oxygen in less than 75 s with a $5 \, l \, min^{-1}$ nitrogen flow if a 15 ml sample is used. This design also prevents the frothing of the solution which can cause problems in sea water if it is vigorously agitated by gas bubbling. These units are available commercially (MPI ElectRoCell, McKee–Pedersen Instruments). The influence of these parameters on the precision and accuracy of lead analyses has been considered in some detail (Interlaboratory lead analysis of standardized sea water, 1974). This report also presents a thorough discussion of the problems involved in maintaining consistently low blank levels (see Section 19.10.5.27).

The flow through cell designed by Lieberman and Zirino (1974) and the elegant digital electronics described by Clem and his co-workers (Clem *et al.*, 1973; Goldsworthy and Clem, 1972) may eventually be combined to give a compact, sensitive and very versatile system for the *in situ* analysis of a number of trace constituents in sea water and for the elucidation of metal complex equilibria in natural samples.

Before full advantage can be taken of these sophisticated instrumental techniques more attention should be paid to the interpretation of the stripping traces themselves. There is no agreed procedure for expressing interference effects, and little detailed work has been done on the causes of these mutual interferences (see however, Bradford, 1972). Where interference effects have been reported there is seldom agreement between authors about the identity of the principal interfering elements, let alone the magnitude of their effects. There is considerable variation in the peak potentials recorded, even for zinc (Table 20.8). These variations may be due in part to the presence of different electroactive complexes in the individual samples. As in polarography, both the strength and the weakness of the anodic stripping technique is its extreme sensitivity to variations in the chemical speciation of the electroactive components. Now that a number of simple analytical procedures have been established and the electrochemical basis of the method is more clearly understood, it should be possible to put this sensitivity to good use.

20.6. Amperometric Titrations

It is possible to extend titration procedures to very low concentration levels if the endpoint is sensed by recording the abrupt changes in the limiting current that occur at a polarized electrode. The electrochemical cell normally consists of a polarized electrode (e.g. DME, HMDE or a platinum electrode) held at a fixed potential with respect to an unpolarizable reference electrode (amperometric titration). Alternatively, two polarized electrodes can be used with a fixed potential applied across the cell (biamperometric titration). The theory of amperometric titrations has been discussed in some detail (Charlot *et al.*, 1962; Meites, 1965; Reilley and Murray, 1963; Stock, 1965) and extensive tables of suitable reaction systems have been prepared (Meites, 1963; Stock, 1965). Stock (1974) has reviewed recent work.

The experimental equipment required is much simpler than that used in polarography or in voltammetric measurements. The current measurements are made only at a single potential. It is not necessary to measure the exact value of the current, only its variation during the course of the titration need be determined. Despite this simplicity, the precision of the analyses can be increased by at least an order of magnitude over that obtainable by direct polarographic and voltammetric measurements. The solution is stirred after the addition of titrant, but is usually allowed to become quiescent before the current measurement is made. To avoid complications caused by alterations in the composition of the supporting electrolyte, the titrant is normally added either as a concentrated solution or *via* constant current coulometry. The sensitivity of the method is limited by the accuracy with which such small quantities of titrant can be added and by the sensitivity of the amperometric procedure used to sense the resulting current change.

If the reagents are generated coulometrically it is possible, for example, to titrate 5×10^{-8} M of EDTA with lead (Pouw *et al.*, 1974) or bismuth (Pouw *et al.*, 1973). Lower levels of naturally occurring complexones in sea water can be estimated if the mercury electrode in an anodic stripping cell is used as the polarized electrode (Clem, 1973, Fig. 20.21). In this instance the individual points on the titration curve represent peak currents measured on separate stripping traces recorded at successive stages in the titration. The theoretical basis of such titrations has been considered in some detail by Shuman and Woodward (1973). In these examples the current measured is that associated with the reversible discharge of the metal cation. Freese *et al.* (1970) demonstrated the possibility of titrating submicrogram quantities of copper (down to 6×10^{-8} M) by following the current changes associated with the discharge of the titrant (triethylene tetramine) at a rotating mercury-coated platinum electrode. They discussed the theory of the method in some detail. An amperometric technique has also been described for the estimation of traces of free chlorine in sea water (Baker, 1969).

As was the case with polarography and stripping voltammetry, the sensitivity of these procedures can be enhanced by applying a modulated voltage across the electrodes. Differential pulse polarography (Myers and

Osteryoung, 1974) and ac polarography (Tanaka and Ogino 1964) have been used for the titration of submicrogram concentrations of metal ions.

Amperometric titrations can also be used to determine species which are not themselves electroactive. Murakami and Hayakawa (1968) have described a method for the determination of sulphate in sea water. Barium chloride is titrated into a sea water sample, adjusted to pH 5·3 to 7·3, and causes the sulphate ion to precipitate as barium sulphate. Potassium chromate is used as an indicator species. When all the sulphate has been precipitated, the excess barium removes chromate ions and causes an abrupt change in the current flowing at the working electrode. A precision of $\pm 0{\cdot}6\%$ was obtained. A somewhat more complicated procedure based on the method of Potter and White (1954) was used by Barkley and Thompson (1960) for the determination of iodine in sea water. The total iodine in solution is converted to be free iodine which is, in turn, allowed to react with an excess of thiosulphate. The amount of iodine present is then determined by titrating the residual thiosulphate with iodate. The endpoint is located by measuring the current flowing between two polarized platinum electrodes. A precision of $\pm 4\%$ was obtained with iodate-iodine at the $50\,\mu g\,l^{-1}$ level. Truesdale and Spencer (1974) improved the precision of the method (to $\pm 1{\cdot}3\%$ at the $30\,\mu g\,l^{-1}$ level) by applying a larger voltage across the electrodes initially and following the subsequent potential changes accompanying the titration.

20.7. Net Current Flow in Stirred Solutions—The Determination of Oxygen

20.7.1. Dropping Mercury Electrodes

Since oxygen is electroactive at a wide range of noble metal substrates in neutral and alkaline solutions, considerable effort has been expended on the design of electrochemical oxygen sensors for continuous monitoring. The general approach is to hold the working electrode at a cathodic potential at which the oxygen reduction generates an appreciable diffusion current. Under favourable conditions the magnitude of this current can be related directly to the partial pressure of oxygen in the solution. The dropping mercury electrode has a continually renewed surface with well defined electrochemical characteristics, and unlike solid electrodes it is not affected by the accumulation of oxide coatings or organic films. Consequently, it has been investigated by a number of workers for the measurement of oxygen in natural waters (see for example Briggs and Knowles, 1958, 1961; Rotthauwe, 1958; Giguère and Lanzier, 1945; Grasshoff, 1963; Føyn, 1965, 1967, 1969).

A suitable potential is applied between the dropping mercury electrode and the reference electrode either from an external source (e.g. see Grasshoff, 1963) or *via* galvanic action between the reference and working electrodes (see e.g. Føyn. 1965). The diffusion current is related to a number of cell parameters by the Ilkovič equation (equation 20.11 with C_x replaced by the oxygen partial pressure P_{O_2}). It depends on the mercury flow rate (*via* the dropping time, t_d, and the drop size, m) and will, therefore, vary if the capillary orifice becomes restricted by salt deposition (Briggs and Knowles, 1958, 1961), or if the height of the mercury in the reservoir varies appreciably when the system is used in the field (Føyn, 1967). The diffusion current and, it is to be hoped, the sensitivity of the analysis, can be increased by using a wide bore capillary and hence increasing m (Føyn, 1965; Briggs and Knowles, 1961). A system with a high mercury flow rate is also less sensitive to vibration than is the conventional fine bore capillary. In practice, these advantages tend to be outweighed by the rapid consumption of purified mercury and the baseline shifts caused by alterations in the head of mercury.

The diffusion coefficient D_x (equation (20.11), varies with the salinity, temperature and pressure of the solution, and these effects must be corrected for before the oxygen partial pressure can be estimated. Other electroactive components in solution might also contribute to the diffusion current, especially in polluted waters, and consequently affect the accuracy of the measurements.

Finally, it must be stressed that oxygen electrodes measure oxygen activity (or partial pressure) rather than oxygen concentration (see van Stekelenburg, 1970). The oxygen activity is the same in sea water and in pure water equilibrated with the same gaseous atmosphere although the solubility of oxygen (i.e. the oxygen concentration observed) in the two solutions will be quite different. The oxygen electrode must, therefore, be calibrated in solutions of known oxygen *concentration* and allowance must be made for the influence of temperature and salinity on this concentration value.

The most complete study of these effects at the dropping mercury electrode in sea water has been made by Grasshoff (1963). The important features of the earlier work have been reviewed by Riley (1965a). Grasshoff (1963) used a two electrode cell with a silver–silver chloride spiral as the counter electrode. A forced drop with a life of 0·25 s was used on the dropping mercury electrode. Its potential was fixed at $-0{\cdot}8$ V with respect to the silver–silver chloride electrode. This rapid drop time resulted in a stable diffusion current that was insensitive to mechanical disturbance of the electrode. The system can therefore be used on board ship. The flow-through polarographic cell had a volume of 50–60 ml. The current recorded by the cell (i_m) is related to the *concentration* of oxygen in the solution (C_{O_2} mg l^{-1}) by the equation

$$C_{O_2} = (i_m - 0{\cdot}160) \,.\, F_T \,.\, F_S. \qquad (20.56)$$

Grasshoff (1963) has tabulated values for F_T, the temperature compensation factor, at 1°C intervals from 0° to 25°C and values for F_S, the salinity compensation factor, from 0–40‰ salinity at 1‰ S intervals. The factor 0·160 corrects the measured diffusion current for the residual and capacitance currents observed at −0·8 V in the absence of oxygen. The current sensitivity of the system is high (0·723 μA (mg l^{-1})$^{-1}$). The analytical procedure is very rapid, requiring about one minute for recording the diffusion current and for noting the temperature of the sample. The reproducibility of the measurements on consecutive aliquots of a single sample was the same as that of the Winkler technique (±0·1 %) and the estimated oxygen concentrations from the two procedures were in good agreement. This would appear to be the most rapid and precise electrochemical procedure for the determination of oxygen concentration in sea water samples. Its main limitations are its rather large appetite for purified mercury and its sensitivity to the presence of unexpected concentrations of other depolarizers.

Føyn (1965, 1967) has adapted a dropping mercury electrode for *in situ* analysis (Fig. 20.22). The dropping mercury electrode is mounted inside a zinc tube and this device is simply lowered over the side of the ship. The potential generated between the zinc and the mercury when the cell is immersed in sea water lies within the range of the second reduction step of oxygen at the mercury electrode (equation (20.51)). The current flowing between the zinc tube and the dropping mercury electrode is therefore proportional to the oxygen partial pressure. The cell is rapid in its response, and is able to resolve fine structure in the oxygen–depth profiles. The high flow rate of mercury used, however, results in a shift in the calibration between the up and down casts, and in addition, the profiles are not compensated for the rather large temperature coefficient of the diffusion current (2·5 % per °C). In practice, the cell can be calibrated by taking a series of discrete water samples for analysis by the Winkler technique in the same position as the continuous profile.

20.7.2. SOLID ELECTRODES

Because of the obvious inconvenience involved in the use of a dropping mercury electrode for prolonged monitoring, attention has been directed toward the development of oxygen electrodes based on a solid substrate. Much higher current sensitivities can be achieved with these electrodes.

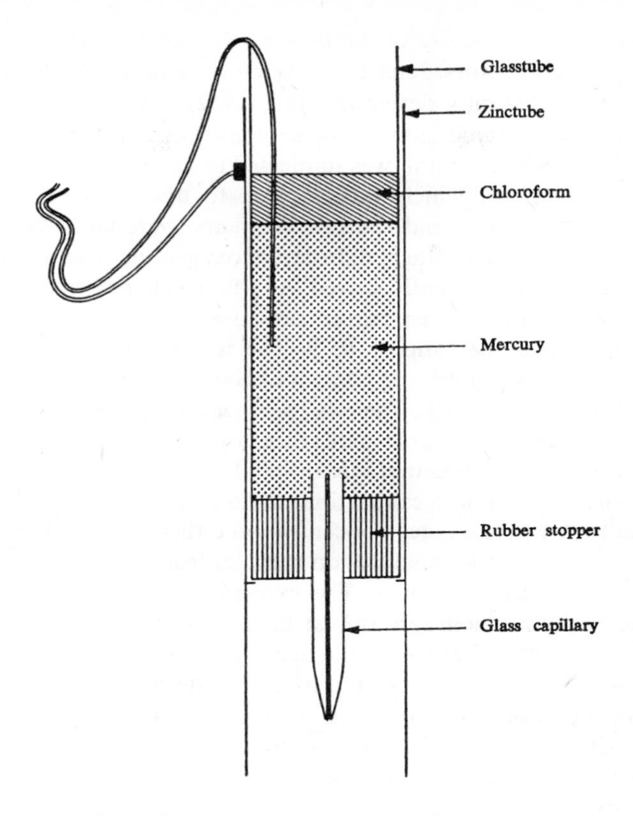

FIG. 20.22. Dropping mercury electrode for the determination of oxygen *in situ*. Reproduced with permission from Føyn (1967).

Giguère and Lanzier (1945) showed that the diffusion current for the discharge of oxygen at a rapidly rotating platinum cathode (0·5 mm diameter, 3 mm long) is about forty times greater than that observed at the dropping mercury electrode. In addition, Ozaki and Suzuki (1964) observed that a mercury coated platinum electrode is sixteen times more sensitive than the dropping mercury electrode for measuring the oxygen tension in flowing solutions. However, these electrodes are readily poisoned by organic compounds in natural samples, giving rise to erratic results, and they are used only in rather special circumstances where their sensitivity, simplicity and rapidity of response are particularly useful.

Apart from its use in high pressure experimental systems (Chandler and

Vidaver, 1971), the main use of the bare noble metal electrode has been in free-falling oxygen sensors. Jeter *et al.* (1972) have described an expendable bathyoxymeter based on a design by Ohle (1952) in which the cathode was a small piece of "gold-plated table ware" (1 cm^2) and the anode is a piece of zinc foil (4 cm^2). When the unit was immersed in sea water the two electrodes formed a galvanic cell which was allowed to discharge through a 30 Ω resistor. If the resistance and the cell geometry were suitably chosen the current was limited by the rate of diffusion of oxygen to the cathode provided that the solution was efficiently stirred. At a flow rate of 3 $m\,s^{-1}$ the current stabilized at constant oxygen partial pressure and was not influenced by variations in pH in the range 5 to 9. The temperature coefficient of the diffusion current was about 3 % per degree Celsius and the salinity was reported to have little effect above 3‰. The cell, together with its loading resistor, was mounted inside an expendable bathythermograph projectile. The voltage drop across the resistor was measured while the projectile fell freely through the water at 6 $m\,s^{-1}$. Smooth calibration curves could be drawn up relating the voltage recorded to the oxygen concentration either from tank experiments, or on the basis of Winkler analyses on samples from standard hydrographic casts taken in the same vicinity. The expendable bathyoxymeter, like the expendable bathythermograph, is best used to give a rapid though coarse picture of the essential features of the depth profile.

Westerberg (1972) studied a cell of ingenious design in which the cathode consisted of numerous platinum bristles connected to a copper wire and set in epoxy resin. The resin surface was polished so that the tips of the bristles were exposed. Since the time constant of the response of a solid electrode to changes in oxygen tension is proportional to its area, it was expected that the individual bristles would exhibit a very rapid response and that the probe would therefore be relatively insensitive to the movement of the solution. The parallel use of a large number of micro electrodes should yield a large total diffusion current. This multiple electrode was held at $-0{\cdot}8$ V with respect to a silver–silver chloride counter electrode. The platinum electrode required 5 min cathodic electrolysis at this potential before it gave a stable reading. The cell was mounted together with a thermistor and a pressure gauge in the *in situ* probe. The instrument showed excellent reproducibility in laboratory tests giving a linear relationship between current and oxygen partial pressure, but the accuracy in the field measurements was rather low (± 10 %). The main source of error appeared to be changes in the boundary layer caused by the oscillation of the probe during its descent. The design does not therefore completely eliminate the sensitivity of the cell to movement of the solution. The procedure is simple and rapid, and a coarse oxygen profile down to 100 m can be obtained in less than five minutes.

20.7.3. MEMBRANE COVERED ELECTRODES

For long term monitoring or for profiling in the deep ocean the noble metal cathode must be protected from stray contamination and from the precipitates of carbonates and hydroxides that tend to accumulate because of the high local pH values associated with the discharge of oxygen (equation (20.51)). Most oceanographic sensors are designed with a gas permeable membrane separating the voltammetric cell from the sample solution (cf Fig. 20.11). The anode is usually wound round the cathode in a concentric spiral and the diffusion layer between the cathode and the membrane is kept as thin as possible. To minimize the sensitivity of the system to shock this diffusion layer is usually stabilized by trapping a disc of filter paper soaked in the cell electrolyte between the membrane and the cathode surface. Although it protects the cathode from contamination, the gas permeable membrane also introduces a number of troublesome characteristics into the cell performance. The rate of diffusion of oxygen through the membrane is very sensitive to variations in the external temperature and pressure. Since the cell actually consumes oxygen during the measurement, the current flowing is also sensitive to changes in the diffusion gradient across the membrane caused by mechanical disturbance of the probe. In addition, the membrane itself admits not only oxygen but also carbon dioxide and water. These two components cause long term drifts in the calibration of the probe. Calibrations must be carried out at least daily if the analytical error is to be kept within 2 or 3 %. These difficulties taken together have given the oxygen membrane electrode a rather poor reputation in oceanographic circles. According to Carpenter (1972) "such devices might have application as thermometers, current meters, photometers or clocks and, in fact, an electrode has been used to sense weak motions in the deeper oceans".

There are many circumstances in which it is worth learning to live with the vagaries of the oxygen electrode because of the convenience of being able to monitor oxygen tension directly and continuously rather than having to use a discrete sampling procedure followed by rather tedious and time consuming chemical analyses. The oxygen electrode is, therefore, useful in studies of plant and animal metabolism, for *in situ* investigations of the fine structure of the oxygen profile in the deep oceans and for *in situ* measurements of the uptake of oxygen by the sediment.

There are also a number of applications where the oxygen electrode, despite its foibles, is likely to yield more accurate results than is the classical Winkler analysis. Talreja *et al.* (1969) showed that in brines containing appreciable amounts of carbonates the effervescence following acidification of the sample led to erratic results if the Winkler procedure was used. The oxygen electrode was more sensitive and reliable under these circumstances. The oxygen electrode is also well suited for work in highly contaminated conditions under which numerous side reactions might affect the reliability of the Winkler technique (McKeown *et al.*, 1967). This is particularly true in the analysis of oxygen in saline waters contaminated with pulp mill waste since components of the sulphite liquor adversely affect the Winkler technique (Felicetta and Kendall, 1965; Moore *et al.*, 1967). The oxygen electrode also gives more reliable results than does the Winkler technique for the determination of oxygen in unfiltered phytoplankton cultures (Solov'ev and Tsvetkova, 1966).

Under conditions in which the Winkler procedure works well, comparative analyses indicate that, with care, the analysis of water samples with the oxygen electrode gives comparable results (Mancy and Jaffe, 1966; Solov'ev and Tsvetkova, 1966; Boutin *et al.*, 1969; Reynolds, 1969). At sea, electrode measurements may agree with on-board Winkler analyses to within 2% (Boutin *et al.* 1969, Fig. 20.23). Detailed *in situ* measurements down to 2000 m indicate a standard deviation of electrode measurements from Winkler analyses of $\pm 0{\cdot}11\,\mathrm{mg\,l^{-1}}$ or about 2·5% (Grasshoff, 1969; Greene *et al.*, 1970; Van Landingham and Greene, 1971). On board ship, the use of the oxygen electrode is simpler and more rapid than the Winkler technique, and does not require the use of a number of reagents that must be carefully prepared, transported and stored (Reynolds, 1969). Thus, there are reasonable grounds for persevering with a description of the structure and function of oxygen membrane electrodes.

The current flowing through the cell may be related to the membrane permeability (π_m) and thickness (d) by the equation

$$i = nFA(\pi_m/d)(P_s - P_c) \tag{20.57}$$

(Greene *et al.*, 1970). A is the effective area of the cathode. The partial pressure of oxygen in the sample and at the cathode are represented by P_s and P_c respectively, F is the Faraday constant, and n (= 4) is the number of equivalents per mole of oxygen reacting. Since the discharge reaction at the cathode proceeds rapidly P_c is effectively zero. The current flowing in the membrane cell is therefore proportional to the partial pressure of oxygen in the sample. Since the solubility of oxygen in sea water is a function of both temperature and salinity (see e.g. Murray and Riley, 1969) it is necessary to convert the information received from the cell from units of oxygen partial pressure to units of oxygen concentration either by prior calibration of the system (e.g. Grasshoff, 1962) or by electronic manipulation of the cell output (Greene *et al.*, 1970; Van Landingham and Greene, 1971). When preparing calibration graphs for oxygen sensors it is important to remember that the partial

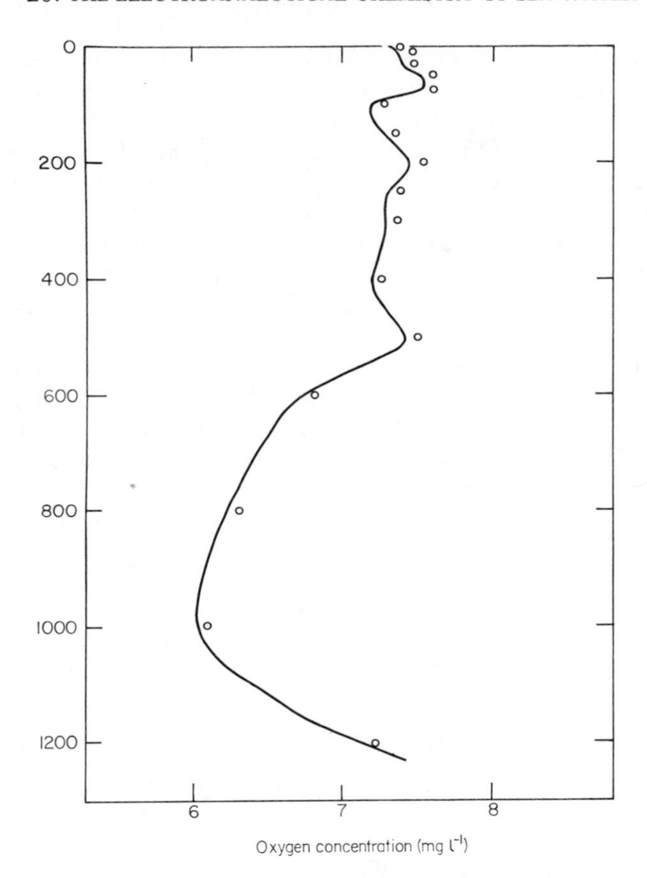

FIG. 20.23. Comparison of oxygen profiles taken at the HYDRALANTE Station III-68 (43° N, 20° W). The solid line indicates the profile as established by Winkler analyses and the circles represent data obtained on-board ship with a Beckman 777 oxygen electrode (Teflon membrane, silver anode, gold cathode). Values estimated on the same profile using an EIL 15 A oxygen electrode (PVC membrane, lead anode, silver cathode) showed a much greater deviation from the Winkler titration (up to 0·7 mg l^{-1} at 1200 m). Data taken from Boutin *et al.* (1969).

pressure of oxygen in the solution should be corrected for variations in the water vapour pressure so that

$$P_s = P_g \times (P_a - VP)/P_a \qquad (20.58)$$

where P_g is the nominal partial pressure of oxygen in the gas calibration mixture, P_a is the barometric pressure and VP is the saturation water vapour pressure of the solution at the particular salinity and temperature.

The difficulties encountered in the design of reliable membrane electrodes centre around the choice of suitable values for π_m and d (equation (20.57)). Both the sensitivity of response ($\mu A(mg\,l^{-1})^{-1}$) and the minimum sample velocity for a specified accuracy are proportional to π_m/d (Greene *et al.*, 1970). The less sensitive the electrode is to agitation, the less sensitive it is to oxygen partial pressure. In addition, the time constant for response to a stepwise change in oxygen partial pressure is proportional to the square of the membrane thickness (d). Any attempt to decrease the sensitivity of the cell to movement of the sample solution by doubling the membrane thickness will result in a fourfold increase in the response time (Greene *et al.*, 1970). The sensitivity of the electrode to flow arises because of depletion of oxygen in the immediate vicinity of the outside surface of the membrane. A sensor with a rapid response will consume a relatively large amount of oxygen and, therefore, will require a correspondingly high rate of exchange of sample solution at the membrane surface.

In practice some compromise must be made between these various demands (Table 20.13). The membrane permeability (π_m, equation 20.57) is

TABLE 20.13
The influence of membrane characteristics on the performance of an oxygen probe[a]

Membrane		Probe performance		
Material	Thickness	Current sensitivity ($\mu A\ (mg\,l^{-1})^{-1}$)	Minimum sample flow for 2% accuracy ($m\ min^{-1}$)	Time 97% response
Teflon	12	0·11	15	3 s
Teflon	24	0·03	4	32 s
Mylar	4	0·002	0	5 min

[a] Greene *et al.* (1970), silver anode, gold cathode.

a logarithmic function of temperature and pressure (Greene *et al.*, 1970; Kanwisher, 1961; Briggs and Viney, 1964). This dependence is summarized by the equation

$$\pi_{T,P} = \pi_S \exp\left(-CP(T_0 - T)/T\right), \qquad (20.59)$$

where $\pi_{T,P}$ is the permeability at temperature T°K and pressure P. π_S is the permeability at 1 atm pressure and at the reference temperature T_0 °K, C is a constant characteristic of the membrane material. For a 12 µm Teflon membrane the output current at constant oxygen tension falls by 50% when the temperature falls from 20°C to 0°C and by 44% when the pressure is increased from 1 atm to 333 atm. These effects must be accounted for in the calibration of the sensor. Alternatively, they can be compensated directly when the output current of the cell is processed (Briggs and Viney, 1964; Greene *et al.*, 1970; Van Landingham and Greene, 1971).

A variety of electrochemical cells has been used as the basis for the oxygen membrane electrode (Table 20.14). Mackereth (1964) described a useful field electrode based on a galvanic cell designed by Mancy *et al.* (1962) with a silver anode and a lead cathode. This cell (see also Duxbury, 1963; Lipner *et al.*, 1964) does not require an external voltage source, and it is therefore a little simpler than the other designs. The cathodic reaction is given by equation 20.51 and the anodic reactions in the saturated potassium bicarbonate electrolyte are

$$Pb \rightleftharpoons Pb^{2+} + 2e^- \qquad (20.60)$$

$$Pb^{2+} + 2OH^- \rightleftharpoons Pb(OH)_2. \qquad (20.61)$$

The cell described by Mackereth (1964) achieved a high current sensitivity (Table 20.14) by using electrodes with a large surface area (equation (20.57)). By this means he was able to circumvent to some extent the impasse created by the conflicting demands of sensitivity to oxygen tension and insensitivity to sample movement. A perforated cylinder of silver with a surface area of 30 cm^2 was used for the cathode. The anode was prepared by cold pressing 150 g of granular lead into an annular perspex mould. The anode was mounted concentrically within the cathode and the two electrodes were shunted by a 100 Ω resistor. A cylindrical polythene sleeve slipped over the outside of the silver cathode acted as the gas-permeable membrane. Using a 70 μm polythene film the sensitivity of the cell (S_t, μA $(mg\,l^{-1})^{-1}$) is related to the temperature (t °C) by the equation,

$$S_t = 6{\cdot}29(1 + 0{\cdot}081t + 0{\cdot}0056t^2). \qquad (20.62)$$

This high current efficiency implies a fairly rapid consumption of the electrode materials. Continuous running in air-saturated water with a 200 μA current flowing will consume lead from the anode at a rate of 0·2 $\mu g\,s^{-1}$. This corresponds to the consumption of only about 13 g of lead in 2 yr or about 8·5% of the total lead available and is not a serious restriction in practice.

A number of improvements have been made on the original design. Automatic compensation for the influence of temperature on π_m (equation (20.57)) has been incorporated within the probe (Briggs and Viney, 1964). The cell can be autoclaved up to 15 $lb\,in^{-2}$ if the original saturated potassium bicarbonate filling solution is replaced by a more dilute potassium carbonate–

TABLE 20.14
Characteristics of representative oxygen probes.

Anode	Cathode	Electrolyte	Membrane	Minimum sample flow rate (m min^{-1})	Applied potential (V)	Response time	Current sensitivity[a] (μA (mg $l^{-1})^{-1}$) in sea water	Reference
Ag/AgCl	Hg[b]	Sea water sample	—	0	−0·8	0·1 s	0·723	Grasshoff (1963)
Zn	Au	Sea water sample	—	180	Galvanic	—	39	Jeter *et al.* (1972)
Ag/Ag_2O	Pt	0·5 N KOH	20–50 μm polythene	0·12	−0·7	90% in 15 s 99% in 2 min	—	Carritt and Kanwisher (1959)
Ag/AgCl	Pt	0·2 M KCl	12 μm polythene	3[c]	−0·8	99% in 15 s	0·28	Grasshoff (1969)
Ag/AgCl	Au	Gelled KCl	12 μm Teflon	15[d]	−0·75	97% in 3 s	0·13	Greene *et al.* (1970) Van Landingham and Greene (1971)
Pb	Ag	Saturated $KHCO_3$	40–70 μm polythene	[e] —	Galvanic	90% in 15 s	30	Mackereth (1964)

[a] at 20°C, 1 atm pressure and 35‰ salinity.
[b] Dropping mercury electrode.
[c] Normal operating speed 10–20 m min^{-1}.
[d] Field tests set the minimum velocity for 2% accuracy at 40 m min^{-1}.
[e] Mackereth (1964) suggested gentle agitation of the electrode during measurement.

bicarbonate buffer. When anoxic waters are encountered hydrogen sulphide may permeate the membrane and interfere with the functioning of the cell (Grasshoff, 1962). This can be avoided by constructing the gas-permeable membrane from two sheets of polythene or Teflon film interleaved with a disc of filter paper soaked in cadmium nitrate solution (Schmid and Mancy, 1969).

Electrodes that require an applied potential use a noble metal cathode (Au or Pt) and a silver anode (Table 20.14). If an alkaline electrolyte is used, the silver anode is coated with silver oxide by pre-electrolysis in a potassium hydroxide solution before the cell is assembled (Carritt and Kanwisher, 1959; Grasshoff, 1962). The anodic reaction when potential is applied may be written as

$$2Ag + 2OH^- \rightleftharpoons Ag_2O + H_2O + 2e^-. \tag{20.63}$$

If chloride electrolytes are used in conjunction with a silver–silver chloride electrode (Grasshoff, 1969; Greene *et al.*, 1970; Van Landingham and Greene, 1971) the anodic reaction is

$$2Ag + 2Cl^- \rightleftharpoons 2AgCl + 2e^-. \tag{20.64}$$

Since hydroxyl ions are continuously produced at the cathode, the cell response will gradually drift as the consequences of reaction 20.63 become apparent at the anode (Carritt and Kanwisher, 1959).

Kanwisher (1961) described the first oxygen profiles obtained by *in situ* electrode measurements in the deep ocean down to 1700 m using a cell with a platinum cathode and silver–silver oxide anode. The effect of temperature on the membrane permeability was corrected within a few percent by adding the output of the cell to that of a thermistor mounted in the probe itself. This thermistor had a temperature coefficient approximately equal in magnitude but opposite in sign to that of the cell. (More elaborate compensating networks have since been described, see e.g. Briggs and Viney, 1964; Greene *et al.*, 1970.) The diffusion current was depressed by increasing pressure (equation 20.59). At 3000 m the response had dropped by 30% and at 10000 m it had dropped by 70%. Down to 3000 m the effect is approximately linear and can be computed out of the cell signal.

Solov'ev has described a cell similar in design to that of Kanwisher, but using a platinum cathode and a silver–silver chloride anode (Solov'ev, 1964, 1967; Solov'ev and Tsvetkova, 1966). He found that the sensitivity of the method approached 0·003 ml $O_2\,l^{-1}$, and that the measurements had an accuracy and a precision comparable to that of the Winkler technique (Solov'ev and Tsvetkova, 1966). This cell has been incorporated by Grasshoff (1969) into a more sophisticated probe that simultaneously measures depth, temperature and oxygen tension to depths of 700 m. The oxygen cell consists of a microplatinum cathode (area 0·5 cm^2) and large silver anode immersed in a potassium chloride solution. The electrolyte space is packed with small glass beads to reduce the sensitivity of the cell to vibration. Pressure compensation is provided by a silicone rubber disc set in the side of the cell. A pressure transducer and two thermistors, one for temperature measurement and one for correcting the output of the oxygen cell, are also mounted in the probe body which is 25 cm in length overall. The signals from the various sensors are transmitted directly through a six-core cable to a deck unit where the temperature and pressure signals are amplified and a temperature correction is applied to the signal from the oxygen cell. *In situ* temperature and oxygen tension are plotted against depth on an X–Y_1–Y_2 recorder, and data sampled at predetermined time intervals are recorded on punched tape for further computation. The oxygen tension measurements are converted into percentage saturation values after the cell has been calibrated by comparison with Winkler titrations. The precision of the resulting oxygen concentration measurements is $\pm 3\%$.

A probe with *in situ* correction for temperature and pressure effects and a direct readout in oxygen concentration (precision and accuracy $\pm 2\%$) has been described (Greene *et al.*, 1970; Van Landingham and Greene, 1971, Table 20.14) and is now commercially available (Beckman Inc., MINOS—Dissolved Oxygen Monitor). The sensor itself (Fig. 20.24) is mounted inside a chamber which is filled with a small quantity of mineral oil which transmits the ambient pressure to the cell electrolyte via a flexible membrane. The potassium chloride electrolyte is set into a gel to reduce the sensitivity of the cell to mechanical shock. A gold cathode is used since this apparently gives a flatter current–potential profile for oxygen discharge (Meyling and Frank, 1962). The current flowing is, therefore, less sensitive to small fluctuations in the applied voltage. Electronics carried within the sensor body use signals from a pressure transducer and a thermistor array to correct for the influence of temperature and pressure on the permeability of the membrane (equation (20.59)) and to compensate for the thermal lag that is observed with sudden changes in the ambient temperature. The monitor output will consequently remain within two per cent of the reading band with $1°\mathrm{C\,s^{-1}}$ transients from -2 to 35°C. An analogue computing element uses the signal from a thermistor to multiply the partial pressure of oxygen (after correction for temperature and pressure effects on the membrane) by the proper function of sample temperature to give a direct readout in oxygen concentration units. For work in sea water the signal is also corrected for the effect of salinity on the oxygen concentration at a reference salinity of 34·3‰ (Greene *et al.*, 1970). In the open ocean, salinity variations between 32‰ and 36‰ will result in an error of less than 0·5% in the computed oxygen concentration.

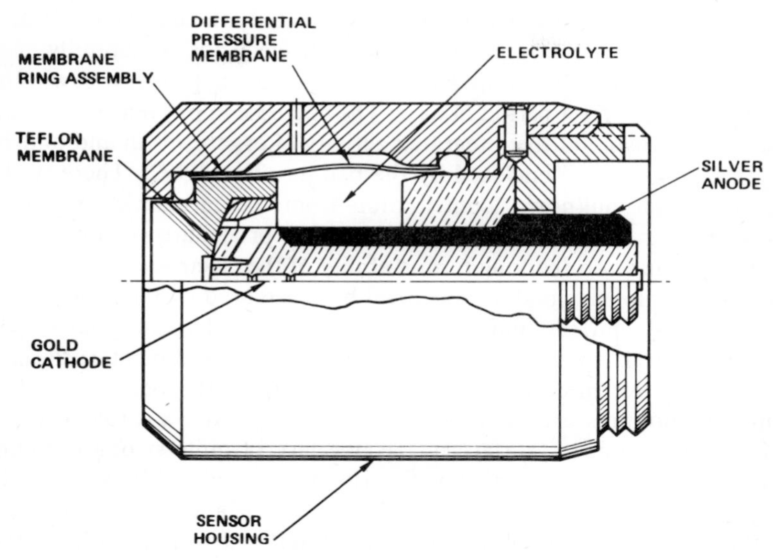

FIG. 20.24. Oxygen sensor used in the Beckman MINOS dissolved oxygen monitor. Reproduced with the permission of Beckman Instruments Inc.

In estuarine work the salinity must be measured along with the oxygen tension and corrections applied using tabulated values of oxygen solubility. The outputs from the temperature, pressure and oxygen sensors are sampled in sequence and converted into a frequency modulated signal which is transmitted up the cable (7500 m length). The working depth of the probe is restricted to 3500 m by the pressure tolerance of the pressure transducer. The oxygen cell itself has been tested to 800 atm (equivalent to 8000 m).

The initial field results (Van Landingham and Greene, 1971) indicated a marked difference in the behaviour of the probe between the upward and the downward traces. With a probe velocity of 40 $\mathrm{m\,min^{-1}}$ eighty per cent of the oxygen concentrations estimated from the downward trace were within $\pm 0{\cdot}1\ \mathrm{ml\,l^{-1}}$ of the values obtained from Winkler titrations on discrete samples taken at the same time. The corresponding figure for the upward trace was 67 %. The probe velocity required for stable readings in the field (40 $\mathrm{m\,min^{-1}}$) was higher than that reported earlier (15 $\mathrm{m\,min^{-1}}$) on the basis of laboratory tests (Greene *et al.*, 1970). Errors in the downward trace were generally negative with respect to the Winkler determinations and those from the upward trace were generally positive. The reasons for these discrepancies are not clear. The measurements obtained indicated that both the oxygen maximum at 300 m and the oxygen minimum at 650 m were subject to short term fluctuations.

Oxygen electrodes have been used *in situ* to study the uptake of oxygen by sediments. A bell jar containing an oxygen electrode is carefully placed on the sediment surface and the gradual decrease in oxygen tension within the jar is monitored over a period of days. The decrease in oxygen concentration is used as a measure of the integrated activity of organisms in and under the portion of the sea bed enclosed by the bell jar (Teal, 1972). There is also a substantial, but unknown, contribution from the inorganic oxidation of organic matter in the sediment column. This procedure obviously places extreme demands on the long term stability of the electrodes.

A rather complex tripod arrangement has been used by Pamatmat and his co-workers (Pamatmat and Fenton, 1968; Pamatmat and Banse, 1969; Pamatmat, 1971) to position two bell jars on the sediment surface. The complexity arises mainly from the need to locate the bell jars gently in position without disturbing the sediment and to provide a stable instrument platform on an uneven surface in the presence of quite strong currents. A television camera, also mounted on the tripod, was used to follow the positioning operation. A lead–silver oxygen cell was used (Mancy *et al.*, 1962) and the water within the bell jars was continuously agitated with a magnetic stirrer bar. A somewhat simpler design has been described by Stein and Denison (1967). Two divers are needed to position the apparatus on the bottom. Edberg and Hofsten (1973) further modified this design and developed a bell jar that can be positioned from a small boat. The controversy over the utility of these *in situ* oxygen uptake measurements is adequately summarized in the discussion following the paper by Teal (1972).

The oxygen electrode has proved its utility, but a number of problems still remain that make it in general a less reliable procedure than the Winkler titration for the measurement of dissolved oxygen concentration. The major difficulty arises from the need to balance the current sensitivity of the probe against its sensitivity to the movement of the external electrolyte. The ideal probe would show a very rapid response to changes in the ambient oxygen tension together with a high current sensitivity and the capability of measurement in slow flowing or stagnant conditions.

The speed of response of the electrode can be increased and the cell design simplified by depositing the cathode material directly on to the gas permeable membrane (Bergman, 1968). Gold or silver coatings can be applied to a thin PTFE membrane (3 to 13 μm) by vacuum evaporation.

In oxygen electrodes with a high current efficiency the sample solution must be rapidly stirred so that oxygen is replaced at the outer surface of the membrane as rapidly as it is consumed by the cell. In this way a constant diffusion gradient is maintained and a stable current output results. If the oxygen at the outer surface is replaced by electrolysis rather than by stirring, then the electrode should function adequately in a nearly static environment. Reilly (1968) suggested that this could be achieved by siting a fine mesh platinum electrode (a minigrid) on the sample side of the membrane. The current released by the reduction of oxygen at the cathode is then fed back, via an operational amplifier, to the minigrid so that oxygen is produced there in an equivalent amount. There should then be no net consumption of oxygen from the solution.

It is also possible to make the output from a conventional oxygen membrane sensor insensitive to the movement of the sample by using a chronoamperometric procedure (Lilley *et al.*, 1969) rather than the conventional constant potential electrolysis. If the polarizing voltage is applied as a pulse rather than as a steady signal then the rate of current decay after the pulse has been applied is given by the Cottrell equation,

$$i = nFAD^{\frac{1}{2}}C/1{\cdot}773t^{\frac{1}{2}}, \qquad (20.65)$$

where i is the current measured after time t, C is the concentration of oxygen initially present at the cathode surface, and the other symbols have the same meanings as in equation (20.57). The current flowing through the cell is related to C and this, in turn is directly proportional to the partial pressure of oxygen in the sample solution (see earlier discussion). It is possible to increase the sensitivity of a Clark oxygen cell (Clark *et al.*. 1953), similar in design to that used by Van Landingham and Greene (1971), fivefold by recording the current flowing 1·2 s after a voltage perturbation rather than by recording the continuous current at constant voltage. Higher sensitivities can be obtained by recording the current at shorter time intervals after the application of the voltage pulse (equation (20.65)). The current flowing is the same in stagnant and in well-stirred environments for periods up to 4·5 s after the application of the voltage pulse. The main drawback of the technique is that the electrode takes 20 min to re-equilibrate completely with the solution after a voltage pulse. However, 99 % equilibration is usually observed within 5 minutes. In its present form the technique is not well suited for continuous profiling, but it should replace the conventional amperometric technique for all long term, intermittent analyses. It is important to note that this technique is not confined to oxygen analysis. Any electroactive species could be determined by a similar procedure if a suitably selective membrane were available.

20.8. Net Current Flow in Stirred Solutions—Electrodeposition

In stripping analysis the trace component is initially concentrated on an electrode by controlled potential electrolysis. The concentration of the element is then estimated by reversing the electrolysis step in a quiescent solution and measuring the resulting stripping characteristics electrochemically (usually by chronoamperometry). The main analytical attraction of this procedure is its high sensitivity which enables measurements of trace metal concentrations to be made in untreated sea water samples and thus opens up the prospect of *in situ* trace metal analysis. The earlier discussion (Section 20.5.4) has emphasized the rather complicated nature of the anodic stripping trace and its dependence on the detailed chemistry of the sample. These problems must be understood more clearly before *in situ* anodic stripping analyses can be used with confidence on a routine basis.

In the meantime it is worthwhile considering the use of electrodeposition procedures for the preconcentration of trace metals for subsequent analysis by other procedures. This may eventually lead to the development of methods for sampling trace metals *in situ* that are less susceptible to contamination than are those using traditional water bottles. The concept of electrolytic preconcentration is not new as Putnam (1953) used an amalgamated zinc–silver couple to collect gold from sea water for analysis. However, only a few other examples can be quoted that may be of interest in marine chemistry.

Brinen and McClure (1972) have described a procedure for the electrodeposition of microgram quantities of lead, cadmium and bismuth at a mercury coated platinum electrode for subsequent analysis by electron spectroscopy (ESCA). Electrodeposition on graphite has also been used as a preconcentration step for the analysis of trace quantities of electroactive metals in biological materials by neutron activation (Mark. 1970; Rottschafer *et al.*, 1972). Vassos *et al.* (1973) deposited chromium, cobalt. copper, mercury, nickel and zinc from water samples onto pyrolytic graphite and analysed the deposit by x-ray spectrometry. Fairless and Bard (1973) used a 30 minute electrolysis time at $-0{\cdot}35$ V (versus the saturated calomel electrode) to deposit copper from natural sea water. The amount of copper deposited was estimated by carbon filament atomic absorption spectrometry after the preliminary evaporation of mercury. A detection limit of $0{\cdot}2\,\mu g\,Cu\,l^{-1}$ was reported. This procedure could probably be extended to the other metals that have been analysed by anodic stripping voltammetry (Table 20.8). These analyses must be calibrated by a standard addition procedure since the electrodeposition stage is not quantitative.

Considerable advantages would result if the electrodeposition stages of these analyses could be carried out *in situ*. Experience with anodic stripping voltammetry has indicated that the reproducibility of analyses between replicate samples is often an order of magnitude worse than that for consecutive analyses on the same sample (e.g. Zirino and Healy, 1972; Rojahn, 1972;

see also Table 20.15 under Zn^{2+}). This discrepancy is apparently caused by contamination introduced during the manipulation of the sample and it can be reduced by *in situ* analysis (Liebermann and Zirino, 1974). A number of difficulties arise in the design of an *in situ* electrolysis cell for the removal of a known and reproducible fraction of trace metal from solution. From equation (20.6) it is clear that such a cell should present a large surface to volume ratio and that the solution must be efficiently stirred to reduce the thickness of the diffusion layer. These conditions are best met by pumping sea water fairly rapidly through the collector electrode which is formed as a tube, a mesh or a labyrinthine thin layer cell. Since stoichiometric deposition is rarely possible under such circumstances, a large volume of sea water must be processed to obtain sufficient material for analysis. The efficiency of the deposition process will depend on the diffusion coefficient of the particular electroactive component (equation (20.6)), and this, in turn, will depend on salinity, temperature and pressure. Where non-stoichiometric deposition is used in analysis the procedure is normally calibrated by the addition of standard 'spikes' of the component being analysed. It is difficult to see how this can be achieved with *in situ* electrodeposition, and some means must be found for calibrating the collection technique under laboratory conditions. A systematic study of these problems and of the practical difficulties of cell design is now underway (H. B. Mark Jr., personal communication).

20.9. Net Current Flow in Stirred Solutions—Coulometry

Constant current coulometry provides a simple and precise method for generating accurately known quantities of chemical reagents either within the sample or in an adjacent vessel for direct addition. If a generating reaction with a 100 % current efficiency is chosen (Lingane, 1958; Charlot *et al.*, 1962; Meites, 1963) then the amount of reagent produced (C) can be calculated directly from the current flowing (I) and the duration of the electrolysis (t)

$$C = It/nF. \tag{20.66}$$

No further calibration is required. Few applications of this elegant procedure have been reported in marine chemistry. Representative examples will be discussed briefly to indicate the possible scope for further development.

Packard and Healy (1968) used a coulometric procedure to standardize the dehydrogenase assay used in the estimation of plankton respiration rates. INT (2-p-iodophenyl-3-nitrophenyl-5-phenyltetrazolium chloride) was reduced quantitatively at a mercury pool cathode to yield formazin which is

TABLE 20.15

A summary of electrochemical methods used for the analysis of sea water.

Component or property	Method	Notes	Precision	Reference
Alkalinity (Total)[a]	Potentiometric titration	Titrate sea water with HCl in closed system	±0·3%	Edmond (1970) Takahashi *et al.* (1970) Almgren *et al.* (1974)
Bi^{3+}	Linear scan ASV[b] with MCGE[c]	Sea water 1 M in HCl	±30%	Gilbert and Hume (1973)
	Linear scan ASV[b] with GCE[d]	Sea water 0·05 M in HCl	±7%	Florence (1974)
Ca^{2+}	Potentiometric titration	Titrate with EGTA and follow with Ca-selective electrode	±0·5% ±1·2%	Whitfield *et al.* (1969) Dyrssen *et al.* (1968)
CO_2 (partial pressure)	Potentiometric (direct)	Using glass electrode covered with gas permeable membrane	±10%	Wheeler (1966) Greene (1968) Moore *et al.* (1962)
CO_2 (total)	Potentiometric titration	Titrate sea water with HCl in a closed system	±0·3% ±0·1%	Edmond (1970) Takahashi *et al.* (1970) Almgren *et al.* (1974b)
Carbonate saturation	Potentiometric (direct)	Measure pH changes accompanying precipitation or dissolution of carbonate mineral	Qualitative	Weyl (1961) Ben-Yaakov and Kaplan (1971b)
Cd^{2+}	Pulse polarography	Preconcentrate on chelating resin. Determine in 0·24 M HCl–0·2 M $CaCl_2$ supporting electrolyte	±3%	Abdullah and Royle (1972)
	Linear scan ASV[b] with GCE[d]	Sea water 0·02 M in HCl	—	Florence (1972)
	ac modulated ASV[b] with	Sea water 0·015 M in HNO_3	±3% (replicates ±21%)	Rojahn (1972)

	with HMDE (Pt)[f]			
	Slow scan subtractive ASV[b] with HMDE (Pt)[f]	Two matched electrodes used in parallel	±30% (replicates)	Zirino and Healy (1972)
Chlorinity	Potentiometric titration with $AgNO_3$	Ag indicator electrode. Ag reference electrode in burette. Dead-stop titration	±0·008%	Hermann (1951)
		Pt indicator electrode Hg/Hg_2SO_4 reference electrode Dead-stop titration	±0·006%	Bather and Riley (1954)
		Differential titration with two matched Ag/AgCl electrodes	±0·01%	Reeburgh and Carpenter (1964)
		Ag indicator electrode and Gran end-point location method (automated)	±0·017%	Jagner and Årén (1970)
Co^{2+}	Pulse polarography	Preconcentrate on chelating resin. NH_4Cl/NH_4OH supporting electrolyte in presence of dimethyl glyoxime	±6%	Abdullah and Royle (1972)
		Preconcentrate by coprecipitation with MnO_2 under the influence of UV light. NH_4Cl/NH_4OH supporting electrolyte in presence of dimethyl glyoxime	±3%	Harvey and Dutton (1973)
Cu^{2+}	Pulse polarography	Preconcentrate on chelating resin. 0·24 M HCl 0·2 M $CaCl_2$ supporting electrolyte	±6%	Abdullah and Royle (1972)
	ac polarography	De-oxygenate sea water with CO_2 to give pH5	—	Odier and Plichon (1971)
	ac modulated ASV[b] with HMDE (Hg)[e]	Sea water 0·015 M in HNO_3	±2% (replicate ±8%)	Rojahn (1973)

TABLE 20.15—*continued*

Component or property	Method	Notes	Precision	Reference
	Linear scan ASV[b] with GCE[d]	Sea water 0·02 M in HCl	—	Florence (1972)
	Rapid scan with HMDE (Pt)	In Dead Sea brine with transfer to 0·5 M NH_4Cl, 0·5 M NH_3 for stripping phase	±15%	Ariel *et al.* (1964)
F^-	Potentiometry (direct)	Add citrate buffer to sea water	±1%	Warner (1971a)
		Adjust sea water to pH 6·6 with HCl	±0·88%	Anfält and Jagner (1971c)
	Standard addition potentiometry	Add citrate buffer to sea water	±1·7%	Warner (1971a)
			±2% (estuarine waters)	Warner (1971b)
		Adjust sea water to pH 6·6 with HCl	±0·59%	Anfält and Jagner (1971c)
	ac polarography	React F^- with zirconium–alizarin S complex and determine free alizarin	±0·6%	Berge and Brügman (1972)
IO_3^-	Pulse polarography	Adjust sea water to pH 3 with strong acid	—	Petek and Branica (1969b)
		Complex zinc by addition of EDTA. Measure at pH 8	±2·5%	Liss *et al.* (1973) Herring and Liss (1974)
	Amperometric titration	Convert to free iodine and back titrate excess of thiosulphate added	±4%	Barkley and Thompson (1960)
	Potentiometric titration at polarised electrodes	Convert to free iodine and back titrate excess of thiosulphate added	±1·3%	Truesdale and Spencer (1974)

I^-	Pulse polarography	Measure iodate concentration of a sea water sample before and after u.v. irradiation	$\pm 1{\cdot}4\ \mu g\ l^{-1}$	Herring and Liss (1974)
K^+	Standard addition potentiometry	Fully automated procedure with Gran end-point location	±0·3%	Anfält and Jagner (1973)
Mg^{2+}	Potentiometric titration	Using Ca^{2+} indicator electrode. Titrate with DCTA to give $(Ca^{2+} + Mg^{2+})$. Subtract Ca^{2+} determined by separate titration	±0·5%	Whitfield *et al.* (1969)
			±1–2%	Dyrssen *et al.* (1968)
		Using pH indicator electrode titrate sea water/40% ethanol mixture with NaOH. Fully automated procedure	±0·08%	Jagner (1967, 1971)
NH_3	Standard addition potentiometry	Adjust sea water to pH 11. Minimum level without distillation $30\ \mu g\ l^{-1}$	±2–5%	Gilbert and Clay (1973)
Ni^{2+}	Pulse polarography	Preconcentrate on chelating resin 0·5 M NH_4Cl/0·5 M NH_3 supporting electrolyte	±3%	Abdullah and Royle (1972)
O_2	Polarography at DME[g]	Analysis time less than 1 min	±0·1%	Grasshoff (1963)
	Polarography with membrane cell	*In situ* probe	±2·5%	Grasshoff (1969)
			±3%	Van Landingham and Greene (1971)
Pb^{2+}	Pulse polarography	Preconcentrate on chelating resin. 0·24 M HCl/0·2 M $CaCl_2$ supporting electrolyte	±6%	Abdullah and Royle (1972)
	Slow scan subtractive ASV[b] with HMDE (Pt)[f]	Two matched electrodes used in parallel	±30% (replicates)	Zirino and Healy (1972)
	ac modulated ASV[b] with HMDE (Hg)[e]	Sea water 0·015 M in HNO_3	±3% (±7% replicates)	Rojahn (1972)

TABLE 20.15—*continued*

Component or property	Method	Notes	Precision	Reference
Pb^{2+}	Linear scan with GCE[d]	Sea water 0·02 M in HCl	—	Liebermann and Zirino (1973)
pH	Potentiometric	Electrodes calibrated with NBS buffers	±0·003 pH	Pytkowicz *et al.* (1966) Takahashi *et al.* (1970)
		Electrodes calibrated on sea water pH scale[h]	±0·003 pH	Almgren *et al.* (1974)
		In situ probes	±0·02 pH	Ben-Yaakov and Kaplan (1968, 1971a) Conti *et al.* (1971) Distèche (1959, 1964) Whitfield (1971b)
Sb^{3+}	Linear scan ASV[b] with MCGE[c]	Sea water 4 M in HCl Determine Sb + Bi and subtract Bi estimated from separate determination	±20%	Gilbert and Hume (1973)
SO_4^{2-}	Potentiometric titration	Electrode sensitive to lead used to follow reaction of $Pb(NO_3)_2$ with SO_4^{2-} after initial separation of halides	±4%	Mascini (1973)
	Amperometric titration	SO_4^{2-} titrated with Ba^{2+} using a chromate indicator	±0·6%	Murakami and Hayakawa (1968)
UO_2^{2+}	Pulse polarography	Preconcentrate UO_2^{2+} by extraction with di-(2 ethyl hexyl)-phosphoric acid. Perchloric acid–sodium tartrate supporting electrolyte	±1·5%	Milner *et al.* (1961)

	Preconcentrate on chelating resin. 0·5 M NH_4Cl/0·5 M NH_3 supporting electrolyte	±3%	Abdullah and Royle (1972)
ac modulated ASV[b] with HMDE (Hg)[e]	Add 1 M acetate buffer to give pH 4·6	±2% (replicate ±16%)	Rojahn (1972)
Linear scan ASV[b] with GCE[d]	Dilute sea water 1/10. Supporting electrolyte 0·004 M HCl, 0·02 M Na acetate (pH 5·7)	—	Florence (1972)
Rapid scan ASV[b] with HMDE (Pt)[f]	Dead sea brines	±2·5%	Ariel and Eisner (1963)
Rapid scan ASV[b] with slow DME[g]	Untreated sea water	±4·5%	Macchi (1964) Bernhard *et al.* (1972)
Slow scan subtractive ASV[b] with HMDE (Pt)[f]	Two matched electrodes used in parallel	±3% (replicates ±50%)	Zirino and Healy (1971)
ASV[b] at tubular MCGE[c]	Flow through system for continual measurement	±9·5%	Liebermann and Zirino (1973)

[a] The carbonate alkalinity can also be estimated if the pH and the total boron concentration are known.
[b] Anodic stripping voltammetry.
[c] Mercury coated graphite electrode.
[d] Glassy carbon electrode.
[e] Hanging mercury drop electrode extracted from mercury filled capillary.
[f] Hanging mercury drop electrode on platinum wire.
[g] Dropping mercury electrode.
[h] Hansson (1973a).

insoluble in aqueous media. The formazin is extracted with an organic solvent and its absorbance determined using a spectrophotometer. A plot of absorbance *versus* the number of coulombs passed through the cell enables the concentration of INT to be calculated. The direct titration procedure works in this instance because the insolubility of formazin prevents the build up of the reduction product in solution. In general, however, side reactions gain prominence when the concentration of the primary reactant falls to a low level and the current efficiency falls below 100 % before the endpoint is reached. The normal procedure in coulometric titrimetry is to generate known quantities of an electroactive agent in the solution and to allow this to react with the substance being analysed, for example the reduction of Fe(III) to Fe(II) can be used to determine the concentrations of Ce(IV) in solution via the reactions

electrolysis $\quad \mathrm{Fe(III)} + e^- \rightleftharpoons \mathrm{Fe(II)}$

secondary reaction $\quad \mathrm{Fe(II)} + \mathrm{Ce(IV)} \rightleftharpoons \mathrm{Ce(III)} + \mathrm{Fe(III)}$

The course of the secondary reaction can be followed potentiometrically by a platinum electrode in this instance, and the endpoint can be located by a suitable Gran extrapolation procedure. Meites (1963) has given details of a large number of analyses that can be performed in this way.

Other analytical procedures can be calibrated by the coulometric generation of reagents. For example, Williams (1968) calibrated gas chromatographic procedures for the determination of nitrogen and oxygen in natural waters by generating known quantities of oxygen from the coulometric electrolysis of a 0·5 M potassium sulphate solution and nitrogen from the electrolysis of a 0·1 M hydrazine sulphate solution (see also Atkinson, 1972). The coulometric generation of fluoride ions (Durst and Ross, 1968) could also be used for the calibration of electrometric methods for the determination of fluoride. Microcoulometric titrations with the generation of hydroxide and hydrogen ions have been used for the determination of dissolved organic carbon in sea water as carbon dioxide (Lyutsarev, 1969) and for the estimation of the total nitrogen content in natural waters as ammonia (Moore and McNulty, 1968; Rhodes and Hopkins, 1971).

Coulometric procedures have also found use as sensitive detectors in gas chromatography (Natusch and Thorpe, 1973) and in liquid chromatography (Johnson and Larochelle, 1973; Takata and Muto, 1973). The analyte may react with an electroactive intermediate (Natusch and Thorpe, 1973) and the chromatogram will indicate the number of coulombs required to regenerate the electroactive component. Alternatively, the analyte may be removed from solution by controlled potential electrolysis and the resultant current flow monitored (Johnson and Larochelle, 1973; Takata and Muto, 1973). The peak area produced in each case will be proportional to the number of coulombs passed and hence to the concentration of the component in the effluent stream. Recent developments in this field have been reviewed by Kissinger (1974).

Microgram quantities of bismuth (Pouw *et al.*, 1973) and lead (Pouw *et al.*, 1974) can be generated coulometrically by the electrolysis of the corresponding amalgams. Such procedures should prove useful aids in the use of polarography and anodic stripping voltammetry for the elucidation of trace metal speciation and for investigating the concentration of metal chelating compounds in sea water (Pouw *et al.*, 1973, 1974; Clem, 1973).

References

Abdullah, M. I., Royle, L. G. and Morris, A. W. (1972). *Nature,* **235**, 158.

Abdullah, M. I. and Royle, L. G. (1972). *Anal. Chim. Acta,* **58**, 283.

Abresch, K. and Claasen, I. (1964). "Coulometric Analysis", Chapman and Hall, London.

Adams, R. N. (1969). "Electrochemistry at Solid Electrodes", 402 pp. Marcel Dekker, New York.

Alberg, W. J. and Hitchman, M. L. (1971). "Ring Disc Electrodes". Oxford University Press, Oxford.

Allam, A. I., Pitts, G. and Hollis, J. P. (1972). *Soil Sci.* **114**, 456.

Allegier, R. J., Halford, B. C. and Juday, C. (1941). *Trans. Wisc. Acad. Sci.* **33**, 115.

Allen, H. E., Matson, W. R. and Mancy, K. H. (1970). *J. Wat. Pollut. Control Fed.* **42**, 573.

Almgren, T., Dyrssen, D and Strandberg, M. (1974). *Deep-Sea Res.* (In press).

Ammann, D., Pretsch, E. and Simon, W. (1972). *Anal. Lett.* **5**, 843

Andersen, N. R. (1973). Convener. "Chemical oceanographic research: present status and future direction". Deliberations of a workshop held at Naval Postgraduate School, Monterey, California, December 11–15, 1972. Arlington, Virginia, Office of Naval Research, ACR-190. 80 pp.

Anderson, D. H. and Robinson, R. J. (1946). *Ind. Eng. Chem.* (Anal.) **18**, 767.

Anfält, T. and Jagner, D. (1969a). *Anal. Chim. Acta,* **47**, 57.

Anfält, T. and Jagner, D. (1969b). *Anal. Chim. Acta,* **47**, 483.

Anfält, T. and Jagner, D. (1970). *Anal. Chim. Acta,* **50**, 23.

Anfält, T. and Jagner, D. (1971a). *Anal. Chim. Acta,* **57**, 177.

Anfält, T. and Jagner, D. (1971b). *Anal. Chim. Acta,* **57**, 165.

Anfält, T. and Jagner, D. (1971c). *Anal. Chim. Acta,* **53**, 13.

Anfält, T. and Jagner, D. (1973a). *Anal. Chim. Acta,* **66**, 152.

Anfält, T. and Jagner, D. (1973b). *Anal. Chem.* **45**, 2412.

Ariel, M. and Eisner, U. (1963). *J. Electroanal. Chem.* **5**, 362.

Ariel, M., Eisner, U. and Gottesfeld, S. (1964). *J. Electroanal. Chem.* **7**, 307.

Armstrong, F. A. J., Williams, P. M. and Strickland, J. D. H. (1966). *Nature,* **211**, 481.

Atkinson, L. P. (1972). *Anal. Chem.* **44**, 885.
Bagg, J. and Rechnitz, G. A. (1973a). *Anal. Chem.* **45**, 1069.
Bagg, J. and Rechnitz, G. A. (1973b). *Anal. Chem.* **45**, 271.
Baker, R. J. (1969). *Ind. Water Eng.* **6**, 20.
Barendrecht, E. (1967). *In* "Electroanalytical Chemistry", Vol. 2 (A. J. Bard, ed.). Edward Arnold, London.
Barić, A. and Branica, M. (1967). *J. Polarogr. Soc.* **13**, 4.
Barić, A. and Branica, M. (1969). *Limnol. Oceanogr.* **14**, 796.
Barica, J. (1973). *J. Fish. Res. Bd. Can.* **30**, 1389.
Barkley, R. A. and Thompson, T. G. (1960). *Anal. Chem.* **32**, 154.
Barker, G. C. and Gardner, A. W. (1973). *J. Electroanal. Chem.* **46**, 150.
Barker, G. C., Gardner, A. W. and Williams, M. J. (1973). *J. Electroanal. Chem.* **42**, App 21.
Bates, R. G. (1973). "Determination of pH, Theory and Practice", 479 pp. John Wiley, London.
Bates, R. G. and Alfenaar, M. (1969). *In* "Ion Selective Electrodes" (R. A. Durst, ed.), pp 191–214. National Bureau of Standards Special Publication No. 314. U.S. Govt Printing Office, Washington, D.C.
Bates, R. G., Staples, B. R. and Robinson, R. A. (1970). *Anal. Chem.* **42**, 867.
Bather, J. M. and Riley, J. P. (1953). *J. cons. perm. int. Explor. Mer.* **18**, 277.
Baumann, E. W. (1971). *Anal. Chim. Acta*, **54**, 189.
Baumann, E. W. (1973). *Anal. Chim. Acta*, **64**, 284.
Ben-Yaakov, S. (1970). *Limnol. Oceanogr.* **15**, 326.
Ben-Yaakov, S. (1972). *Geochim. Cosmochim. Acta*, **36**, 1395.
Ben-Yaakov, S. (1973). *In* "Marine Electrochemistry", pp 111–123 (J. B. Berkowitz, R. A. Horne, M. Banus, P. L. Howard, M. J. Pryor and G. C. Whitnack, eds.). Electrochem. Soc., Princeton, N.J. 406 pp.
Ben-Yaakov, S. and Kaplan, R. (1968a). *Rev. Sci. Instrum.* **39**, 1133.
Ben-Yaakov, S. and Kaplan, I. R. (1968b). *J. Ocean. Tech.* **2**, 25.
Ben-Yaakov, S. and Kaplan, I. R. (1968c). *Limnol. Oceanogr.* **13**, 688.
Ben-Yaakov, S. and Kaplan, I. R. (1969). *Limnol. Oceanogr.* **14**, 874.
Ben-Yaakov, S. and Kaplan, I. R. (1971a). *Mar. Tech. Soc. J.* **5**, 41.
Ben-Yaakov, S. and Kaplan, I. R, (1971b). *J. Geophys. Res.* **76**, 722.
Ben-Yaakov, S. and Kaplan, I. R. (1973). *J. Mar. Res.* **31**, 79.
Ben-Yaakov, S. and Kaplan, I. R. (1973b). *In* "Marine Electrochemistry", pp 98–108 (J. B. Berkowitz, R. A. Horne, M. Banus, P. L. Howard, M. J. Pryor, G. C. Whitnack, eds.). Electrochem. Soc. Princeton, N.J. 406 pp.
Ben-Yaakov, S. and Ruth, E. (1974). *Limnol. Oceanogr.* **19**, 144.
Ben-Yaakov, S., Ruth, E. and Kaplan, I. R. (1974). *Deep-Sea Res.* **21**, 229.
Berge, H. and Brügmann, L. (1972). *Beitr. Meeresk.* **29**, 115.
Bergmann, I. (1968). *Nature*, **218**, 266.
Berner, R. A. (1963). *Geochim. Cosmochim. Acta*, **27**, 563.
Bernhard, M., Branica, M. and Piro, A. (1972). "Zinc in Sea Water I: Distribution of Zn in the Ligurian Sea and Gulf of Taranto under Special consideration of the sampling procedure". (In press).
de Bethune, A. J. and Swerdeman-Loud, N. A. (1964). *In* "Encyclopedia of Electrochemistry" (C. A. Hampel, ed.), pp 414–426. Reinhold, New York.
Bewers, J. M., Miller, G. R. Jnr, Kester, D. R. and Warner, T. B. (1973). *Nature Phys. Sci.* **242**. 142.

Bond, A. M. (1973a). *Anal. Chem.* **45**, 2027.
Bond, A. M. (1973b). *Talanta*, **20**, 1139.
Bond, A. M. and Canterford, J. H. (1971). *Anal. Chem.* **43**, 228.
Bond, A. M. and Canterford, D. R. (1972a). *Anal. Chem.* **44**, 721.
Bond, A. M. and Canterford, D. R. (1972b). *Anal. Chem.* **44**, 1803.
Bos, M. and Meersherck (1972). *Anal. Chim. Acta*, **61**, 185.
Bos, P. and Van Dalen, E. (1973). *J. Electroanal. Chem.* **45**, 165.
Boutin, C., Grimaldi, J. P., Pencalet, J. P. and Revel, J. (1969). *Cah. Oceanogr.* **21**, 555.
Bradford, W. L. (1972). "A Study on the Chemical Behaviour of Zinc in Chesapeake Bay Water using Anodic Stripping Voltammetry", 103 pp. Tech. Rep. Chesapeake Bay Inst., No. 76.
Bradford, W. L. (1973). *Limnol. Oceanogr.* **18**, 757.
Brainina, Kh. Z. (1971). *Talanta*, **18**, 513.
Brainina, Kh. Z. and Sapozhnikova, É. Ya. (1966). *Russ. J. Anal. Chem.* **21**, 1192.
Brand, M. J. D. and Rechnitz, G. A. (1969). *Anal. Chem.* **42**, 1172.
Brand, M. J. D. and Rechnitz, G. A. (1970). *Anal. Chem.* **41**, 616.
Branica, M. (1970). *In* "Reference Methods for Marine Radioactivity Studies", pp 243–259. Tech. Report Ser. No. 118, IAEA, Vienna.
Branica, M. and Petek, M. (1969). *Rapp. Comm. int. Mer Médit.* **19**, 767.
Branica, M., Petek, M., Barić, A. and Jeftić L. (1969). *Rapp. Comm. int. Mer Médit.* **19**, 929.
Branica, M., Piro, A. and Bernhard, M. (1972). "Zinc in Sea Water II: Determination of Physico-chemical States of Zinc in Sea Water". (In press).
Breck, W. G. (1972). *J. Mar. Res.* **30**, 121.
Breck, W. G. (1973). *J. Mar. Res.* **31**, 83.
Brewer, P. G. and Spencer, D. W. (1970). Woods Hole Oceanographic Institution, Technical Report No. 70–21.
Brewer, P. G., Spencer, D. W. and Wilkniss, P. E. (1970). *Deep-Sea Res.* **17**, 1.
Breyer, B. and Bauer, H. H. (1963). "Alternating Current Polarography and Tensammetry". Interscience, New York.
Briggs, R. and Knowles, G. (1958). *Analyst*, **83**, 304.
Briggs, R. and Knowles, G. (1961). *Analyst*, **86**, 603.
Briggs, R. and Viney, M. (1964). *J. Scient. Instrum.* **41**, 78.
Brinen, J. S. and McClure, J. E. (1972). *Anal. Lett.* **5**, 737.
Brinkman, A. A. M. (1968). "Theory of Pulse Polarography". Drukkerij-Bronder-Offset, Rotterdam.
Broecker, W. S. (1971). *Quat. Res.* **1**, 188.
Brooks, R. R., Presley, B. J. and Kaplan, I. R. (1967). *Talanta*, **14**, 809.
Browning, D. R. (1969) ed. "Electrometric Methods", 131 pp. McGraw-Hill, London.
Buffle, J. (1972). *Anal. Chim. Acta*, **59**, 439.
Buffle, J., Parthasarathy, N. and Monnier, D. (1972). *Anal. Chim Acta*, **59**, 427.
Carlson, R. M. and Paul, J. L. (1968). *Anal. Chem.* **40**, 1292.
Carlson, R. M. and Paul, J. L. (1969). *Soil. Sci.* **108**, 266.
Carpenter, J. H. (1972). *In* "Analytical Chemistry: Key to Progress on National Problems" (W. W. Meinke and J. K. Taylor eds), pp 403–404. National Bureau of Standards Special Publication 351. U.S. Govt Printing Office, Washington D.C.
Carpenter, J. H. and Manella, M. E. (1973). *J. Geophys. Res.* **78**, 3621.
Carr, P. W. (1971). *Anal. Chem.* **43**, 425.

Carr, P. W. (1972). *Anal. Chem.* **44**, 453.
Carritt, D. E. and Kanwisher, J. W. (1959). *Anal. Chem.* **31**, 5.
Carter, J. M. and Hume, D. N. (1972). *Chem. Anal. (Warsaw)*, **17**, 747.
Chandler, M. T. and Vidaver, W. (1971). *Rev. Sci. Instr.* **42**, 143.
Charlot, G., Badoz-Lambling, J. and Trémillon, B. (1962). "Electrochemical Reactions", 376 pp. Elsevier, Amsterdam.
Christian, G. D. (1969). *J. Electroanal. Chem.* **23**, 1.
Clark, L. C. Jr, Wolf, R., Granger, D. and Taylor, Z. (1953). *J. Appl. Physiol.* **6**, 189.
Clem, R. G. (1971). *Anal. Chem.* **43**, 1853.
Clem, R. G. (1973). *McKee-Pedersen Instruments. Application Notes*, **8** (i).
Clem, R. G. and Goldsworthy, W. W. (1971). *Anal. Chem.* **43**, 918.
Clem, R. G., Jakob, F., Anderberg, D. H. and Ornelas, L. D. (1971). *Anal. Chem.* **43**, 1938.
Clem, R. G., Litton, G. and Ornelas, L. D. (1973). *Anal. Chem.* **45**, 1306.
Coleman, D. M., Van Atta, R. E. and Klatt, L. N. (1972). *Envir. Sci. Technol.* **6**, 452.
Connors, D. N. and Park, K. (1967). *Deep-Sea Res.* **14**, 481.
Connors, D. N. and Weyl, P. K. (1968). *Limnol. Oceanogr.* **13**, 39.
Conti, U. (1972). "Study of an Underwater Environmental Monitor". Ph.D. Thesis, University of California.
Conti, U. and Wilde, P. (1972). *Mar. Tech. Soc. J.* **6**(2), 17.
Conti, U., Wilde, P. and Richards, T. L. (1971). Offshore Tech. Conf., Dallas, Paper 1456.
Cooper, L. H. N. (1937). *J. Mar. Biol. Ass. U.K.* **22**, 167.
Copeland, T. R., Christie, J. H., Osteryoung, R. A. and Skogerboe, R. K. (1973). *Anal. Chem.* **45**, 2171.
Covington, A. K. (1966). *Electrochim. Acta*, **11**, 959.
Covington, A. K. (1969). *In* "Ion Selective Electrodes" (R. A. Durst, ed.), pp 107–190. National Bureau of Standards Special Publication 314. U.S. Government Printing Office, Washington, D.C.
Covington, A. K. (1974). C.R.C. *Crit. Rev. Anal. Chem.* **4**, 356.
Cox, R. A., Culkin, F. and Riley, J. P. (1967). *Deep-Sea Res.* **14**, 203.
Crow, D. R. (1969). "Polarography of Metal Complexes", 203 pp. Academic Press, London.
Culberson, C., Pytkowicz, R. M. and Hawley, J. E. (1970). *J. Mar. Res.* **28**, 15.
Culkin, F. C. and Cox, R. A. (1966). *Deep-Sea Res.* **13**, 789.
Davey, E. W., Morgan, M. J. and Erickson, S. J. (1973). *Limnol. Oceanogr.* **18**, 993.
Davies, J. E. W., Moody, G. J. and Thomas, J. D. R. (1972). *Analyst*, **97**, 87.
Davis, H. M. and Rooney, R. C. (1962). *J. Polarogr. Soc.* **8**, 25.
Delahay, P. (1954). "New Instrumental Methods in Electrochemistry", 437 pp. Interscience, New York.
Distèche, A. (1959). *Rev. Sci. Instrum.* **30**, 474.
Distèche, A. (1964). *Bull. Inst. Océanogr.* **64**, 1320.
Distèche, A. and Dubuisson, M. (1960). *Bull. Inst. Oceanogr.* **57**, 1174.
Donadey, G. (1969). Ph.D. Thesis, Faculté des Sciences, L'Université de Paris. 88 pp.
Donadey, G. and Rosset, R. (1972). *Rapp. Comm. int. Mer. Medit.* **20**, 657.
Durst, R. A. (1969a) ed. "Ion-selective Electrodes", 452 pp. National Bureau of Standards Special Publication 314. U.S. Government Printing Office, Washington D.C.
Durst, R. A. (1969b). *Mikrochim. Acta*, **3**, 611.

Durst, R. A. and Ross, J. W. Jr (1968). *Anal. Chem.* **40**, 1343.
Duxbury, A. C. (1963). *Limnol. Oceanogr.* **8**, 483.
Dyrssen, D. (1965). *Acta Chem. Scand.* **19**, 1265.
Dyrssen, D. and Jagner, D. (1966). *Anal. Chim. Acta*, **35**, 407.
Dyrssen, D. and Sillén, L. G. (1967). *Tellus*, **19**, 113.
Dyrssen, D., Jagner, D. and Wengelin, F. (1968a). "Computer Calculations of Ionic Equilibria and Titration Procedures", 250 pp. Almqvist and Wiksell, Stockholm.
Dyrssen, D., Jagner, D. and Johansson, H. (1968b). Report on the Chemistry of Seawater, V. Dept Anal. Chem., University of Gothenberg, Sweden.
Dyrssen, D., Novikov, Y. and Uppstrom, L. (1972). *Anal. Chim. Acta*, **60**, 139.
Edberg, N. and Hofsten, B. (1973). *Wat. Res.* **7**, 1285.
Edmond, J. M. (1970). *Deep-Sea Res.* **17**, 737.
Edmond, J. M. (1972). *Proc. R. Soc. Edinburgh (B)*, **72**, 371.
Eisenman, G. (1967). Ed. "Glass Electrodes for Hydrogen and Other Cations". Edward Arnold, London. 582 pp.
Eisner, U., Rottschafer, J. M., Berlandi, F. J. and Mark, H. B. Jr (1967). *Anal. Chem.* **39**, 1466.
Ellis, W. D. (1973). *J. Chem. Educ.* **50**, A131.
Erhardt, M. (1969). *Deep-Sea Res.* **16**, 393.
Ewing, G. W. (1969). *J. Chem. Ed.* **46**, A717.
Fairless, C. and Bard, A. J. (1973). *Anal. Chem.* **45**, 2289.
Felicetta, V. F. and Kendall, D. R. (1965). *TAPPI*, **48**, 362.
Fitzgerald, W. F. (1970). Ph.D. Thesis, Mass. Inst. of Technol., Cambridge, Mass.
Flato, J. B. (1972). *Anal. Chem.* **44**, 75A.
Fleet, B. and Ho, A. Y. W. (1973). *Talanta*, **20**, 793.
Florence, T. M. (1970). *J. Electroanal. Chem.* **27**, 273.
Florence, T. M. (1972). *J. Electroanal. Chem.* **35**, 237.
Florence, T. M. (1974). *J. Electroanal. Chem.* **49**, 255.
Florence, T. M. and Farrar, Y. J. (1973). *J. Electroanal. Chem.* **41**, 127.
Føyn, E. (1965). *In* "Progress in Oceanography" (M. Sears ed.), Vol. 3, pp 137–144. Pergamon Press, London.
Føyn, E. (1967). *In* "Chemical Environment and the Aquatic Habitat" (E. Gottermann and E. Clymo, eds) pp 127–132. N.V. Noord-Hollandsche Uitgevers Maatschappij, Amsterdam. 322 pp.
Føyn, E. (1969). *In* "Chemical Oceanography" (R. Lange, ed.), pp 126–129. Universitets Forlaget, Copenhagen.
Frant, M. S. and Ross, J. W. Jr (1968). *Analyt. Chem.* **40**, 1169.
von Fraunhofer, J. A. and Banks, C. H. (1972). "Potentiostat and its Applications", 254 pp. Butterworths, London.
Freese, F., Jasper, H. J. and Den Boef, G. (1970). *Talanta*, **17**, 945.
Fuoss, R. M. (1968). *Rev. Pure and Appl. Chem.* **18**, 125.
Garrels, R. M. (1967). *In* "Glass Electrodes for Hydrogen and other Cations" (G. Eisenman, ed.), pp 344–361. Marcel Dekker, New York.
Gertz, K. H. and Loeschke, N. H. (1958). *Naturwiss.* **45**, 160.
Gibbard, H. F. Jr (1973). *J. Electrochem. Soc.* **120**, 624.
Gieskes, J. M. (1967). *Kiel. Meeresforsch.* **23**, 75.
Gieskes, J. M. (1970). *Limnol. Oceanogr.* **15**, 329.
Giguère, P. A. and Lanzier, L. (1945). *Canad. J. Res.* **B23**, 76.
Gilbert, T. R. (1971). Ph.D. Thesis, Mass. Inst of Technol., Cambridge, Mass.

Gilbert, T. R. and Clay, A. M. (1973a). *Anal. Chem.* **45**, 1757.
Gilbert, T. R. and Clay, A. M. (1973b), *J. Electrochem. Soc.* **120**, C254 (abstract only).
Gilbert, T. R. and Hume, D. N. (1973). *Anal. Chim. Acta*, **65**, 451.
Goldsworthy, W. W. and Clem, R. G. (1971). *Anal. Chem.* **43**, 1718.
Goldsworthy, W. W. and Clem, R. G. (1972). *Anal. Chem.* **44**, 1360.
Gran, G. (1950). *Acta. Chem. Scand.* **4**, 559.
Gran, G. (1952). *Analyst*, **77**, 661.
Grasshoff, K. (1962). *Kieler Meersforsch.* **18**, 151.
Grasshoff, K. (1963). *Kieler Meeresforsch.* **19**, 8.
Grasshoff, K. (1969). *Kieler Meeresforsch.* **25**, 133.
Greene, M. W. (1968). *Mar. Sci. Instr.* **4**, 116.
Greene, M. W., Gafford, R. D. and Rohrbaugh, D. G. (1970). *In* "Marine Technology", Vol. 2, pp 1485–1502. Marine Technology Society, Washington D.C.
Greenhalgh, R. and Riley, J. P. (1963). *Nature,* **197**, 371.
Griffiths, G. H., Moody, G. J. and Thomas, J. D. R. (1972). *Analyst*, **97**, 420.
Guilbault, G. G. and Brignac, P. (1969). *Anal. Chem.* **4**, 1136.
Gutknecht, W. F. and Perone, J. P. (1970). *Anal. Chem.* **42**, 906.
Hahn, P. B., Wechter, M. A., Johnson, D. C. and Voigt, A. F. (1973). *Anal. Chem.* **45**, 1016.
Hansson, I. (1973a). *Deep-Sea Res.* **20**, 479.
Hansson, I. (1973b). *Deep-Sea Res.* **20**, 461.
Hansson, I. (1973c). *Acta Chem. Scand.* **27**, 931.
Hansson, I. (1973d). *Acta Chem. Scand.* **27**, 924.
Hansson, I. and Jagner, D. (1973). *Anal. Chim. Acta*, **65**, 363.
Harrison, D. E. F. and Melbourne, K. V. (1970). *Biotechnol. Bioeng.* **12**, 633.
Harvey, B. R. and Dutton, J. W. R. (1973). *Anal. Chim. Acta*, **67**, 377.
Harwell, K. E. (1954). *Anal. Chem.* **26**, 616.
Hawley, J. E. and Pytkowicz, R. M. (1973). *Mar. Chem.* **1**, 245.
Hayes, J. W., Glover, D. E., Smith, D. E. and Overton, M. W. (1973). *Anal. Chem.* **45**, 277.
Headridge, J. B. (1969). "Electrochemical Techniques for Inorganic Chemists", 124 pp. Academic Press, New York.
Hecht, F., Korkisch, J., Patzak, R. and Thiard, A. (1956). *Mikrochim. Acta*, **9**, 1283.
Hermann, F. (1951). *J. cons. perm. int. Explor. Mer.* **17**, 223.
Herman, H. B. and Rechnitz, G. A. (1974). *Science, N.Y.* **184**, 1074.
Herring, J. R. and Liss, P. S. (1974). *Deep-Sea Res.* **21**, 777.
Heyrovsky, J. and Kuta, J. (1966). "Principles of Polarography". Academic Press, New York.
Hills, G. J. (1965). *Talanta*, **12**, 1317.
Hills, G. J. and Ovenden, P. (1966). *Adv. Electrochem.* **4**, 185.
Hindman, J. C., Anderson, L. J. and Moberg, E. G. (1949). *J. Mar. Res.* **8**, 30.
Holm-Hansen, O., Taylor, F. J. R. and Barsdate, R. J. (1970). *Mar. Biol.* **7**, 37.
Hseu, T. M. and Rechnitz, G. A. (1968). *Anal. Chem.* **40**, 1054.
Hubbard, A. T. (1973). *Crit. Rev. Anal. Chem.* **3**, 201.
Hubbard, A. T. and Anson, F. C. (1970). *In* "Electroanalytical Chemistry" (A. J. Bard, ed.), Vol. 4, Chapter 2. Edward Arnold, London.
Huderová, L. and Štulík, K. (1972). *Talanta*, **19**, 1285.
Hulanicki, A. and Trojanowicz, M. (1973). *Talanta*, **20**, 599.

Ingle, S. E., Culberson, C. H., Hawley, J. E. and Pytkowicz, R. M. (1973). *Mar. Chem.* **1**, 295.
Ingman, F. and Still, E. (1966). *Talanta*, **13**, 1431.
"Interlaboratory Lead Analysis of Standardised Samples of Sea Water" (1974). *Mar. Chem.* **2**, 69.
Ingri, N., Kakolowicz, W., Sillén, L. G. and Warnqvist, B. (1967). *Talanta*, **14**, 1261.
Isbell, A. F. Jr, Pecsok, R. L., Davies, R. H. and Purnell, J. H. (1973). *Anal. Chem.* **45**, 2363.
Ishibashi, M., Fujinaga, T., Izutsu, K., Yamamoto, T. and Tamura, H. (1961). *Rec. Oceanogr. Wks Jap. N.S.* **6** (i), 106.
Ives, D. J. G. and Janz, J. G. (1961). "Reference Electrodes: Theory and Practice", 651 pp. Academic Press, New York.
Jackson, J. (1948). *Chem. and Ind. (London)*, **1948**, 7.
Jagner, D. (1967). *Report on Chemistry of Sea Water, III.* Dept Anal. Chem., University of Gothenburg, Sweden.
Jagner, D. (1970). *Anal. Chim. Acta*, **50**, 15.
Jagner, D. (1971). "A Computer Treatment of Theoretical and Practical Aspects of Titration Procedures". Gothenburg Dissertation in Science No. 25. University of Gothenburg.
Jagner, D. and Årén, K. (1970). *Anal. Chim. Acta*, **52**, 491.
Janata. J. and Mark, H. B. Jr. (1967). *Anal. Chem.* **39**, 1896.
Jasinski, R. and Trachtenberg, I. (1973). *Anal. Chem.* **45**, 1277.
Jasinski, R., Trachtenberg, I. and Andrychuck, D. (1974). *Anal. Chem.* **46**, 364.
Jensen, C. R., Van Gudy, S. D. and Stolzy, L. M. (1965). *Soil Sci. Soc. Am. Proc.* **29**, 631.
Jeter, H. W., Føyn, E., King, M., Gordon, L. I. (1972). *Limnol. Oceanogr.* **17**, 288
Johansson, A. and Pehrsson, L. (1970). *Analyst.* **95**, 652.
Johnson, D. C. and Larochelle, J. (1973). *Talanta*, **20**, 959.
Kambara, T. (1966). "Modern Aspects of Polarography". Plenum Press, N.Y.
Kanwisher, J. (1961). *Mar. Sci. Instr.* **1**, 334.
Kečkeš, S. and Pucar, Z. (1966). Annual Report 1966, IEA Research. Contract No. 201/R2/RB, 287 pp. Institute "Ruder Boškvić" Rovinj and Zagreb.
Keller, H. E. and Osteryoung, R. A. (1971). *Anal. Chem.* **43**, 342.
Kemula, W. and Kublik, Z. (1963). *Adv. Anal. Chem. Instr.* **2**, 123.
Kissinger, P. T. (1974). *Anal. Chem.* **46**, 19R.
Korkisch, J., Thiard, A. and Hecht, F. (1956). *Mikrochim. Acta*, **9**, 1422.
Koryta, J. (1972). *Anal. Chim. Acta*, **61**, 329.
Koske, P. H. (1964). *Kiel. Meeresforsch*, **20**, 138.
Kozlovskii, M. T. and Zebreva, A. I. (1972). *Progr. Polarogr.* **3**, 157.
Kramer, J. R. (1961). Proc. 4th Conference on Great Lakes Research. Publ. 7, pp. 27–56. Great Lakes Res. Div. Instr. Sci. and Techn. University of Michigan.
Kunkler, J. L., Koopman, F. L. and Swenson, F. A. (1967). U.S. Geol. Surv. Prof. Paper 575-B, pp B250–253.
Lagrange, P. and Schwing, J. P. (1970). *Anal. Chem.* **42**, 1844.
Laitinen, H. (1967). *In* "Trace Characterisation" (W. W. Meinke and B. F. Scribner, eds), pp 75–120. NBS Monograph 100. U.S. Govt. Printing Office, Washington, D.C.
Lang, H. N. (1973). Ph.D. Thesis, Institut de Mathématiques et Sciences Physique, L'Université de Nice.

Laser, D. and Ariel, M. (1974). *J. Electroanal. Chem.* **49**, 123.
Lee, W. H. (1969). *In* "Electrometric Methods" (D. R. Browning, ed.), pp 1–35. McGraw-Hill, London. 131 pp.
Lerman, A. and Shatkay, A. (1968). *Earth Planet. Sci. Letters*, **5**, 63.
Leyendekkers, J. V. (1973). *Mar. Chem.* **1**, 75.
Leyendekkers, J. V. and Whitfield, M. (1970). *Anal. Chem.* **42**, 444.
Leyendekkers, J. V. and Whitfield, M. (1971). *Anal. Chem.* **43**, 322.
Lieberman, S. H. and Zirino, A. (1974). *Anal. Chem.* **46**, 20.
Light, T. S. and Fletcher, K. S. (1967). *Anal. Chem.* **39**, 70.
Light, T. S. and Swartz, J. L. (1968). *Anal. Lett.* **1**, 825.
Lilley, M. D., Story, J. B. and Raible, R. W. (1969). *J. Electranal. Chem.* **23**, 425.
Lingane, J. (1958). "Electroanalytical Chemistry", 2nd Edition, 669 pp. Interscience, New York.
Lipner, H., Witherspoon, L. R. and Champeaux, V. C. (1964). *Anal. Chem.* **36**, 204.
Liss, P. S., Herring, J. R. and Goldberg, E. D. (1973). *Nature Phys: Sci.* **242**, 108.
Loveland, J. W. (1963). "Conductometry and Oscillometry. *In* "Treatise on Analytical Chemistry". (I. M. Kolthoff and P. J. Elving, eds), Part I, Vol. 4, p. 2569. Interscience, New York.
Lyman, J. (1972). *Proc. R. Soc. Edinburgh* (*B*), **72**. 381.
Lyman, J. and Fleming, R. H. (1940). *J. Mar. Res.* **3**, 134.
Lyutsarev, S. V. (1969). *Gidrokhim. Mater.* **49**, 207.
McCallum, C. and Midgley, D. (1973). *Anal. Chim. Acta*, **65**, 155.
MacIntyre, F. (1970). *Scient. Amer.* **235**(5), 164.
McKee, R. G. (1970). *Anal. Chem.* **42**, 91A.
McKeown, J. J., Brown, L. C. and Gove, G. W. (1967). *J. Wat. Pollut. Control Fed.* **39**, 1323.
Macchi, G. (1965). *J. Electroanal. Chem.* **9**, 290.
Mackereth, F. J. H. (1964). *J. Scient. Instrum.* **41**, 38.
Maienthal, E. J. and Taylor, J. K. (1968). *In* "Trace Inorganics in Water" (R. B. Baker, ed.), pp 172–182. Am. Chem. Soc. Adv. Chem. Ser. 73. Am. Chem. Soc., Washington, D.C.
Malijković, D. and Branica, M. (1971). *Limnol., Oceanogr.* **16**, 779.
Mancy, K. F. and Jaffe, T. (1966). U.S. Public Health Serv. Publ. 999-WP-37, 94 pp.
Mancy, K. H., Okun, D. A. and Reilley, C. N. (1962). *J. Electroanal. Chem.* **4**, 65.
Mangelsdorf, P. C. and Wilson, T. R. S. (1971). *J. Phys. Chem.* **75**, 1418.
Manheim, F. T. (1961). *Stockholm Contributions in Geology*, **8**, 27.
Marinenko, G. and Taylor, J. K. (1966). *J. Res. Natl. Bur. Std.* **70C**, 1.
Marinenko, G. and Taylor, J. K. (1967). *Anal. Chem.* **39**, 1568.
Marinenko, G. and Taylor, J. K. (1968). *Anal. Chem.* **40**, 1645.
Mark, H. B. Jr (1970). *J. Pharm. Belg.* **25**, 367.
Mark, H. B. Jr, Eisner, U., Rottschafer, J. M., Berlandi, F. J. and Mattson, J. S. (1969). *Environ. Sci. Technol.* **3**, 165.
Mascini, M. (1973). *Analyst*, **98**, 325.
Matson, W. R. (1968). Ph.D. Thesis, Mass. Inst. of Technol., Cambridge, Mass.
Matson, W. R. and Roe, D. K. (1967). *Anal. Instr.* **4**, 19.
Matson, W. R., Roe, D. K. and Carrit, D. (1965). *Anal. Chem.* **37**, 1594.
Mattson, J. S., Mark, H. B. Jr and MacDonald, H. C. Jr (1972) eds. "Computers in Chemistry and Instrumentation", Vol. 2, Electrochemistry, 466 pp. Marcel Dekker, New York.

Meites, L. (1963) Ed. "Handbook of Analytical Chemistry". McGraw-Hill Inc., New York.
Meites, L. (1965). "Polarographic Techniques", Second Edition, 752 pp. Wiley Interscience, New York.
Merril, C. R. (1961). *Nature*, **192**, 1087.
Meyling, A. H. and Frank, G. H. (1962). *Analyst*, **87**, 684.
Midgley, D. and Torrance, K. (1972). *Analyst*, **97**, 626.
Millero, F. J. (1974). *Ann. Rev. Earth Planet. Sci.* **2**, 101.
Milner, G. W. C. (1957). "Principles and Applications of Polarography and Other Voltammetric Processes', 729 pp. Longmans, London.
Milner, G. W. and Phillips, G. (1967). "Coulometry in Analytical Chemistry". Pergamon Press, London.
Milner, G. W. C., Wilson, J. D., Barnett, G. A. and Smale, A. A. (1961). *J. Electroanal. Chem.* **2**, 25.
Minear, R. A. and Murray, B. B. (1973). *In* "Trace Metals and Metal–Organic Interactions in Natural Waters" (P. C. Singer, ed.), pp 22–29. Ann Arbor Science, Ann Arbor, Michigan.
Moody, G. J. (1973). "IUPAC International Symposium on Selective Ion–Sensitive Electrodes". Butterworths, London. (Also published in *Pure and Appl. Chem.* **36**(4), 1973).
Moody, G. J. and Thomas, J. D. R. (1971). "Selective Ion-Sensitive Electrodes" 140 pp. Merrow Publishing Co., Watford.
Moody, G. J. and Thomas, J. D. R. (1972). *Talanta*, **19**, 623.
Moore, G. W., Roberson, C. E. and Nygren, H. D. (1962). *U.S. Geol. Surv. Proj. Paper, No 450-B*, 83.
Moore, M. G., Rieck, R. H. and Moore, W. A. (1967). *J. Wat. Pollut. Control Fed. Res. Suppl.* **39**, R64.
Moore, R. T. and McNulty, J. A. (1968). *Anal. Instrum.* **6**, 319.
Morel, F. and Morgan, J. (1972). *Envir. Sci. Technol.* **6**, 58.
Morgenthaler, L. P. (1968). "Basic Operational Amplifier Circuits for Analytical Chemical Instrumentation". McKee-Pederson Instruments, Danville, California.
Morris, J. C. and Stumm, W. (1967). In: "Equilibrium Conditions in Natural Water Systems". *Am. Chem. Soc. Adv. Chem. Series No. 67* (W. Stumm, ed.), p. 270. Am. Chem. Soc., Washington, D.C.
Murakami, S. and Hayakawa, M. (1968). *Kobe Daigaku Kyoikugakubu Kenkyu Shuroku*, **1968**, 11.
Murray, C. N. and Riley, J. P. (1969). *Deep-Sea Res.* **16**, 311.
Muzzarelli, R. A. A. and Sipos, L. (1971). *Talanta*, **18**, 853.
Myers, D. J. and Osteryoung, J. (1974). *Anal. Chem.* **46**, 356.
Nakagawa, G. and Tanaka, M. (1972). *Talanta*, **19**, 559.
Natusch, D. F. S. and Thorpe, T. M. (1973). *Anal. Chem.* **45**, 1184A.
Naumann, R. and Weber, C. (1971). *Fresenius' Z. Anal. Chem.* **253**, 111.
Neeb, R. and Saw, D. (1966). *Z. Anal. Chem.* **222**, 200.
Negus, L. E. and Light, T. S. (1972). *Instr. Tech.* **19**, 23.
Neiman, E. Ya. and Brainina, Kh. Z. (1973). *Zh. Anal. Khim.* **28**, 886.
Odier, M. and Plichon, V. (1971). *Anal. Chim. Acta*, **55**, 209.
Ogata, N. (1972). *Japan. Analyst*, **21**, 780.
Ohle, W. (1952). *Vom Wasser*, **19**, 99.
Osteryoung, R. A. and Christie, J. H. (1974). *Anal. Chem.* **46**, 351.

Ozaki, T. and Suzuki, J. (1964). *Japan. Analyst.* **13**, 107.
Packard, T. T. and Healy, M. L. (1968). *J. Mar. Res.* **26**, 66.
Pamatmat, M. M. (1971). *Limnol. Oceanogr.* **16**, 536.
Pamatmat, M. M. and Banse, K. (1969). *Limnol. Oceanogr.* **14**, 250.
Pamatmat, M. M. and Fenton, D. (1968). *Limnol. Oceanogr.* **13**, 537.
Park, K. (1964). *Science,* **146**, 56.
Park, K. (1969). *Limnol. Oceanogr.* **14**, 179.
Park, K. and Curl, H. C. Jr (1968). *Nature,* **205**, 274.
Park, K., Weyl, P. K. and Bradshaw, A. (1964). *Nature,* **201**, 1283.
Parthasarathy, N., Buffle, J., Monnier, D. (1972). *Anal. Chim. Acta,* **59**, 447.
Perone, S. P. and Brumfield, A. (1967). *J. Electroanal. Chem.* **13**, 124.
Perone, S. P. and Davenport, K. K. (1966). *J. Electroanal. Chem.* **12**, 269.
Perrin, D. D., Sayce, I. G. and Sharma, V. S. (1967). *J. Chem. Soc.* **A 1967**, 1755.
Petek, M. and Branica, M. (1969a). *Rapp. Comm. int. Mer. Médit.* **19**, 765.
Petek, M. and Branica, M. (1969b). *Thalassia jugosl.* **5**, 257.
Peterson, M. H. and Groover, R. E. (1972). *Materials Protection and Performance,* **11**(5), 19.
Piro, A., Bernhard, M., Branica, M. and Verzi, M. (1973). *In* "Radioactive Contamination of the Marine Environment". IAEA, Vienna.
Platford, R. F. (1965a). *J. Fish. Res. Bd, Can.* **22**, 885.
Platford, R. F. (1965b). *J. Mar. Res.* **23**, 55.
Platford, R. F. and Dafoe, T. (1965). *J. Mar. Res.* **23**, 63.
Pommer, A. M. (1967). *In* "Glass Electrodes for Hydrogen and Other Cations" (G. Eisenmann, ed.), pp. 375. Marcel Dekker, N.Y.
Portmann, J. E. and Riley, J. P. (1966a). *Anal. Chim. Acta,* **34**, 201.
Portmann, J. E. and Riley, J. P. (1966b). *Anal. Chim. Acta,* **35**, 35.
Potter, E. C. and White, J. F. (1957). *J. Appl. Chem.* **7**, 309.
Pouw, Th. J. M., Den Boef, G. and Hannema, U. (1973). *Anal. Chim. Acta,* **67**, 427.
Pouw, Th. J. M., Den Boef, G. and Hannema, U. (1974). *Anal. Chim. Acta,* **68**, 243.
Proctor, C. M. (1956). *Trans. Amer. Geophys. Un.* **37**, 31.
Pungor, E. (1965). "Oscillometry and Conductimetry". Pergamon Press, New York.
Pungor, E. and Havas, J. (1966). *Acta Chim. Hung.* **50**, 77.
Pungor, E. and Toth, K. (1964). *Mikrochim. Acta,* **2**, 656.
Pungor, E., Toth, K. and Havas, J. (1966). *Mikrochim. Acta,* **4**, 689.
Putnam, G. L. (1953). *J. Chem. Ed.* **30**, 576.
Pytkowicz, R. M. (1969). *Geochem. J.* **3**, 181.
Pytkowicz, R. M. and Fowler, G. A. (1967). *Geochem. J.* **1**, 169.
Pytkowicz, R. M. and Kester, D. R. (1971). *Oceanogr. Mar. Biol. Ann. Rev.* **9**, 11.
Pytkowicz, R. M., Kester, D. R. and Burgener, B. C. (1966). *Limnol. Oceanogr.* **11**, 417.
Pytkowicz, R. M., Distéche, A. and Distéche, S. (1967). *Earth Planet. Sci. Lett.* **2**, 430.
Randles, J. E. B. (1948). *Trans. Faraday Soc.* **44**, 327.
Rechnitz, G. A. (1963). "Controlled Potential Analysis". Pergamon, New York.
Rechnitz, G. A., Lin, Z. F. and Zamochnick, S. B. (1967). *Anal. Lett.* **1**, 29.
Rechnitz, G. A., Fricke, G. H. and Mohan, M. S. (1972). *Anal. Chem.* **44**, 1098.
Reeburgh, W. S. and Carpenter, J. H. (1964). *Limnol. Oceanogr.* **9**, 589.
Reilley, C. N. (1968). *Rev. Pure and Appl. Chem.* **18**, 137.
Reilley, C. N. and Murray, R. W. (1963). *In* "Treatise on Analytical Chemistry" (I. M. Kolthoff and P. J. Elving, eds), Part 1, Volume 4, pp 2163–2232. John Wiley. New York.

Reilley, C. N., Everett, G. W. and Johns, R. H. (1955). *Anal. Chem.* **27**, 483.
Reinmuth, W. H. (1961). *Anal. Chem.* **33**, 185.
Reynolds, J. F. (1969). *J. Water Pollut. Contr. Fed.* **41**, 2002.
Rhodes, D. R. and Hopkins, J. R. (1971). *Anal. Chem.* **43**, 630.
Riley, J. P. (1965a). *In* "Chemical Oceanography" (J. P. Riley and G. Skirrow, eds), Vol. 2, pp 295–424. Academic Press, London.
Riley, J. P. (1965b). *Deep-Sea Res.* **12**, 219.
Riley, J. P. and Chester, R. (1971). "Introduction to Marine Chemistry", 465 pp. Academic Press, London.
Riley, J. P. and Tongudai, M. (1967). *Chem. Geol.* **2**, 263.
Ringbom, A. (1963). "Complexation in Analytical Chemistry". Interscience, New York.
Robbins, C. W., Carter, D. L. and James, D. W. (1973). *Soil Sci. Amer. Proc.* **37**, 212.
Robinson, R. A. and Bates, R. G. (1973). *Anal. Chem.* **45**, 1666.
Robinson, R. A., Duer, W. C. and Bates, R. G. (1971). *Anal. Chem.* **43**, 1862.
Roe, D. K. and Toni, J. E. A. (1965). *Anal. Chem.* **37**, 1503.
Rojahn, T. (1972). *Anal. Chim. Acta,* **62**, 438.
Rooney, R. C. (1963). *J. Polarogr. Soc.* **9**, 45.
Rooney, R. C. (1966). *Chem. and Ind.* **1966**, 875.
Ross, J. W. Jr (1969). *In* "Ion-selective Electrodes" (R. A. Durst, ed.), pp 57–71. NBS Special Publication 314, Nat. Bur. Stand. (US), U.S. Govt Printing Office, Washington, D.C.
Ross, J. W., Riseman, J. H. and Krueger, J. A. (1973). *Pure Appl. Chem.* **36**, 73.
Rossotti, H. (1969). "Chemical Applications of Potentiometry". D. Van Nostrand, London.
Rotthauwe, H. W. (1958). *Kieler Meeresforsch,* **14**, 48.
Rottschafer, J. M., Boczkowski, R. J. and Mark, H. B. Jr (1972). *Talanta,* **19**, 163.
Ruzicka, J., Hansen, E. H. and Tjell, J. Chr. (1973). *Anal. Chim. Acta,* **67**, 155.
Sayce, I. G. (1968). *Talanta,* **15**, 1397.
Scarano, E., Bonicelli, M. G. and Forina, M. (1970). *Anal. Chem.* **42**, 1470.
Schmid, M. and Mancy, K. H. (1968). *Chimia,* **23**, 398.
Schmidt, H. and von Stackelberg, M. J. (1963). "Modern Polarographic Methods", 3rd ed. Academic Press, New York.
Schroeder, R. R. (1972). *Comput. Chem. Instrum.* **2**, 263.
Schultz, F. A. (1971a). *Anal. Chem.* **43**, 502.
Schultz, F. A. (1971b). *Anal. Chem.* **43**, 1523.
Seitz, R., Jones, R. Klatt, L. N. and Mason, W. D. (1973). *Anal. Chem.* **45**, 840.
Seitz, W. R. (1970). Ph.D. Thesis, Mass. Inst. of Technol., Cambridge, Mass.
Severinghaus, J. W. and Bradley, A. F. (1958). *J. Appl. Physiol.* **13**, 515.
Shain, I. (1961). *Anal. Chem.* **33**, 1966.
Shain, I. (1963). *In* "Treatise on Analytical Chemistry" (I. M. Kolthoff and P. J. Elving eds), Part I, Vol. 4, pp 2533–2568. John Wiley and Son, New York.
Shu, F. R. and Guilbault, G. G. (1972). *Anal. Lett.* **5**, 559.
Shuman, M. S. and Woodward, G. P. Jr (1973). *Anal. Chem.* **45**, 2032.
Siegel, A. (1971). *In* "Organic Compounds in Aquatic Environments" (S. D. Faust and J. V. Hunter, eds), pp 275–280. Marcel Dekker Inc., New York.
Siegerman, H. and O'Dom. G. (1972). *Amer. Lab.* **4**, 59.
Siever, R. (1965). *J. Geol.* **73**, 39.
Sillén, L. G. (1967a). *Science,* **156**, 1189.

Sillén, L. G. (1967b). *In* "Equilibrium Concepts In Natural Water Systems" (W. Stumm, ed.), Adv. Chem. Ser. 67, 45–56. Am. Chem. Soc., Washington.
Smith, J. D. and Redmond, J. D. (1971). *J. Electroanal. Chem.* **33**, 169.
Smith, M. J. and Manahan, S. E. (1973). *Anal. Chem.* **45**, 836.
Smith, W. H. and Hood, D. W. (1964). *In* "Ken Sugawara Festival Volume, Recent Researches in the Fields of Hydrosphere, Atmosphere and Nuclear Geochemistry" (Y. Miyake and T. Koyama, eds), pp 185–202. Maruzen, Tokyo.
Solorzano, L. (1969). *Limnol. Oceanogr.* **14**, 799.
Solov'ev, L. G. (1964). *Okeanologiya*, **4**, 149.
Solov'ev, L. G. (1967). *Trudy Inst. Okeanol. Akad. Nauk. SSSR*, **83**, 63, 177.
Solov'ev, L. G. and Tsvetkova, A. A. (1966). *Oceanology*, **6**, 444.
Springer, J. S. (1970). *Anal. Chem.* **42**, 22A.
Srinivasan, K. and Rechnitz, G. A. (1969). *Anal. Chem.* **41**, 1203.
Srna, R. F., Epifanio, C., Hartman, M., Pruder, G. and Stubbs, A. (1973). College of Marine Sciences, University of Delaware, Report No. DDCA-SG-14-73, 20 pp.
Stein, J. E. and Denison, J. G. (1967). *Adv. Wat. Pollut. Res.* **3**, 181.
van Stekelenburg, G. J. (1970). *J. Electroanal. Chem.* **28**, 222.
Stock, J. T. (1965). "Amperometric Titrations". Interscience, New York.
Stock, J. T. (1974). *Anal. Chem.* **46**, 1R.
Strickland, J. D. H. and Parsons, T. R. (1972). "A Practical Handbook of Sea Water Analysis", Bulletin 167 (second edition) 310 pp. Fisheries Research Board of Canada, Ottawa.
Stumm, W. (1967). *Adv. Water Pollut. Res.* **1**, 283.
Sybrandt, L. B. and Perone, S. P. (1971). *Anal. Chem.* **43**, 382.
Szekeilda, K. H. (1969). *J. Cons. Perm. Int. Explor. Mer*, **32**, 318.
Takahashi, T., Weiss, R. F., Culberson, C. H., Edmond, J. M., Hammond, D. E., Wong, C. S., Li, Y.-H. and Bainbridge, A. E. (1970). *J. Geophys. Res.* **75**, 7648.
Takata, Y. and Muto, G. (1973). *Anal. Chem.* **45**, 1864.
Talreja, S. T., Bhalala, B. G. and Rao, P. S. (1969). *Salt Res. Ind.* **6**, 82.
Tanaka, N. and Ogino, H. (1964). *J. Electroanal. Chem.* **7**, 332.
Taylor, G. R. (1961). *In* "pH Measurement and Titration" (G. Mattock, ed.) Heywood, London.
Taylor, J. K., Maienthal, E. J. and Marinenko, G. (1965). *In* "Trace Analysis—Physical Methods" (G. H. Morrison, ed.) pp 377–433. Interscience, New York.
Teal, J. M. (1972). *In* "Barobiology and Experimental Biology of the Deep Sea" (R. W. Brauer ed.), pp 212–222. University of North Carolina.
Thomas, R. F. and Booth, R. L. (1973). *Env. Sci. Tech.* **7**, 523.
Thompson, H. and Rechnitz, G. A. (1972). *Chem. Instr.* **4**, 239.
Thompson, M. E. (1966). *Science*, **153**, 866.
Thompson, M. E. and Ross, J. W. (1966). *Science*, **154**, 1643.
Tikhonov, M. K. and Shalimov, G. A. (1965). *Gidrofiz. i Gidrokhim. Issled., Akad. Nauk Ukr. S.S.R.* **1965**, 133.
Truesdale, V. W. and Spencer, C. P. (1974). *Mar. Chem.* **2**, 33.
Tseung, A. C. C. and Bevan, H. L. (1973). *J. Electroanal. Chem.* **45**, 429.
Tsunogai, S. (1971). *Deep-Sea Res.* **18**, 913.
Van Breemen, N. (1972). *Geochim. Cosmochim. Acta*, **37**, 101.
Van Duin, P. J. (1972). *TNO-Nieuws*, **27**, 457.
Van Landingham, J. W. and Greene, M. W. (1971). *Mar. Tech. Soc. J.* **5**(4), 11.
Vassos, B. H., Hirsch, R. F. and Letterman, H. (1973). *Anal. Chem.* **45**, 792.

Velghe, N. and Claeys, A. (1972). *J. Electroanal. Chem.* **35**, 229.
de Vries, V. T. (1965). *J. Electroanal. Chem.* **9**, 448.
Walker, A. C., Bray, U. B. and Johnston, J. (1927). *J. Am. Chem. Soc.* **49**, 1235.
Wallace, W. J. (1973). "The Development of the Chlorinity/Salinity Concept in Oceanography", 240 pp. Elsevier, Amsterdam.
Warner, T. B. (1969a). *Anal. Chem.* **41**, 527.
Warner, T. B. (1969b). *Science*, **165**, 178.
Warner, T. B. (1971a). *Deep-Sea Res.* **18**, 1255.
Warner, T. B. (1971b). *Water Res.* **5**, 459.
Warner, T. B. (1972a). *Mar. Techn. Soc. J.* **6**, 24.
Warner, T. B. (1972b). *J. Geophys. Res.* **77**, 2728.
Warner, T. B. and Bressan, D. J. (1973). *Anal. Chim. Acta*, **63**, 165.
Wedborg, M. (1972). "Answers to Problems in Computer Calculations of Ionic Equilibria and Titration Procedures". Dept. Anal. Chem., University of Gothenburg.
Westerberg, H. (1972). *Meddn. Havsfiskelab. Lysekil, No. 126.*
Weyl, P. K. (1961). *J. Geol.* **69**, 32.
Wheeler, E. A. (1966). *Proc. Conf. Electronic Engineering in Oceanography* (Southampton, UK), IERE Conf. Proc. No. 8, pp 1–4.
Whitfield, M. (1969). *Limnol. Oceanogr.* **14**, 547.
Whitfield, M. (1970). *Electrochim. Acta*, **15**, 83.
Whitfield, M. (1971a). "Ion-selective Electrodes for the Analysis of Natural Waters", 130 pp. AMSA Handbook No. 2. Australian Marine Association, Sydney.
Whitfield, M. (1971b). *Limnol. Oceanogr.* **16**, 829.
Whitfield, M. (1973). *Mar. Chem.*, **1**, 251.
Whitfield, M. (1975). "Sea Water as an Electrolyte Solution". *In* "Chemical Oceanography" (J. P. Riley and G. Skirrow, eds), second edition, Vol. I, Chapter 2.
Whitfield, M. and Leyendekkers, J. V. (1969a). *Anal. Chim. Acta*, **45**, 383.
Whitfield, M. and Leyendekkers, J. V. (1969b). *Anal. Chim. Acta*, **46**, 63.
Whitfield, M. and Leyendekkers, J. V. (1970). *Anal. Chem.* **42**, 444.
Whitfield, M., Leyendekkers, J. V. and Kerr, J. D. (1969). *Anal. Chim. Acta*, **45**, 399.
Whitnack, G. C. (1961). *J. Electroanal. Chem.* **2**, 110.
Whitnack, G. C. (1966). *In* "Polarography 1964", Proc. 3rd Int. Congr. Southampton (ed. G. J. Hills), Vol. I, pp 641–651. MacMillan and Co. Ltd.
Whitnack, G. C. (1973). *In* "Marine Electrochemistry", pp 342–351 (J. B. Berkowitz, R. A. Horne, M. Banus, P. L. Howard, M. J. Pryor, G. C. Whitnack, eds). Electrochem. Soc., Princeton, N.J. 406 pp.
Whitnack, G. C. and Sasselli, R. (1969). *Anal. Chim. Acta*, **47**, 267.
Wilde, P. and Rogers, P. W. (1970a). *Rev. Sci. Instr.* **41**, 356.
Wilde, P. and Rogers, P. W. (1970b). *Mar. Techn. Soc. 6th Annual Meeting*, **2**, 1445.
Wilkniss, P. E., Warner, T. B. and Carr, R. A. (1971). *Marine Geol.* **11**, M39.
Williams, W. G. (1968). M.S. Thesis, Massachusetts Institute of Technology. 60 pp.
Wilson, T. R. S. (1971). *Limnol. Oceanogr.* **16**, 581.
Windom, H. L. (1971). *Limnol. Oceanogr.* **16**, 806.
Yoshimori, T. and Tanaka, T. (1971). *Anal. Chim. Acta*, **55**, 185.
Yoshimori, T. and Tanaka, T. (1973). *Anal. Chim. Acta*, **66**, 85.
Zakharov, M. S. and Trushina, L. F. (1966). *Zh. Anal. Khim.* **21**, 145.
Zieglerová, L., Štulik, K. and Doležal, J. (1971). *Talanta*, **18**, 603.
Zirino, A. and Healy, M. L. (1970). *Limnol. Oceanogr.* **15**, 956.

Zirino, A. and Healy, M. L. (1971). *Limnol. Oceanogr.* **16**, 773.
Zirino, A. and Healy, M. L. (1972). *Envir. Sci. Technol.* **6**, 243.
Zirino, A., Lieberman, S. and Healy, M. L. (1973). *In* "Marine Electrochemistry" (J. B. Berkowitz, R. A. Horne, M. Banus, P. L. Howard, M. J. Pryor and G. C. Whitnack eds.), pp 319–332. Electrochem. Soc., Princeton, N.J. 406 pp.
Zobell, C. E. (1946). *Bull. Amer. Ass. Petrol Geol.* **30**, 477.

Chapter 21

Extraction of Economic Inorganic Materials from Sea Water

W. F. McILHENNY

The Dow Chemical Company, Freeport, Texas, U.S.A.

21.1. Introduction

From its beginning, the chemical industry has been closely associated with the ocean. Sodium chloride, sodium carbonate, bromine, magnesium, its salts and those of potassium were first recovered in industrial quantities from sea water. Marine algae were, for a period, the major industrial source for iodine, bromine and potash. One of the first genuinely physico-chemical processes attempted by man was the manufacture of salt from the sea by solar evaporation. The recovery of salt, Glauber's salt and Epsom salt from sea water in New England by John Sears was the first chemical manufacturing operation in the Western hemisphere, and the first patent issued in North America was for a monopoly on sea salt manufacture.

In addition to the use of sea water as a major source of raw materials, the chemical industry uses the ocean as an artery of commerce and as a heat sink. Chemical raw materials in large amounts are obtained from marine beaches, the floors of bays and estuaries, beneath the surface of the continental shelf, and marine plants and animals. The total value of materials derived from the World Ocean by the chemical production industry certainly exceeds a billion dollars per annum.

The present scale of production of chemicals produced from sea water itself is considerable. More than 324 million metric tons of materials, with a value of $500 million are recovered annually from sea water or from the bitterns remaining after the recovery of salt. In order of decreasing value, salt, magnesium metal, water, magnesium compounds and bromine are currently recovered industrially (see Table 21.1). In addition, gypsum and potassium

TABLE 21.1
Production of chemicals from sea water

Commodity	Total annual world production	Annual production from sea water	% of total	Annual value of sea water production
	(metric tons)	(metric tons)		($)
Salt (NaCl)	138 874 000	39 100 000[a]	29	280 000 000
Magnesium (Mg)	201 000	121 000[b]	60	77 000 000
Water (H_2O)	338 000 000	284 000 000	84	75 000 000
Magnesium compounds (MgO)	10 900 000[a]	660 000[a, b]	6	48 000 000
Bromine (Br)	132 000	36 000[c]	27	20 700 000
Total		324 000 000		500 700 000

[a] Estimated
[b] Includes magnesium from dolomitic lime
[c] Includes sea salt bittern

compounds are recovered in lesser amounts and heavy water has been produced on an industrial scale. The locations of some of the major production centres for seawater chemicals are shown in Fig. 21.1.

The literature on the recovery, or potential recovery, of the dissolved constituents of sea water is extensive (Christensen, *et al.*, 1967; Tallmadge, *et al.* 1964; Armstrong and Miall, 1946; McIlhenny, 1967) and processes have been proposed or developed for the extraction of all of the major components, and many of the minor and trace elements.

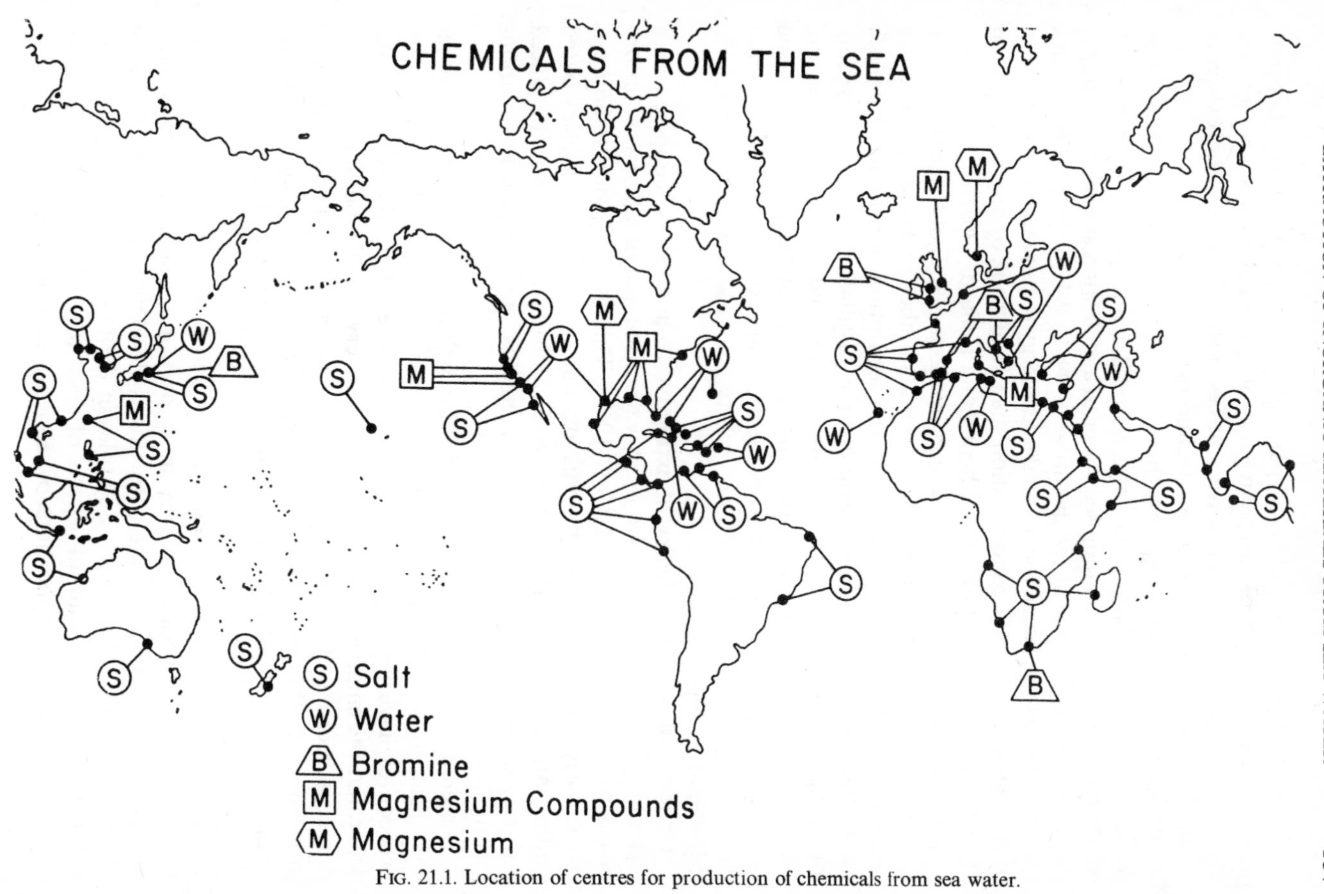

FIG. 21.1. Location of centres for production of chemicals from sea water.

The feasibility and viability of particular recovery processes depends primarily on the economic attractiveness of sea water as a source of raw materials, compared to alternative sources, rather than on the existence of suitable recovery and processing technology.

The energy required to remove from sea water salts other than those present in relatively large amounts, is small compared with that which would be expended in pumping sea water to a collection site or circulating it through process equipment (of the order of 0·04 kWh ton^{-1}). Indeed, it has been estimated that, because of capital and energy requirements, the recovery of any elements lower in concentration than strontium, or any combination of them, cannot be profitable (McIlhenny and Ballard, 1963).

A survey made by Christensen and co-workers (1967) of brine processing technology, showed that a relatively small number of basic methods are being used or have been proposed for the separation of products from sea water; these include, adsorption, evaporation, distillation, solvent extraction, ion exchange, precipitation, electrolysis, flotation and oxidation. In general, the initial separation step does not produce a marketable product, and one or more subsequent processing steps are necessary before a saleable product can be produced having sufficient purity and a suitable physical form (McIlhenny, 1966).

Understanding of the theory of extractive technology is hampered by a lack of knowledge of the behaviour of water itself and of aqueous solutions of electrolytes (Gillam and McCoy, 1966). This lack of understanding is particularly restrictive when attempts are made to set up an industry based on the handling of sea water and brine. As a result, reliance must be placed on past experience, laboratory measurements, and empirical equations (McIlhenny, 1966).

The emphasis of the following discussion will be on those processes that have been used industrially, or which are commercially viable. Separation procedures have been proposed for virtually everything that exists in sea water, but the emphasis has not been on recovery for sale, but for analytical purposes because of the need to eliminate interference in the analysis or processing of other materials. Both Christensen *et al.* (1967) and Tallmadge *et al.* (1964) have published excellent reviews of proposed minor element recovery processes.

Actual operating data, reaction equilibra and process kinetics are, as with most industrial processes, for the most part, not available for sea water processing. A notable exception to this is desalination and the Office of Saline Water (1971) has nearly 800 titles in the Research and Development report series dealing with all aspects of desalting including both the basic theory and practical applications of saline water technology.

21.2. WATER

Although water has been produced from sea water for many centuries, it is only within the past two decades that the production of potable water from the ocean has become an established technology.

There is no deficiency of fresh water on the earth nor will there be in the foreseeable future, since the lakes, rivers and ice caps contain about a billion gallons of water for each inhabitant of the earth. More than 300 million gallons of water falls as rain on the earth's surface each year per head of the population. Water is available—but not always where we choose to live, or to establish our industries, and localized water imbalances arise because of rapidly increasing population, increased residence in arid areas, increased agricultural irrigation, and pollution of existing supplies. Most of the water in the hydrosphere is contained in the ocean and adjacent to it lie many of the arid areas, having climates and potentially fertile soils favourable to colonization. In these areas desalination of ocean water is a convenient means of supplying the local water demand.

The technology of desalination is determined by the nature of both the water itself and its dissolved components. Because of the need for production at low cost (\$0·26 m^{-3} at a rate of 4×10^6 m^3 day^{-1}) processes having a high cost efficiency are required. As a result, novel processes based on sophisticated thermodynamic cycles have been devised, and low cost constructional materials have been developed and used in carefully designed large plants.

Practical desalting is directed towards the removal of only sufficient dissolved salt to make the remaining solution usable for the required purpose. In the United States, desalting equipment is generally designed to yield a product containing less than 500 ppm of total dissolved solids, in order to comply with the recommended potable water standard of the U.S. Public Health Service. Specialized uses, such as steam generation, laboratory usage, or soft drink bottling often require a better quality water.

In all desalting processes energy must be expended to separate a less saline from a more saline solution. The minimum energy input required can be calculated by considering a hypothetical simple desalting process (Spiegler, 1962). The theoretical minimum work for desalting sea water (salinity 34·3‰, at 25°C) is 0·70 kWh m^{-3} (2·65 kWh/1000 U.S. gal). This minimal energy required increases as the amount of water removed from the solution increases. It also increases with increasing temperature. Typical minimum energy requirement for the production of fresh water from sea water for various degrees of water removal are shown in Table 21.2. These figures are based on the assumption that each step is carried out reversibly.

TABLE 21.2

Minimum energy for desalting sea water (*Stoughton and Litzke, 1965*)
(values in kWh ton^{-1} water produced)

Temp °C	Per cent recovery 0	25	50	75	100
25	0·71	0·82	0·99	1·36	3·1
50	0·77	0·89	1·08	1·49	3.4
75	0·83	0·96	1·16	1·60	3·7
100	0·88	1·01	1·23	1·69	3·9
125	0·92	1·07	1·30	1·78	4·1
150	0·96	1·11	1·35	1·84	4·3
175	1·00	1·15	1·39	1·90	4·5
200	1·03	1·18	1·43	1·93	4·7

In practice it is, of course, impossible to attain these minimum energy requirements since infinitely slow operation with no losses and 100% efficiency is an unrealistic ideal, and it appears that about four times the thermodynamic minimum is about the best that can be attained for practical purposes (Murphy *et al.*, 1956).

Desalination would be possible if regions of different concentration could be established within the solution as a result of, for example, differential transport, or if in some way an equilibrium or a concentration gradient could be achieved. At present, there is no means of accomplishing this, and all existing desalination processes use either differentiation into two phases or segregation into two solutions by making use of a membrane process.

A convenient classificiation of desalination processes is shown in Table 21.3. Usable water may be produced by the removal of either salt or water from a saline solution, and energy may be supplied in several different ways. Hybrid processes using a combination of these simple processes have been proposed, but have not been extensively investigated.

Of the 748 land based desalting plants having an output of 25000 gal day^{-1}, or larger, in operation or under construction in January of 1970, 253 use sea water as the feed material (O'Shaughnessy, 1970). The total daily production capacity for the sea water-based plants is 778000 tons of fresh potable water or about 84% of the total amount of desalted water produced. The locations of the twenty-five largest sea water desalting plants and the processes used are shown in Table 21.4. It may be seen that all of the largest plants use a form of distillation, and that most of them employ multistage flash distillation.

Prerequisites of all processes for the recovery of fresh water from sea water are that energy be consumed and that either a phase change of the

TABLE 21.3
Desalination processes

Constituent removed	Phase in which transported	Primary form of energy supplied	Process
Salts	Solid	Chemical	Ion-exchange
		Thermal	Thermally reversible resins
		Electrical	Reversible electrodes
Salts	Liquid	Thermal	Solvent extraction
Salts	Gas		None known
Water	Solid	Thermal	Absorption freezing
		Thermal	Hydrate formulation
		Thermal	Absorption
		Mechanical	Compression freezing
		Mechanical	Compression freezing
Water	Liquid	Mechanical	Reverse osmosis
		Electrical	Electrodialysis
		Electrical	Transport depletion
		Chemical	Osmionic
		Thermal	Solvent extraction
Water	Gas	Mechanical	Compression distillation
		Thermal	Distillation

water takes place, or a diffusion of the water relative to the dissolved salts occurs. In either case transport of water must be made against a finite resistance. Thus, in each of the four principal methods of desalination—distillation, freezing, electrodialysis and reverse osmosis—transfer occurs, under the action of a potential driving force, against a resistance. The driving force is related to temperature difference for both distillation and freezing, to electrical potential for electrodialysis and to pressure difference for reverse osmosis.

Some generalizations about the desalination process can be made (Simpson and Silver, 1963). If ϕ is the flux density of the process, γ is the amount of material transport necessary per unit time to produce unit mass of water, and η is the efficiency of the device producing the work for the transfer process, then the total transfer area required for the production device will be γ/ϕ. The work, per unit mass of water produced is $\phi\rho\gamma/\eta$.

In general, the cost of producing water can be separated into three terms, the magnitude of the first two being proportional to the transfer area, and to the work dissipated by the flux respectively; the third term is a constant for a given process.

$$\text{Cost} = K_A(\gamma/\phi) + K_E\left(\frac{\phi\rho\gamma}{\eta}\right) + K_C$$

TABLE 21.4
Largest sea water desalting plants

Daily capacity tons	Location	Process
28932	Terneuzen, Netherlands	MSF
28365	Rosarita, Mexico	MSF
27230	Abu Dhabi, Oman	MSF
19984	Las Palmas, Canary Islands	MSF
18910	Jidda, Saudi Arabia	MSF
18153	Shauiba, Kuwait	MSF
18153	Shuwaikh, Kuwait	MSF
18153	Shuwaikh, Kuwait	MSF
18153	Shuwaikh G., Kuwait	MSF
16640	Valetta, Malta	MSF
13615	Shuaiba, Kuwait	MSF
13237	Shevchenko, USSR (Caspian Sea)	VTE
13237	Shevchenko, USSR (Caspian Sea)	VTE
13237	Shevchenko, USSR (Caspian Sea)	VTE
13010	Mundo Nobo, Curacao	MSF
10166	Balashi, Aruba	MSF
9909	Key West, Florida	MSF
9833	San Diego, California	MSF
9644	Brindisi, Italy	MSF
9455	Virgin Islands	MSF
9455	Penuelas, Puerto Rico	MSF
9077	Shuwaikh E, Kuwait	MSF
9077	Shuwaikh F, Kuwait	MSF
9077	Shuwaikh B, Kuwait	MSF
9077	Bahamas	MSF
8509	Guantanamo, Cuba	MSF

MSF = Multistage flash distillation
VTE = Long tube vertical multiple effect distillation

The identification of the parameters for particular desalination processes is shown in Table 21.5. Vapour compression is considered separately from other distillation processes because the furnished energy is mechanical and the thermal efficiency is that of the compressor. In freezing and hydrate processes, two refrigeration cycles must be considered—one pumping heat from the freezer to the melter and the other abstracting the residual heat from the freezer using the sea water as a heat sink.

In general, a good process uses a low grade energy source; the transfer takes place through an inexpensive, low resistance surface and uses a phase change requiring minimum energy. No process completely satisfies all of these requirements.

TABLE 21.5
Sea water desalination processes. Identification of general process parameters with particular processes (after Simpson and Silver, 1963)

Process	Amount of property transferred to produce unit mass of water (γ)	Flux density (ϕ)	General resistance (ρ)	Efficiency of device producing work for transfer process (η)
Distillation	$\lambda v V/T$	$U\Delta T/RT$	T/U	1
Vapour compression distillation	λ_v/T	$U\Delta T/T$	T/U	η_b
Freezing	$\lambda_f + C_w(T_d - T)/T$	$U_1\Delta T_1/T$	T/U_1	η_b
	and $C_w(1+x)\Delta T_p - C_w(T_d - T)/T$	$U_2\Delta T_2/T$	T/U_2	η_b
Electrodialysis	F	i	r	η_c
Reverse osmosis	v	w	$(1/P)+(\pi/w)$	η_p

21.2.1. SCALE

A particularly troublesome feature associated with the design of sea water desalting plants is the precipitation of insoluble crystalline materials. This may be due to a simple concentration effect or because of the occurrence of chemical reactions (e.g. loss of CO_2 causing $CaCO_3$ precipitation. The deposited material, if dense and adherent is known as "scale". The economic and operational handicaps caused by scale are so extremely severe that practical plants must be designed to operate under conditions in which no scale is formed, or alternatively some method of controlling its formation must be employed.

Scale can form only from supersaturated solutions. In addition, nucleation sites must be available for crystalline growth to occur and there must be sufficient contact time between the nucleation site and the supersaturated solution. The rate of scale formation depends on both the degree of supersaturation and on the rate at which the solution passes the crystallization site.

Two particularly troublesome scales are encountered in sea water desalting units. The alkaline scales, $Mg(OH)_2$ and $CaCO_3$ are produced as a result of shifts in the carbon dioxide equilibria, and supersaturation due to the increased ionic concentration. A more intractible type of scale, calcium sulphate, forms solely because its solubility is exceeded and it is deposited as one of its three crystalline forms. It is, indeed, the limiting factor in the design of all sea water desalting equipment as it restricts the temperature to which sea water may be heated, and the concentration of salts that may be reached in the residual brines.

Normal surface sea water, with a total alkalinity of about 2·4 meq l^{-1} and a pH of about 8·1 is generally saturated or even supersaturated with respect to $CaCO_3$, and it cannot therefore be concentrated appreciably without the precipitation of $CaCO_3$ occurring. The shift in the carbonate equilibria accompanying evaporation and temperature rise (and loss of CO_2) may be summarized by

$$2HCO_3^- \rightleftharpoons CO_3^{2-} + CO_2 \uparrow + H_2O$$

The increased carbonate ion concentration leads to precipitation of calcium carbonate

$$Ca^{2+} + CO_3^{2-} \rightleftharpoons CaCO_3 \downarrow$$

An additional reaction which occurs, viz.

$$CO_3^{2-} + H_2O \rightleftharpoons 2OH^- + CO_2 \uparrow$$

leads to the production of hydroxyl ions and consequently the precipitation of $Mg(OH)_2$.

$$Mg^{2+} + 2OH^- \rightleftharpoons Mg(OH)_2 \downarrow$$

It has been found that in practice, $CaCO_3$ is the principal scale formed at temperatures below 85°C, but that $Mg(OH)_2$ deposits at higher temperatures (McCutchan and Sieder, 1969). The reason for this is not clear, but it is probable that the loss of carbonate ion and the concominant production of hydroxyl ion shown in the penultimate equation above is the dominant reaction at elevated temperatures; under these conditions carbonate formation is therefore not appreciable.

The solubilities of the alkaline scales in sea water and sea water concentrates at various temperatures are reasonably well known (Langelier, 1950; Badger and Assoc., 1959) and equilibrium diagrams are available. The alkaline scales are acid soluble and may be removed by washing the scaled surface with a solution having a low pH. It is usual to lower the pH of the feed sea water by the addition of acid and to remove the liberated CO_2 by vacuum-degassing or by inert gas stripping (Kellog, 1965)

$$H^+ + HCO_3^- \rightleftharpoons CO_2 \uparrow + H_2O$$

As may be seen in Fig. 21.2, (Dow, 1965), it is desirable to lower the pH to slightly below 4 so as to convert as much as possible of the HCO_3^- to molecular CO_2. There is only a marginal advantage in lowering the pH further. The pH of the decarbonated solution is normally adjusted to between 6 and 8 at which it is least corrosive to the desalination plant.

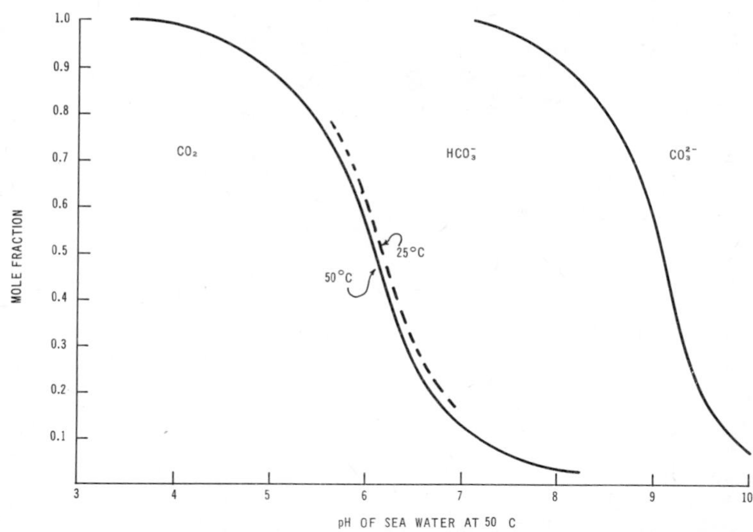

FIG. 21.2. Distribution of total CO_2 in sea water of chlorinity 15‰ at 50°C (Dow, 1965). A perpendicular erected at any pH will indicate by its intersection with the curves the relative mole fraction in each state.

The liberated gaseous CO_2 is undesirable in many processes for several reasons; (i) it blankets condensation surfaces and impedes thermal transfer processes (ii) it dissolves in the pure water product, lowering the pH and producing a corrosive solution and (iii) the gas pockets to which it gives rise interfere with hydraulic flow in membrane systems. Decarbonators are usually operated as separate pieces of equipment, and are frequently combined with deoxygenators in a single unit (deaerator). Oxygen removal is desirable in order to protect the materials used in the construction of the desalination plant against corrosion at elevated temperatures and salinities.

The solubility products of each of the three crystalline forms of $CaSO_4$ in sea water and sea water concentrates may be described as functions of the ionic strength (I), the solubility product (K°_{sp}), at zero ionic strength, the limiting Debye–Hückel slope (S) and some empirical parameters, by the following equation developed by Marshall *et al.* (1964), and further modified by Marshall and Slusher (1968) and Yeatts and Marshall (1972).

$$\log K_{sp} = \log K^{\circ}_{sp} + \frac{8S(I)^{\frac{1}{2}}}{1 + A_{sp}I^{\frac{1}{2}}} + B_G I - C_G I^2.$$

Values for the empirical parameters A_{sp}, B_G, C_G in which the subscripts G, H and A refer to gypsum, hemihydrate and anhydrite, respectively for various temperatures are given in Table 21.6 (Marshall and Slusher, 1968).

TABLE 21.6

Parameters for the prediction of solubility product constants for calcium sulphate from 0–300 °C in sea water and other solutions. (Adapted from Marshall and Slusher, 1968).

Temp °C	Solubility product constants (− log K_{sp}°)			Modified Debye–Hückel slope* (S)	A_{sp}	B_G	C_G
	(G)	(H)	(A)				
0	4·466	3·682	3·818	0·4875	1·450	0·0880	0·0234
25	4·373	3·734	4·192	0·5080	1·500	0·0194	0·0134
50	4·409	3·933	4·539	0·5337	1·544	0·0000	0·0108
75	4·514	4·236	4·884	0·5645	1·575	0·0000	0·0068
100	4·610	4·614	5·240	0·6006	1·594	0·0000	0·0020
125	4·792	5·044	5·617	0·6422	1·600	0·0000	0·0000
150	4·920	5·512	6·020	0·6900	1·600	0·0000	0·0000
175	5·025	6·007	6·453	0·7451	1·600	0·0000	0·0000
200	5·098	6·520	6·917	0·8097	1·600	0·0000	0·0000
250	5·130	7·576	7·941	0·9848	1·600	0·0000	0·0000
300	4·996	8·648	9·093	1·2870	1·600	0·0000	0·0000

* Deby–Hückel slopes for 1–1 electrolytes multiplied by square root of density of water to convert for use with molal units

Key: (G) $CaSO_4 \cdot 2H_2O$
(H) $CaSO_4 \cdot \frac{1}{2}H_2O$
(A) $CaSO_4$

The results of the application of the theoretical treatment are shown diagrammatically in Fig. 21.3. This shows, as a function of temperature, the limits of solution stability of concentrated sea waters if precipitation of $CaSO_4$ in one of its forms is to be avoided. It is believed that the equations and constants are accurate to within $\pm 3\%$ for gypsum, for the range 0° to 110°C; for hemihydrate and anhydrite the accuracy limits are within $\pm 6\%$ at temperatures up to 200°C. These limits are adequate for engineering calculations.

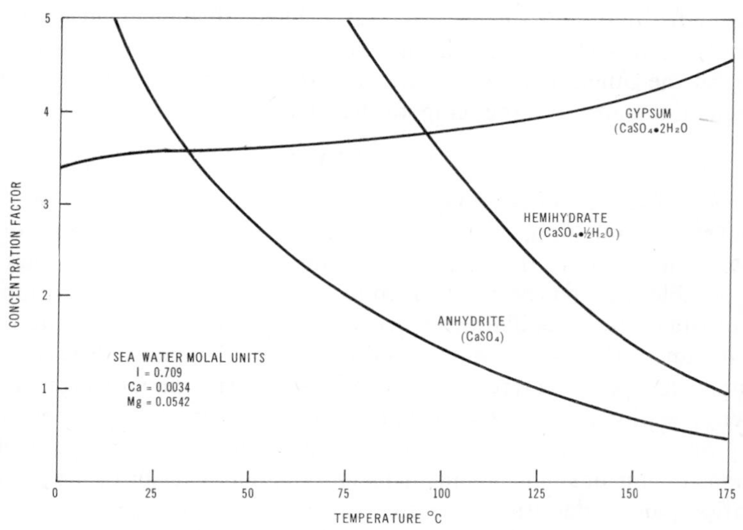

FIG. 21.3. Limits of solution stability of $CaSO_4$ in its various crystalline forms in sea water and concentrated sea water at temperatures up to 175°C (Marshall and Slusher, 1968).

21.2.2. DISTILLATION

During distillation only the solvent is vaporized and the condensed water is salt free. However, the large endothermicity of the vaporization process demands that large amounts of energy must be supplied. Distillation is the oldest desalting process. Aristotle was aware that the vapour from a salt solution was drinkable when condensed, and Julius Caesar used a crude distillation apparatus to supply water for the Roman armies besieging Alexandria. The first recorded use of a shipboard desalter was on a Spanish galleon in the Manila trade in 1606, although water was certainly distilled on ships many years earlier. The extensive use of shipboard desalting units followed the installation of steam engines and boilers, and evaporators were quite common by the 1880's. A land based distillation unit, based on shipboard experience was erected in Curaçao in the early 1930's.

Many types of distillation processes have been proposed, some of considerable complexity. Almost all large distillation plants use either multistage flash, multiple effect falling film or vapour compression as the distillation process.

Two important parameters relevant to the design and evaluation of distillation processes are the performance ratio (the amount of distillate produced per unit of heat input) and the heat transfer area required. In all practical distillation processes, heat is transferred through a surface (usually the wall of a tube) from condensing steam on one side to evaporating saline water on the other. The rate of heat transfer depends on the heat transfer coefficient, the area and the temperature driving force, viz,

$$Q = UA\Delta T$$

The net resistance to heat flow is the sum of the thermal resistances of the evaporating water film, the scale layer, the metal wall and the steam condensate film. Because the magnitude of the net resistance is so important, considerable effort has been expended to minimize its individual components.

Multistage flash distillation is, as was shown in Table 21.4, the process used in almost all large sea water desalting production plants. The equipment is simple, the operation is relatively reliable and the manufacturing techniques and engineering design are sufficiently well established to allow dependable, easily operated units to be produced. The present technical level of flash distillation, a process little known 15 years ago, is a result of the increasing attention paid to desalting processes during the last decade.

A typical commercial multistage flash process flow sheet is shown in Fig. 21.4. Raw sea water is pumped through the tubes of a heat rejection

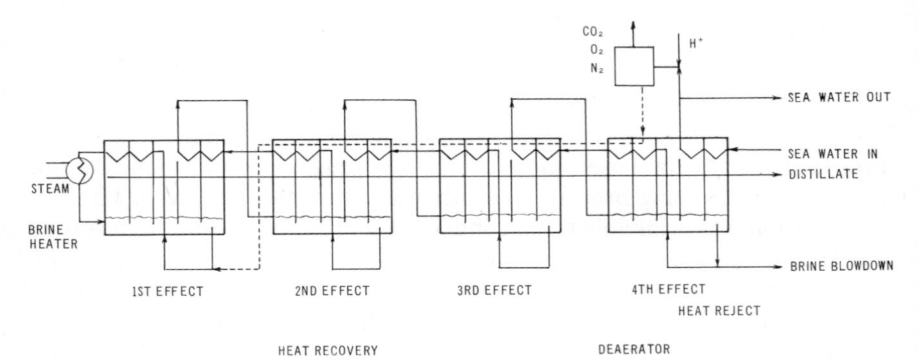

FIG. 21.4. Multistage flash distillation flowsheet.

section and is heated by vapour condensation on the tube exterior. Roughly one quarter of the sea water is acidified, degassed in a vacuum tray degasifier and mixed with the recirculated brine. The balance is discharged. The brine and sea water mixture is circulated through the condenser tubes, being heated in each stage to successively higher temperatures by the flashed water vapour which condenses on the outside of the tubes. The brine is heated to its maximum temperature in the brine heaters by externally supplied steam.

The hot brine enters the first flash chamber which is maintained at a slightly reduced pressure. A part of the saline water undergoes flash vaporization, rises to the condenser bank, is cooled, condensed and collected in the product water trough. The flashing process is continued through the succeeding stages, each at a lower temperature and pressure. A vacuum is maintained in the final stages. Part of the concentrated brine is discharged to the sea, together with the cooling water, but the major proportion is recirculated to mix with fresh incoming feed sea water.

On the basis of a number of simplifying assumptions equations have been derived for the estimation of both the performance ratio of a multistage flash process and the necessary heat transfer area (Silver, 1966). The performance ratio, R, (the number of pounds of distillate produced per 1000 BTU of heat input—corresponding to 1·8 g per 100 cal) is given by (Porter, 1967)

$$R = \frac{1000}{\lambda_v}\left(\frac{1 - e^{-a}}{a}\right)\frac{n}{j}$$

where

$$a \equiv \frac{C_w(\Delta t_b)}{\lambda_v}$$

and n and j are the number of heat recovery and heat rejection stages respectively.

The heat transfer area required is

$$A = w_d \frac{n\lambda_v}{U\Delta t_b} \ln\left(\frac{n}{n - R}\right).$$

As the number of stages is increased the heat transfer surface required will decrease. The optimum number is decided by the economic balance between the cost of subdividing the plant into stages and the cost benefit of the decreased surface. The multistage flash system has several disadvantages. A large quantity of brine must be recycled to conserve thermal energy. Because the conventional flash plant has only a single effect, divided into incremental stages, the brine cannot be highly concentrated and relative to the amount of water produced, a large amount of brine must be discarded. In addition, considerable energy is required to recycle the brine.

In a recently patented innovation the brine heater is replaced with a half-stage heater–evaporator. The same amount of steam is used as in a conventional heater, but the half-stage vapours contain about one-third of the heat supplied. This gives an additional amount of condensate, and a corresponding thermal efficiency gain (Othmer, 1970). It has been suggested that in the vapour-reheat process the condenser tube surface can be eliminated by absorbing the flashed vapours in a spray of product water. In this way large capital saving is possible, and because narrower temperature differentials can be achieved, more stages can be accommodated within a given temperature range than in a conventional system. The flows are such that cooler water from a successive stage is used to condense the vapour. The heat must be recovered in an additional exchanger.

If the evaporating and condensing vessels are so arranged that the vapour produced in one vessel is condensed in a succeeding vessel where the latent heat of vaporization is used to generate a second supply of vapour, each vessel is termed an effect. Since the latent heat required to evaporate a unit mass of water from a saline solution is very nearly the same as the latent heat liberated in condensing a unit mass of steam, each transfer of energy will produce somewhat less than an additional unit mass of product water. The coupling of a series of effects produces a multiple effect evaporator. The initial vaporization of water is accomplished by the input of thermal energy—usually by the condensation of steam produced in an external generation unit.

Clearly, as more effects are added, more water is produced per unit of energy. It is also evident that additional capital investment is required for each effect and that there must exist a point at which the cost of steam saved no longer balances the cost of additional equipment.

Multiple effect evaporators which have been used for many years by the chemical process industry are reliable, have good heat transfer coefficients, and are thermodynamically attractive. The multiple-effect distillation process has the advantages that fewer stages (or effects) are required for the same production, that a more concentrated brine can be produced and that a smaller volume of sea water is handled (Porter, 1967).

The flowsheet for a typical multiple-effect process is shown in Fig. 21.5. The process flows can be arranged in a feed forward or feed backward manner. Because of the limitations of scale formation, sea water evaporators are usually designed so that the most dilute solution—sea water—is the hottest, and the most concentrated solution is the coolest. The incoming feed water is preheated by cross-exchanging with the hot brine and the hot product water.

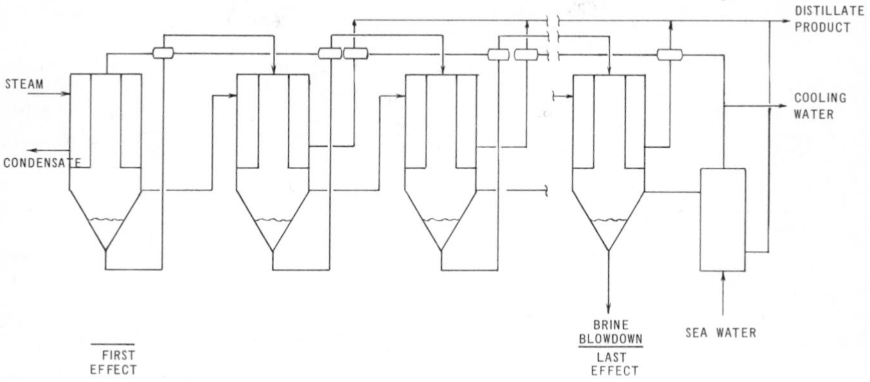

FIG. 21.5. Multiple effect distillation flowsheet.

Equations have been developed for the estimation of the performance ratio and the required heat transfer area in multiple effect distillation (Silver, 1966).

The performance ratio of a multiple-effect distillation unit is

$$R = \frac{1000n}{\lambda_v}$$

as a working approximation $R \simeq n$.
The total heat transfer area required is given by

$$A = w_d \frac{1000(n + 1)}{(t_s - t_{sw})} \left[\frac{n}{U_e} + \frac{f(k)}{U_c}\right]$$

where $f(k) = 1 + \frac{1}{2}k + \frac{1}{3}k^2 + \frac{1}{4}k^3 + \ldots$

and $k = \dfrac{t_{bh} - t_{sw}}{t_d - t_{sw}}$

Since both the multi-stage flash and multi-effect distillation processes have advantages, several combinations of them have been designed and operated. Thus, the multiple effect–multi-stage process, shown in Fig. 21.6 possesses a combination of these advantages. In this concept a group of stages provides the heat input for a further group. Most of the structural simplicity of flash designs can be retained, while still keeping the greater efficiency of multiple effects. The brine concentrations can be higher in the later effects than is possible with a conventional flash evaporator.

The vapour compression evaporator

In a vapour compression evaporator, when the vapour produced by boiling

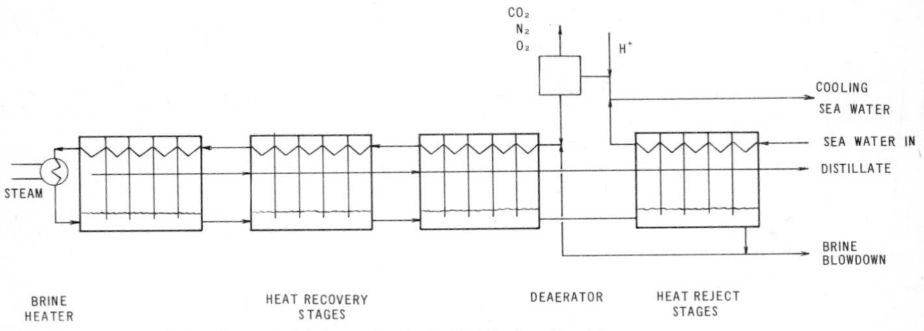

FIG. 21.6. Multiple effect, multiple stage flash distillation flowsheet.

sea water is compressed, its temperature (and energy content) is increased. The additional energy furnished by the compressor is sufficient to produce the designed quantity of product water (Fig. 21.7).

The work done by an ideal adiabatic compressor is

$$W_c = \frac{k}{k-1} P_1 V_1 \left[\left(\frac{P_2}{P_1} \right)^{k/(k-1)} - 1 \right]$$

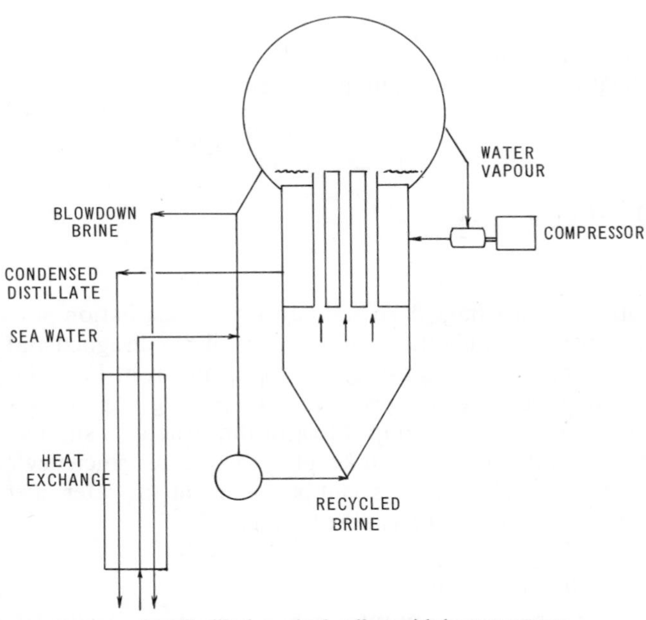

FIG. 21.7. Vapour compression distillation, single effect with heat recovery.

For sea water vapour compression, the compression ratio, r, (P_2/P_1) is usually close to unity, and the energy required is, therefore (Spiegler, 1962)

$$W_c = P_1 V_1 \frac{r^2 - 1}{2r} = RT_1 \frac{r^2 - 1}{2r}$$

Vapour compression is attractive because it is possible to evaporate the condensed water with a relatively small input of energy. For a given performance ratio, the total surface requirement in a vapour compression plant is approximately 25% less than that required for a flash plant. However, in order to secure the necessary heat transfer rate, the boiling solution is usually pumped rapidly past the heat transfer surface; this necessitates the input of considerable pumping energy.

Vapour compression evaporators require no source of thermal energy as mechanical compression is sufficient. Energy for the compressors can be supplied by a variety of prime movers, although the lack of compressors of sufficiently large size has limited vapour compression plants to small units. Since compressors can be driven by diesel or gasoline engines, the energy source can be a readily available petroleum product. There are available many commercial vapour compression evaporators; often these are portable or skid mounted units having capacities of up to 400 m^3 day^{-1}. Vapour compression units are often used on ship-board service where flexibility of operation is desirable.

Solar evaporators

Solar radiation is an attractive source of energy for desalting, especially in arid areas. Most of the inhabited parts of the world receive solar radiation at an average daily rate of 50 to >760 cal cm^{-2}. The average rate for the United States is about 406 cal cm^{-2} (Löf, 1966). Many devices for the utilization of solar energy have been built and tested. A large glass solar still was built in 1872 in the nitrate mining area of Northern Chile. This machine still, which had an area of about 4000 m^2, supplied the water requirements of 180 mules and was used for many years.

Typical solar stills are shown in Figs. 21.8 and 21.9. The solar energy is absorbed in the sea water contained in a shallow basin, mounted on a thermally insulated base. Condensation from the warm moisture laden air occurs on the transparent cover.

Although the quantity of solar energy, over a large area is quite large, it is of low intensity, and intermittent. Solar stills are most useful for the production of relatively small amounts of water where comparative simplicity of operation is important. In these circumstances, the absence of moving parts and of large pressure or temperature differentials makes them attractive. However, the capital investment required is relatively high. The output

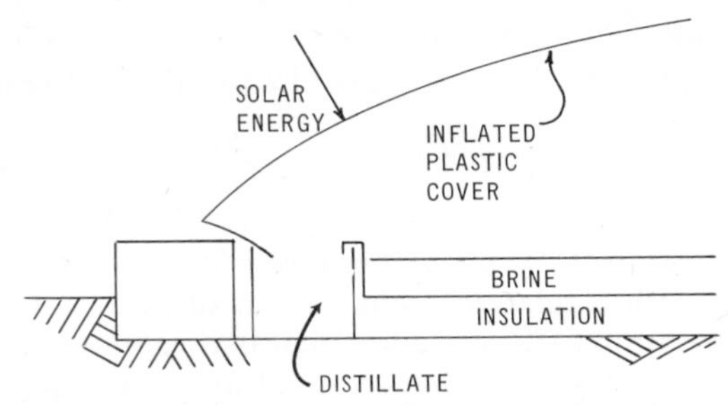

FIG. 21.8. Inflated plastic film solar distillation unit.

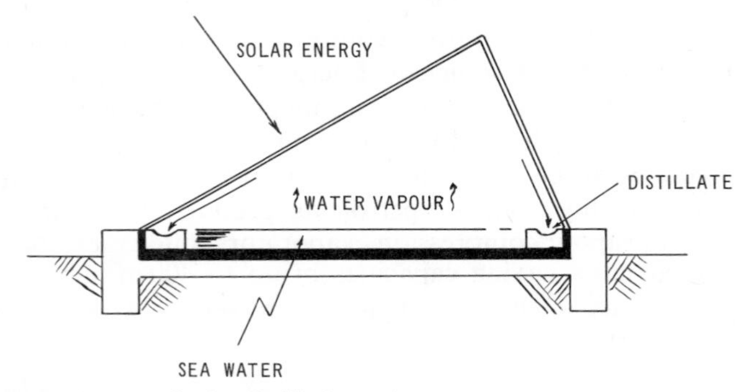

FIG. 21.9. Tilted glass panel solar distillation unit.

depends on the amount of solar radiation and on the efficiency of the system. An energy balance for a unit area still is (Löf *et al.*, 1961)

$$w_d T_c - T_{sw}) = I(1 - \Sigma r) - h_0(T_c - T_a) - \sigma \Sigma_c (T_c^4) - T_r^4) - w_b(T_b - T_{sw}) - \Sigma L.$$

The rate of evaporation depends on the temperature difference between the brine and the cover. In turn this depends on the rate of air circulation and the absolute humidity difference at the two surfaces. The absorption of the incident solar energy is increased by arranging the cover so that it lies in a plane at an angle as nearly as possible equal to the latitude. The rate of evaporation is

$$w_d = h_i(T_b - T_c) \Big/ \frac{M_a}{M_w} (T_{a_c} - T_{a_b}) C_a F$$

where M_a/M_w is the mass ratio of the air circulating to water condensed, and F is a correction allowing for departure from vapour pressure equilibrium.

Two types of solar still covers are in general use today—glass sheet and inflated plastic. The inflated plastic still developed by the World Church

Service (Fig. 21.8) uses a polyvinyl fluoride cover kept inflated by a slight air pressure; surface treatment of the interior of the plastic cover so as to render it wettable facilitates transfer of the condensate to the distillate trough. The brine pan is lined with butyl rubber and insulated with 2·5 cm layer of sawdust or vermiculite. Single stills have plastic spans as wide as 3 m and are up to 60 m long.

The second type of solar still is exemplified by that designed at the Technical University of Athens (Fig. 21.9). A sloping glass cover is supported at an angle of 12° by an aluminium frame. Auxiliary tubular condensers containing flowing sea water are added to increase the condensing surface. The basins are constructed with concrete walls and are lined with butyl rubber sheeting on a tamped earth or levelled sand bottom. Black plastic netting which floats on the surface of the brine increases the amount of the incident solar energy which is absorbed. A pilot unit operating on the island of Symi with an average daily radiation flux of $6{\cdot}46 \times 10^6$ cal m^{-2} produced 5760 g m^{-2} day^{-1}.

21.2.3. OTHER PROCESSES

Freezing processes

Separation of water from sea water by freezing to form a salt-free crystalline ice is attractive for several reasons. The latent heat of fusion is only about one-seventh of that of vaporization. Scales are not generally formed, and the rate of corrosion of constructional materials is greatly diminished. Freezing processes operate much closer to ambient temperatures than do evaporation processes and so require less protection against loss of energy to the environment. A principal disadvantage of all freezing processes is the difficulty of separating the residual brine from the ice crystals. All workable freezing processes produce small ice crystals, although crystals larger than 1 mm in size are desirable (Othmer, 1970).

The theoretical work requirement of the freezing process is that necessary to transfer the latent heat of freezing from the brine to the pure water at the temperature of freezing (Wiegandt, 1960)

$$W_F = \frac{\Delta H \Delta T_{\text{fpd}}}{T}$$

Several freezing cycles have been developed. Since the process does not lend itself to multi-stage or multi-effect operation freezing units are single staged. All current methods use similar mechanisms for forming ice and for separating it from the brine (Snyder, 1966).

In the vacuum-freezing vapour compression process (Fig. 21.10) water itself is used as the refrigerant. Sea water is sprayed into a vacuum chamber

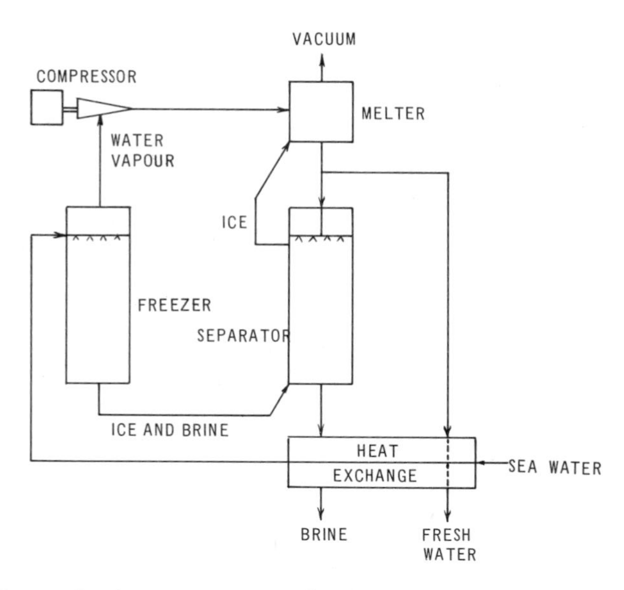

FIG. 21.10. Vacuum-freezing vapour compression desalination flowsheet.

where part of the water is evaporated, cooling the sea water to form a slurry of brine and ice. The vapour produced is compressed and discharged to the melting unit, where the ice, after separation from the adhering brine, is melted to form the product water. The process is simple and requires little additional equipment. The compressor must be large because of the great volume of the low-temperature water vapour that must be handled. The only sizeable sea water desalting plant using a freezing process is located in Israel; this uses the vacuum freezing-vapour compression process.

Other freezing processes include the use of materials such as lithium bromide to absorb the evolved vapour. The absorbed water is later driven off by heating, is condensed and the dewatered absorbent recycled. Auxiliary refrigeration is required. In the direct refrigeration method, the refrigerant, which must be immiscible with the seawater, and must not form solid hydrates is in direct contact with sea water. *Iso*-butane is a suitable solvent as it has a boiling point (−0·5°C at atmospheric pressure) very near the freezing point of sea water. In this process when liquid *iso*-butane is mixed with sea water, it boils, abstracting heat from the solution and forming ice. The *iso*-butane vapour is compressed and the liberated heat is used to melt the ice after it has been separated from both the *iso*-butane and the brine.

Hydrate formation-processes. In a variation of the freezing process, solid hydrates are deliberately produced and separated from the brine. They are then broken down to produce the parent compound and liquid water. Propane, for example, forms a clathrate with 17 molecules of water for each molecule of the hydrocarbon. Clathrates (in which the guest molecule occupies a hole in the host water structure) are also formed with fluorinated hydrocarbons. The operation of hydrate processes is very similar to that of direct injection freezing. Properties of four typical hydrate forming compounds are shown in Table 21.7.

TABLE 21.7
Hydrate forming compounds (Barduhn and Hu, 1960)

Hydrate former	Critical conditions for decomposition of hydrate		Heat of formation	
	Temp./°C	press/atm.	kcal mol^{-1}	moles of water/ moles of hydrate
CH_2ClF	17·7	1·8	21	8
CH_3Br	14·4	0·5	20	8
$CHCl_2F$	8·9	1·0	33	17
$CH_3CH_2CH_3$	5·0	4·9	26	17

Solvent extraction processes. In a somewhat similar manner, certain liquid organic compounds are able to extract water from a saline solution at one temperature, and release it at another.

Amines, particularly secondary and tertiary alkyl amines, possess satisfactory temperature-solubility relationships (see Fig. 21.11 and Davison, 1958; Texas A & M, 1960). Their heats of solution are about 25 cal g^{-1} (relatively small compared with the heats of fusion or vaporization of water). Solvent extraction of sea water with amines requires that magnesium be removed by softening because the high alkalinity of the solvents, can cause precipitation of magnesium salts.

Membrane processes. Membrane processes, in which the water solvent is separated from the dissolved salt by transportation of either salt or water through a semipermeable membrane, are less well developed than are other desalination processes. In electrodialysis, there are two separatory membranes which are cationic and anionic exchangers respectively. Under the influence of an electrical potential gradient, ions of the appropriate charge but not the water molecules are transported through the membrane. In this

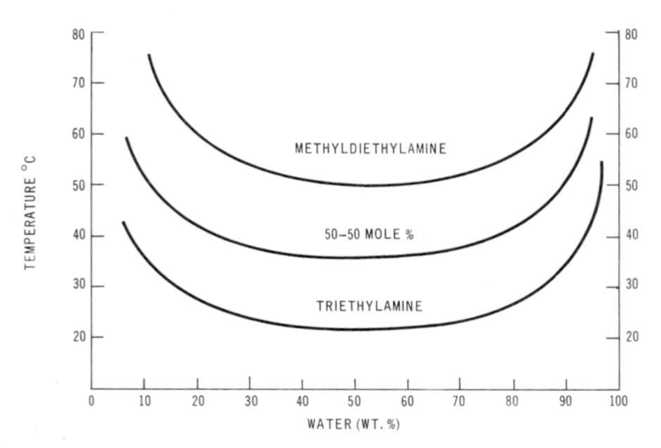

FIG. 21.11. Phase equilibria of tertiary amines usable as water extraction solvents.

way the solution between the two membranes becomes increasingly more dilute. Since the energy consumed is proportional to the concentration of the salts in the saline solution, cost limitations usually confine electrodialysis to the desalination of brackish water. However, the process provides an economic means for the preparation of concentrated salt solutions from the sea water and several comparatively large plants are in operation. Electrodialysis is more fully covered in Section 21.3.1.

In recent years, membranes have been developed which will allow fresh water to be produced from sea water in a single stage. For this remarkable achievement it is necessary that the membrane should withstand a bursting pressure of more than 24·8 atm. (the osmotic pressure of seawater), have a salt rejection of greater than 99·5 %, and allow an appreciable water flux.

In hyperfiltration (usually, but somewhat incorrectly termed reverse osmosis) desalting is accomplished by forcing a salt solution under pressure through a membrane which passes water more readily than it does ions. Its attractiveness lies in its potential extreme simplicity (Fig. 21.12). The flux of water through a membrane is, to a first approximation, proportional to the effective pressure. If the solution is not stirred (and if there is no convection) and the system is operated at a reasonable transmission rate, a rapid salt accumulation occurs to such an extent that transmission of water through the membrane ceases at the imposed pressure. This phenomenon which is termed concentration polarization, is an important factor to be taken into consideration in the design of plants depending on reverse osmosis (Johnson, *et al*, 1966; Strathmann, 1968).

The total energy needed for a hyperfiltration process comprises the energies required for the salt separation, for compensation of the concentra-

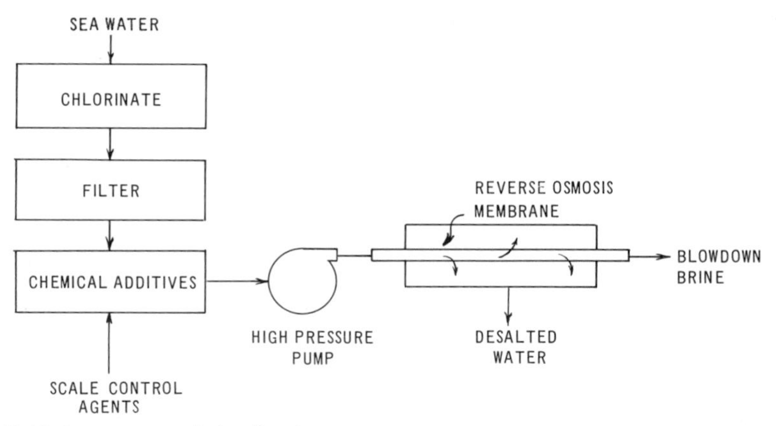

FIG. 21.12. Reverse osmosis desalination.

tion polarization and for maintenance of a useful water flux. With present membranes the total energy consumption is several times larger than that required for the separation of salts alone.

Data on the osmotic pressure of sea salt solution have been published by Stoughton and Lietzke (1965)—see Appendix Table 37. Because the osmotic pressure increases disproportionately rapidly with increasing salt concentration it is likely that sea water reverse osmosis methods will be limited to low recoveries.

The mechanism for the transfer of water through a membrane and the reasons for salt rejection are not well understood, but it probably involves the differences in transport rates of the solvent and solute and the equilibrium distributions of these substrates in the organic polymer–water–salt system (Johnson *et al.*, 1966; Merten, 1965). Membranes to be used for desalting brackish water consist of two layers—a 0·5 μm dense active surface layer which is essentially non-porous supported on a 30 μm porous base consisting of hollow fine fibres (Merten, 1965). Various types of organic polymeric films have been successfully used as reverse osmosis membranes; these include cellulose acetate, sulphonated polystyrene, polyvinyl alcohol and substituted polyamides. At present, only cellulose triacetate has sufficient strength and rejection power to function as a reverse osmosis membrane for the desalination of sea water.

As with other desalting processes, the determination of the active membrane area is of great practical importance. The membrane flux for water (J_1) depends on the system driving force (Hittman Assoc., 1970) according to

$$J_1 = K_1[(P_f - P_p) - (\pi_f - \pi_p)]$$

The membrane water flux also depends on the feed-water temperature. It is usual to correct the flux to a standard temperature (25°C). The active membrane area can then be found from the expression

$$A = w_d/J_1.$$

The rate of water transport and the absolute amount of salt transferred with the water limits the design of reverse osmosis units because the salinity of the product is usually the limiting design condition. The product salinity can be estimated from

$$c_p = c_w(1 - \eta).$$

The effect of concentration polarization can be allowed for by including, as a measure of polarization, the factor $(c_w/\bar{c})$ where $\bar{c}$ is the bulk salinity of the feed solution. Then

$$c_p = \frac{c_w}{\bar{c}}(1 - \eta)\,\bar{c} = c_{pf}(1 - \eta)\,\bar{c}$$

and the actual salt transport is

$$J_2 = B(c_w - c_p)$$

However, the non-linear pressure dependencies of K_1 and B complicate plant design. Detailed programmes have been worked out for calculating the designs of reverse osmosis units; these necessitate the use of complex mathematical relationships for the application of which large computers is desirable.

21.3. Chemical Recovery Processes

21.3.1. SALT

Sea salt is produced in commercial quantities in about sixty countries, and in small amounts for local consumption in another thirty. In fact, sea salt has been, or is now being produced, in virtually every country with a sea coast. Although salt is a dietary necessity, only a small part of that produced is actually used in food. Sodium compounds play a large part directly or indirectly, in most things used by man. The chemical usages for sodium compounds are so extensive that salt is a mainstay of the chemical industry. Thus, about two-thirds of the salt used in the United States is consumed by the chemical industry. About half of this is employed for the electrolytic production of chlorine and caustic soda and most of the remainder is used in the production of soda ash (Kaufman, 1960).

21.3.1.1. *Solar salt*

Almost certainly the first widely used salt was obtained from the ocean by solar evaporation. Probably, shore-dwelling tribes noticed the natural deposition from evaporating tidal pools and then deliberately impounded sea water to produce similar evaporation and crystallization. The process as used today is virtually unchanged from its development thousands of years ago. Thus, in all solar processes, sea water is spread in a series of shallow basins where it evaporates, gradually increasing in concentration until the solution becomes saturated with sodium chloride and salt begins to crystallize. Other precipitating salts, turbidity from the initial sea water, and organic matter from biological growth are sources of contamination. The purity of the product depends on the technical sophistication and can be adjusted to meet the requirements of the consumer.

There are usually three sets of ponds in the overall evaporation sequence: concentrating ponds, pickle ponds and crystallizing ponds (Fig. 21.13). Most of the water is evaporated in the concentrating ponds, which proportionately occupy the largest area. Individual ponds are often 2–4 × 10^6 m^2 in size (Kaufman, 1960). The operation is designed to obtain maximum water evaporation, minimum leakage of concentrated solutions and minimum contamination of the product by other precipitating compounds. Each set of basins has a specific purpose which is reflected in its design.

In the concentrating ponds, the incoming sea water at a density of about 1·027 g l^{-1}, is held until the density of the solution is increased by evaporation to 1·16. Mud and silt settle, and iron hydroxides and calcium carbonate precipitate in the initial ponds, which can be uneven in depth and which generally follow the ground contours. Most of the water contained in the incoming sea water is evaporated in these ponds. The concentrated sea water is further evaporated in a subsequent series of lime ponds (pickle ponds) in which gypsum ($CaSO_4 \cdot 2H_2O$) precipitates. The brine is held here until a specific gravity of 1·21 is reached. These ponds can also be uneven and follow land contours (Ver Planck, 1958). Dyes are sometimes added to increase absorption of the solar radiation.

The brine then is moved to a series of carefully constructed harvesting ponds located near the storage area. These are usually rectangular with flat or gently sloping floors, and have suitably shaped retaining levees. Salt begins to precipitate as large crystals which settle to the pond floor. The evaporation of the final liquor is stopped at a density of about 1·26 so as to prevent deposition of magnesium sulphate and magnesium—potassium double salts. The residual liquor, termed bittern, is often further processed to produce magnesium salts or bromine. The evaporation cycle is continued

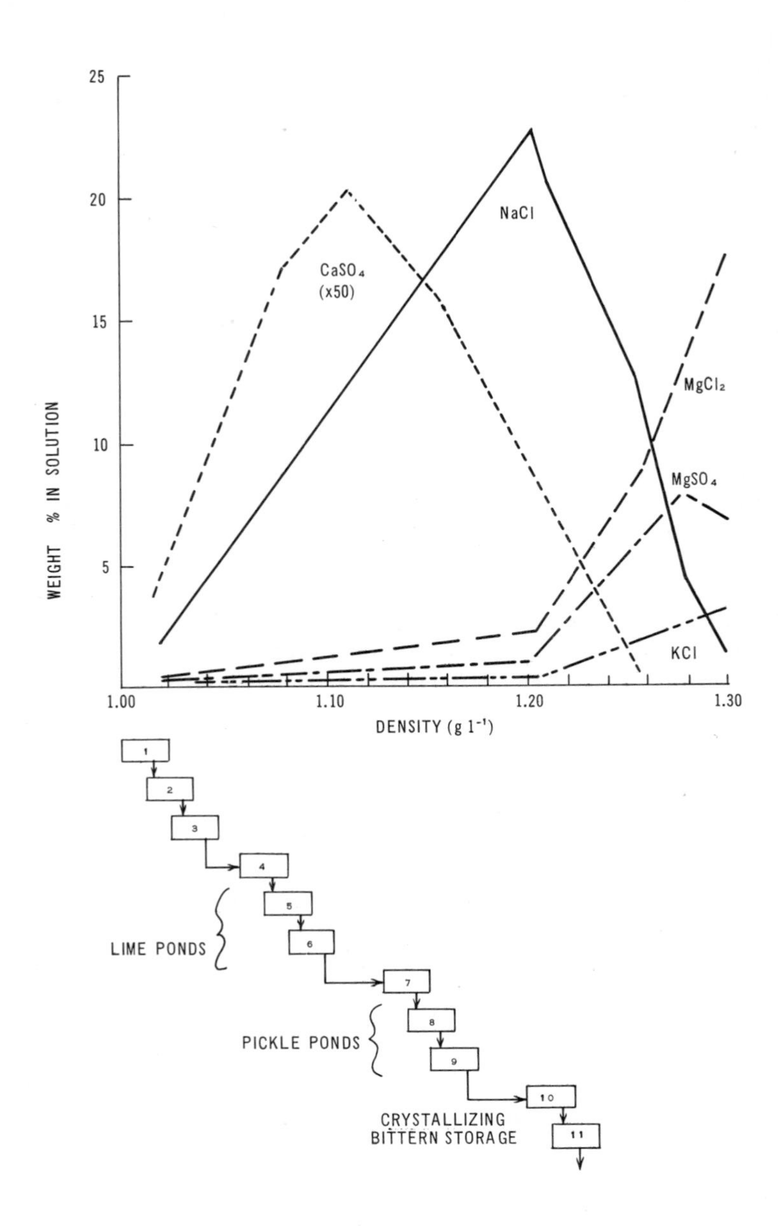

FIG. 21.13. Typical pond operation and solution compositions, solar sea salt production.

until a sufficient depth of salt has accumulated. The pond is then drained and the salt harvested by large, specially designed harvesters and conveyed to a stockpile. The raw salt is washed and sold in bulk, dried and bagged, or further processed by recrystallization to obtain purer grades.

To be economically viable a solar salt plant must be either located near a market or have access to cheap and convenient transport to distant markets. The necessity for coastal operation means that deep draught shipping is accessible, and modern plants are usually located so as to take advantage of this. The terrain must be suitable—preferably a low-lying tidal area with an impervious clay soil. The number of sites satisfying these various requirements is limited. In the selection of the site for the plant it is obviously of prime importance that the evaporation rate or, more accurately, the net excess of evaporation over precipitation must be sufficiently high. Net evaporation rates range from 380 cm y^{-1} in arid, tropical countries to the practical minimum of ca. 50 cm y^{-1}. However, a value of 100 to 120 cm y^{-1} is usually considered to be the lower limit.

Many salt producing areas have alternating annual periods of drought and rainfall. Where this is so, production is intermittent with an intensive harvesting period. The time which elapses between salt entering in sea water and leaving the plant can be as much as two years.

The rate of evaporation of water from brine or from sea water depends on the difference between the vapour pressure of the brine and the atmos-

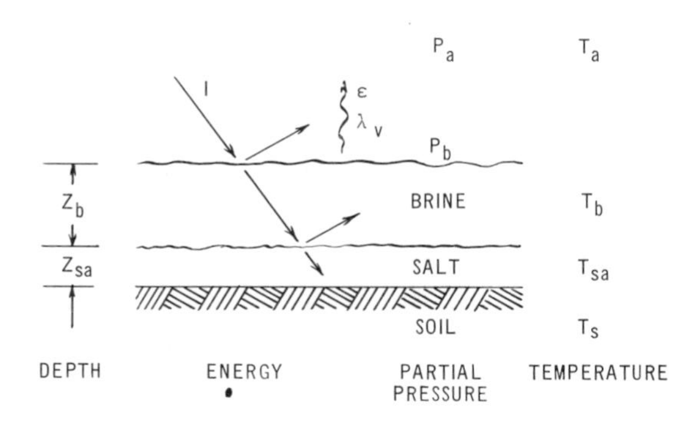

FIG. 21.14. Mass and energy balance considerations for solar sea salt pond operation.

pheric partial pressure of water vapour and on the evaporation coefficient. The vapour pressure of the brine is a function of the temperature of its and salt concentration. The partial pressure of the water vapour in the air depends on the local meteorological conditions.

A convenient form of the evaporation equation (Bloch *et al.*, 1951) is:

$$\frac{dE}{dt} = f_w(P_b - P_a) = f_w(i + jT_b - P_a)$$

An energy balance for unit surface area relates the absorbed solar radiation, the back radiation from the brine, the energy required for evaporation of the water, the thermal energy exchanges between the brine and the walls and the bottom of the pond and between the brine and the air and the energy involved in changing the temperature of the salt and the air (Fig. 21.14). The heats of crystallization and concentration can be ignored because they are small compared with the other terms. Thus,

$$I = y + zT_b + \frac{k_{sa}}{Z_{sa}}(T_b - T_s) + h_w(T_b - T_a) + \left(Z_bC_b + \frac{Z_{sa}C_{sa}}{2}\right)\frac{dT_b}{dt} + \lambda_v \frac{dE}{dt}$$

If certain assumptions are made (viz, a linear relationship between P_b and T_b, and a constant ratio of h_w to f_w), this simplifies to

$$K\frac{dT_b}{dt} + l(t)\,T_b + m(t) = 0$$

where $K = Z_bC_b + \dfrac{Z_{sa}C_{sa}}{2}$,

$$l(t) = \frac{k_{sa}}{Z_{sa}} + \lambda_v j f_w + z + h_w$$

and

$$m(t) = \lambda_v F(i - P_a) - h_w T_a - \frac{K_{sa}T_s}{Z_{sa}} + y - I$$

In practice, the temperature of the brine and, therefore, the rate of evaporation will vary periodically because of variations in the intensity of the incident solar radiation. For this reason Bloch, *et al.* (1951) replaced $l(t)$ with a diurnal time average $\bar{l}$ and used an expression for $m(t)$ to take account of the periodic (time) dependencies of P_a, T_a, T_s, I, h_w, and f_w. The temperature of the brine is given by:

$$T_b = \sum_{n=0}^{\infty} - \frac{a_n \cos \dfrac{2n\pi}{s}(t - \theta_n - \theta_n^1)}{\left\{\bar{l}^2 + \dfrac{4n^2\pi^2}{s^2}\left(Z_bC_b + \dfrac{Z_{sa}C_{sa}}{2}\right)^2\right\}^{\frac{1}{2}}},$$

where $\theta_n^1 = \dfrac{s}{2n\pi} \tan^{-1} \dfrac{2n\pi}{ls} \left(Z_b C_b + \dfrac{Z_{sa} C_{sa}}{2} \right)$.

This shows the temperature of the brine to depend on the depth of the brine layer.

The daily evaporation per unit area can be found by integrating the equation

$$E_D = \int_0^s f_w (i - P_a)\, dt + j \int_0^s f_w T_b \, dt.$$

The first integral depends only on the meteorological conditions; the second depends on the wind velocity and the brine temperature.

The incident solar energy is not absorbed completely by a transparent salt solution, and much (up to 30%) is reflected by the white salt deposits. Addition of dyes such as 2-Naphthol Green increases the energy absorption and reduces the loss by reflection to a small percentage.

Although a knowledge of the phase chemistry (see Fig. 21.15 and Borchert, 1965) of the solids precipitating from evaporating sea water is of value in the operation of a solar salt plant, it is more important for the subsequent

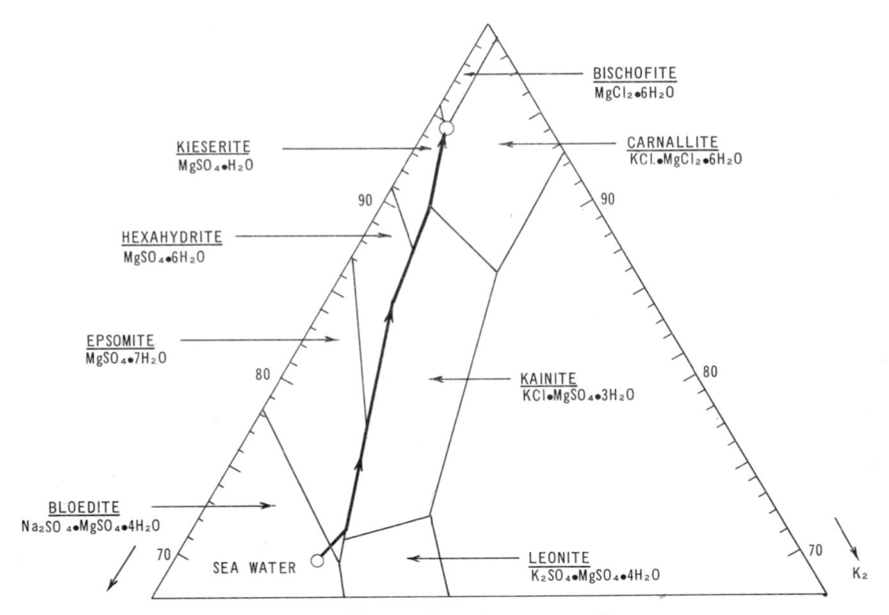

FIG. 21.15. Quinary salt system Mg^{2+}—Na^+—K^+—Cl^-—SO_4^{2-} at 25°C showing sea water evaporation path (Stewart, 1963).

processing of the bitterns. It usually suffices in salt manufacture to prevent contamination of the salt by precipitates. Two of the minor constituents of sea water deposit early in the evaporation sequence. The removal of iron and calcium carbonate are essentially complete by the time the brine reaches densities of 1·05 and 1·115 respectively. Calcium sulphate crystallizes first as gypsum, when the brine density is about 1·264; at later stages it is deposited as anhydrite. Saturation with respect to sodium chloride is attained at a density of about 1·214 and it continues to crystallize as the brine is evaporated further. Some bromide coprecipitates with chloride in the salt lattice. About 79% of the salt present can be recovered from raw sea water before contamination becomes excessive (Garrett, 1965; Usiglio, 1849; Stewart, 1963).

21.3.1.2. *Electrodialysis*

Historically, salt in Japan was produced by the evaporation of sea water. Because of the temperate climate and rainfall it is not possible to use direct solar evaporation and small amounts of salt were produced by ingenious arrangements of sand beds and trickle-film towers. For about twenty years, substantial quantities of salt have been produced in Japan from sea water concentrates manufactured in electrodialysis equipment. The total capacity of electrodialytic salt plants now exceeds 260000 tons year^{-1}, which amounts to about one-quarter of the national salt consumption.

An electrodialysis stack consists of multiple pairs of ion exchange membranes (Fig. 21.16), one membrane being permeable to the anions and the other to the cations. When an electrical potential gradient is imposed on the stack, alternate compartments become enriched and depleted in ions. A similar flow system can be used for the desalination of sea water. The principal difference is that, for the manufacture of salt, the concentration of salts remaining in the depleted stream is not as important as is the concentration in the enriched stream, and the amount of salt being transported across a unit area of membrane is considerably greater than in electrodialytic desalting. The dilute stream is discarded at a relatively high concentration, no attempt being made to recover fresh water. No sea water is fed to the concentration compartment in order to maintain as high a concentration in the brine stream, as is technically feasible.

The formation of scale is a potential problem as it is in any sea water concentration apparatus. In electrodialytic concentration plants two control methods are used (Tsunoda, 1965). In one, the concentration of Ca^{2+} ion is kept low by recycling a part of the stream from which calcium is removed, into the concentration compartment. This ensures that the solubility product of gypsum is not exceeded. In the other, a special membrane with a low perm-selectivity for SO_4^{2-} is used to increase the Cl^-/SO_4^{2-} ratio in the concentrate.

A typical module of an electrodialytic salt plant has 1500 pairs of membranes, each with an effective membrane area of $1\ m^2$, divided into six stacks. The current density is $3{\cdot}65\ A\ dm^{-2}$ at 620 V with a membrane spacing of about 0·75 mm. A brine concentrate containing about $118\ g\ l^{-1}$

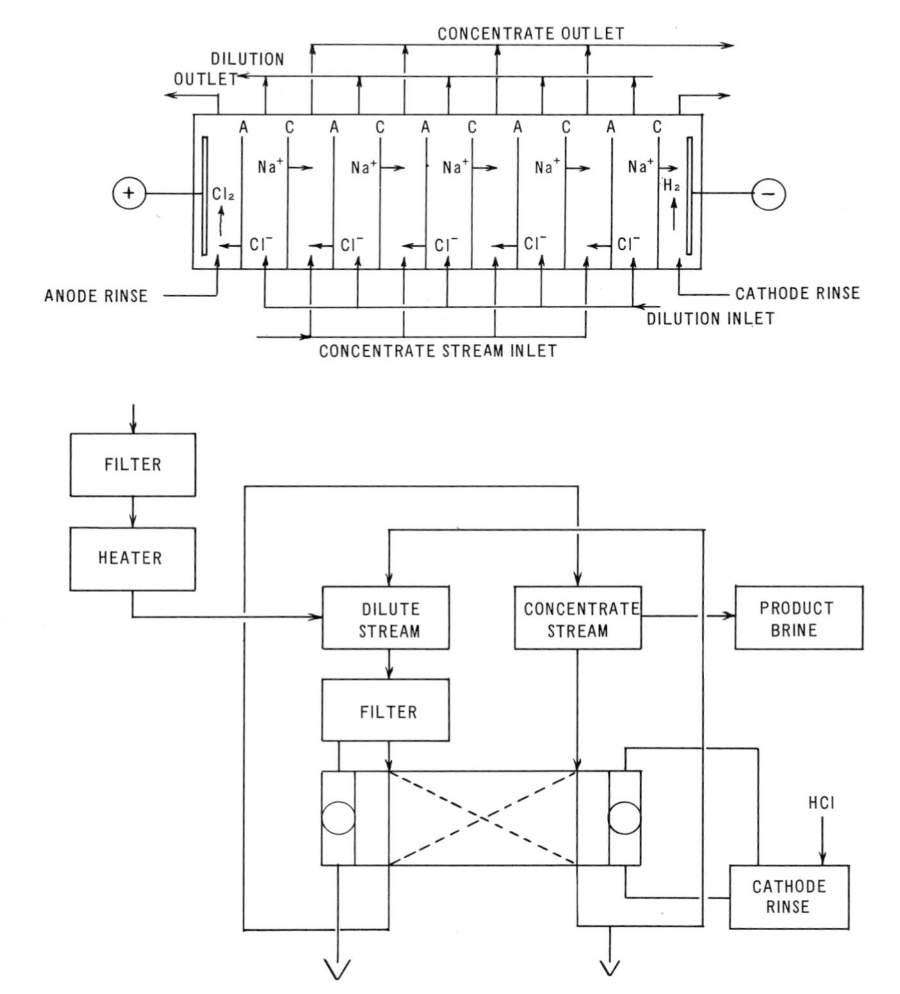

FIG. 21.16. Electrodialysis salt concentration; stack operation and flow arrangement.

chloride can be attained (Tsunoda, 1965). The overall current efficiency is 73% for Na^+ and 85% for Cl^-.

Typically, the membranes are divinylbenzene cross-linked polystyrene with sulphonic acid or quaternary ammonium exchange groups (Tsunoda, 1965); often these are reinforced with PVC cloth (Yamane *et al.*, 1969), or other materials. The exchange capacity is 1·8 to 2·8 meq g^{-1} at 25°C.

Working specifications required for the design of electrodialysis plants include the concentration and composition of the product brine, and its production rate. In addition, information is needed on the electrical requirements and current efficiency. The latter has an important bearing on the capacity of an electrodialysis unit. It is related to the production rate, and brine composition by

$$\eta_c = \frac{(w_b)(N_b)(1{\cdot}62)}{I_n}$$

There is a limiting current density above which the process cannot be operated. This is reached when the ions which are moved through the membrane by the potential gradient cannot be replenished rapidly enough at the membrane surface. This phenomenon, concentration polarization, is the result of the depletion of ions on one side and the accumulation of ions on the other. The limiting current for a cation permeable membrane is (Shaffer and Mintz, 1966)

$$I_{\text{lim}} = \frac{nFDc}{(T_{\text{mt}} - T_{\text{st}})\delta}.$$

In practice considerably less current is used since under conditions of very high current densities, an appreciable proportion of the current is carried by the H^+ and OH^- ions. This leads to a fall in the efficiency of the process and moreover may adversely affect the pH values of the process streams.

The electrical characteristics of the cell can be described in terms of the cell resistance factor and the cell potential. The potential drop is the most important electrical parameter. Changes in it occur as a result of membrane fouling or stack polarization. The potential drop is linearly related to the energy consumption per unit weight of salt concentrated, and is given by

$$E = E_0 + R\left(\frac{i}{C_{\text{do}}}\right).$$

Differences in ionic transport through the membrane lead to differences in the rates of concentration of particular ions. Therefore, in an electrodialytic sea water concentration process, the ionic ratios in the brine produced differ from those in the raw sea water. Generally there is a depletion in calcium and in sulphate in the brine relative to the raw feed.

The overall efficiency is the product of those individual stages in the overall processes (the electrode reactions, the total stack resistance and polarization and the efficiency associated with manifold leakage, perm-selectivity and water transport). Thus,

$$\eta = \eta_e \eta_R \eta_i \eta_m \eta_s \eta_w.$$

The first and fourth terms are primarily functions of stack design, the second depends upon the membrane thickness and on the operating conditions and the third term depends primarily upon the operating conditions. The last two terms depend upon the properties of the membrane selected.

21.3.2. MAGNESIUM

Both magnesium metal and its compounds are in considerable commercial demand. Magnesium compounds are employed by the cement, rubber, textile, metal, chemical and construction industries. Basic magnesia refractories are vitally necessary in the steel industry. Magnesium is the eighth most abundant element in the earth's crust and is the third highest in concentration in sea water. Although, at 1300 ppm, the concentration in sea water is much lower than it is in competitive commercial magnesium ores (dolomite and brucite), high-purity ores are rare, and sea water has been for more than forty years a major source of magnesium compounds, and for thirty years the major source of magnesium metal.

21.3.2.1. *Magnesium compounds*

All commercial processes for the utilization of magnesium in sea water, whether for the production of magnesium metal, or its compounds, begin with the precipitation of magnesium hydroxide from sea water by the addition of hydroxyl ion. In practice, the alkali used is almost always a slaked lime produced by calcining limestone (or oyster shell) or a mixed alkali formed by the calcination of dolomite ($CaCO_3 \cdot MgCO_3$) or dolomitic limestone. Caustic soda is sometimes used when it is available at a reasonable price. If dolomite is the source of alkalinity, about 40% of the precipitated magnesium ion is furnished by the dolomite and 60% by the sea water.

Magnesium hydroxide is a sparingly soluble compound having a thermodynamic solubility product for the process.

$$Mg^{++} + 2OH^- \rightleftharpoons Mg(OH)_2 \downarrow$$

of $1{\cdot}255 \times 10^{-11}$ at 25°C. The value of the solubility product decreases with temperature increase (Strelez, *et al*, 1953). At the ionic strength of sea water the hydroxide is slightly more soluble than in fresh water (Langelier, *et al.*,

1950). Magnesium hydroxide crystallizes in uniaxial hexagonal platelets. The precipitate has a high specific surface area with an abundance of exposed OH^- rich faces. For this reason it is an excellent sorbent for many of the trace elements in sea water (Boegelin and Whaley, 1967). Boron in particular is strongly chemisorbed at pH values below 10·5 (Schambra, 1944).

Basic problems in the precipitation of magnesium hydroxide are its slow settling and poor filtration characteristics. Most of the art in magnesium production lies in the production of a precipitate which settles rapidly and filters easily; this can be achieved in several ways, including reduction in the activity of the precipitant (Mastin, 1938), recirculation of the slurry (Robinson, *et al*., 1946) and the addition of organic compounds (Kaliforschungs-Anstalt, 1938).

Magnesium oxide, the magnesia of commerce, is produced in many grades. It is almost unique among materials with a simple crystalline structure in the way in which physical properties depend on the method of production. Thus, active magnesium oxide is produced by calcination at a relatively low temperature (400–500°C) in a multiple hearth or rotary furnace. Its principal uses are in the production of Sorel cement (magnesium oxychloride or oxysulphate), as a chemical neutralizer, and in the manufacture of magnesia insulation (Comstock, 1964; Boegelin and Whaley, 1967). The unreactive oxide is produced by heating the active oxide to an elevated temperature (up to 2800°C, the melting point of pure MgO) for extended periods. This heat treatment orders the crystalline structure and leads to the production of an oxide with an extremely low hydration rate. Periclase is pure crystalline magnesium oxide produced for use in the manufacture of high temperature crucibles, and as an insulation for heating elements (Havighorst and Swift, 1965; Williams, 1965).

The thickened, washed hydrate is often mixed with selected metal oxides (e.g. CaO, Al_2O_3, Fe_2O_3, SiO_2) in order to lower the melting point sufficiently to permit grain recrystallization and particle sintering when it is calcined at temperatures of 1700–2500°C. The product is a dense unreactive material with an MgO content of about 90%. It is used in the production of basic refractory bricks and shapes, and as a ramming mix for metallurgical furnaces (Comstock, 1964; Havighorst and Swift, 1965; Williams, 1965).

A pretreatment is usually employed to lower the amount of calcium carbonate coprecipitating with the magnesium hydroxide and thus increase the purity of the product. This can be achieved in two ways. In the first of these, a controlled amount of hydroxyl ion is added to the feed sea water so that the increased pH leads to the precipitation of most of the carbonate present along with part of the magnesium as magnesium hydroxide. At a pH of about 10, the carbonate content is lowered to about 35 ppm, less than

10% of the magnesium being lost in the pretreatment (Meschter, 1961). Significantly better carbonate removal is obtained if the pH is lowered by acid addition and the molecular CO_2 is removed by steam or vacuum stripping. Wash waters are also often treated to minimize carbonate levels in the subsequent steps (Anon, 1966, 1967).

In a typical sea water magnesia plant (Fig. 21.17) the incoming sea water is screened and treated to prevent calcium contamination of the magnesium hydroxide product. Calcined limestone or dolomite is added in a water slurry to the sea water in a reactor. Magnesium hydroxide is precipitated and the resultant slurry is allowed to settle in a series of thickeners. The spent sea water is discharged and the concentrated magnesium hydroxide is washed with a countercurrent stream of water and filtered (Havighorst and Swift, 1965).

High purity magnesium compounds for chemical and pharmaceutical uses are produced from sea water by a process (Fig. 21.18) in which the initially precipitated slurry of magnesium hydroxide is carbonated with boiler stack gas to produce a slurry of magnesium carbonate. The temperature and magnesium content are adjusted and additional carbon dioxide is added. Basic magnesium carbonate ($3MgCO_3 \cdot Mg(OH)_2 \cdot 4H_2O$) precipitates and is dried and calcined. The basic carbonate does not sorb impurities to the same extent as does the hydroxide, and on dehydration the open crystalline lattice structure produces an oxide with a high surface area and great chemical activity which owes its properties to its lattice (Comstock, 1964; Shreve, 1967). The reactions involved are:

$$Mg(OH)_2 + 2H_2O + CO_2 \xrightarrow[52-58^\circ C]{} MgCO_3 \cdot 3H_2O \downarrow$$

$$Mg(OH)_2 + 3(MgCO_3 \cdot 3H_2O) \xrightarrow[100^\circ C]{steam} 3MgCO_3 \cdot Mg(OH)_2 \cdot 4H_2O \downarrow + 5H_2O.$$

21.3.2.2. *Magnesium metal*

Magnesium metal is only produced from sea water in two large plants, which use somewhat different processes. In both, magnesium hydroxide, precipitated from sea water, is converted to magnesium chloride; electrolysis of this produces magnesium metal and chlorine. The principal differences between the two processes lie in the type of cell feed employed and the method of preparation of the magnesium chloride feed. In the IG-MEL process (Heroya, Norway), an anhydrous cell feed is produced by the direct chlorination of MgO, this being produced by the calcination of $Mg(OH)_2$ precipitated from sea water. In the Dow process (Freeport, Texas) a hydrous magnesium chloride feed containing about 25% water is used.

The IG-MEL process was developed in Germany in about 1896 and the Dow process in the United States in about 1915. Magnesium was produced

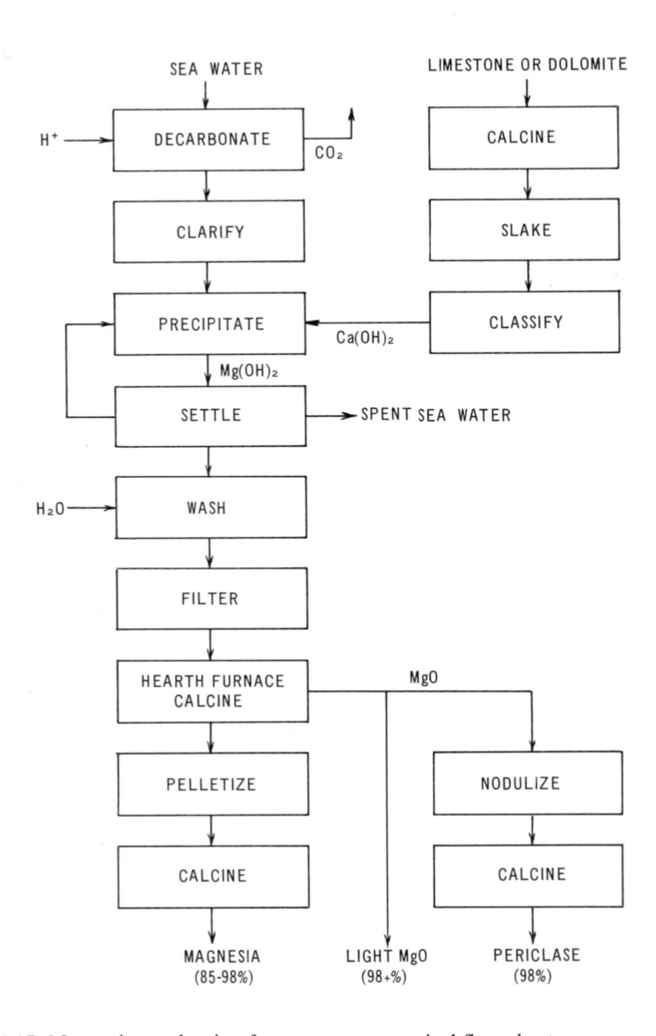

FIG. 21.17. Magnesia production from sea water, typical flow sheet.

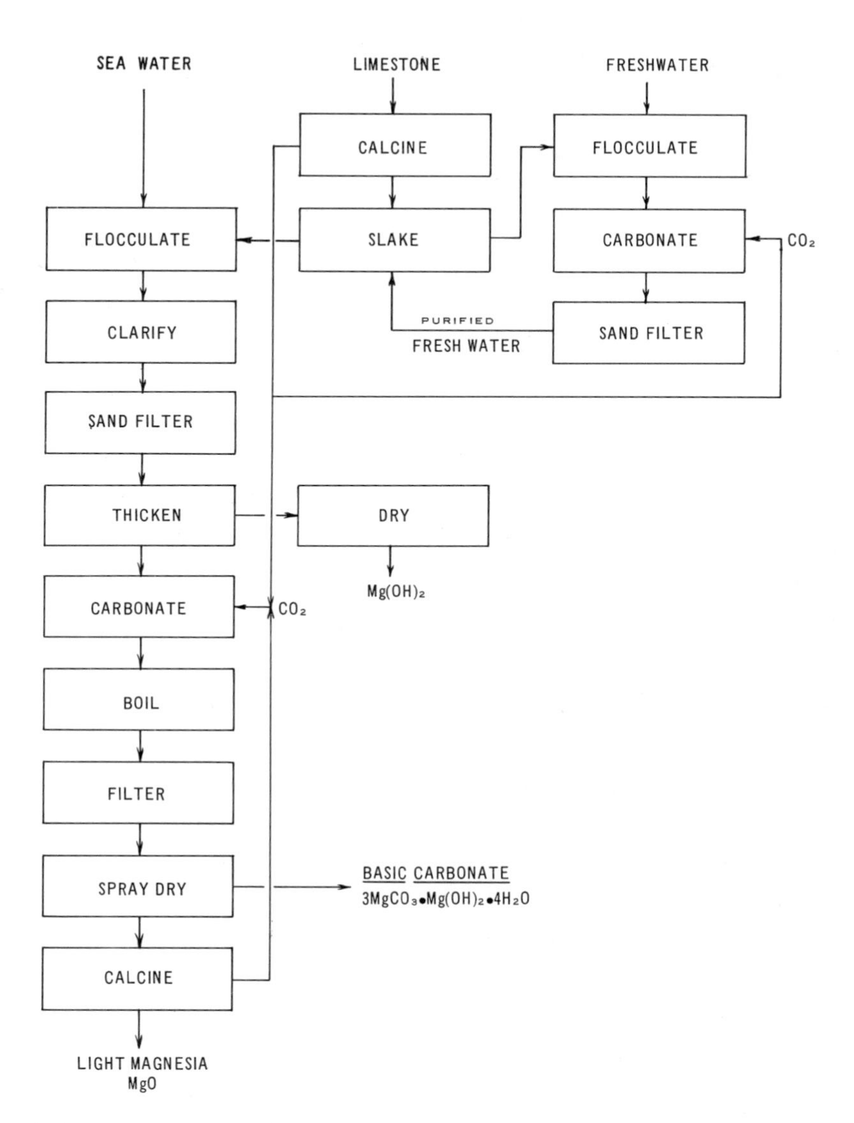

FIG. 21.18. High purity magnesium chemical production from sea water.

from sea water about the same time in the United Kingdom and in the United States just before World War II, and the processes have co-existed for more than thirty years and between them now produce the bulk of the world's magnesium (Hanawalt, 1964; Ball, 1956).

The initial separation of magnesium from sea water by precipitation with hydroxyl ion addition is exactly similar to that required for the production of magnesia. However, it is essential to prevent the adsorption of boron by the $Mg(OH)_2$ when it is precipitated from sea water as it interferes with the operation of a magnesium cell (Schambra, 1944). The effect of the stoichiometry of the hydroxyl addition on boron adsorption is shown in Fig. 21.19 (Robinson, *et al.*, 1946). Up to the equivalence point, boron, probably present as $B(OH)_3$, is strongly sorbed on the $Mg(OH)_2$. As excess alkali is added, $B(OH)_4^-$ ions are formed and are released from the surface (Gross, 1967).

Processes for the production of magnesium after its initial separation from sea water are, of course, independent of the source of magnesium and operate equally well for both brine, and terrestrial derived ores. The processes are described below.

In the IG-MEL process—see Fig. 21.20—(Ball, 1956; Emley, 1966; Roberts, 1960), the reactive MgO produced by the calcination of $Mg(OH)_2$ is reacted with carbon (from coal or peat) and chlorine at about 1000°C to produce molten anhydrous $MgCl_2$.

$$MgO + C + Cl_2 \rightarrow MgCl_2 + CO$$

and

$$2MgO + C + 2Cl_2 \rightarrow 2MgCl_2 + CO_2.$$

The $MgCl_2$ is tapped from the chlorinators and fed to rectangular flat bottomed steel shells, with carbon anodes and steel cathodes. The cell is operated without external heat and the Mg metal which is produced is removed periodically from the surface of the cell bath (Emley, 1966).

The Dow process—see Fig. 21.21—has been described in detail in the literature (Gross, 1967; Schambra, 1944; Hunter, 1944; Rave, 1953). A cell feed having the approximate composition $MgCl_2$. 0·5–1·0 H_2O is produced by the neutralization of $Mg(OH)_2$ (from sea water) with HCl followed by several sequential dehydration steps. Electrolysis is accomplished in a large steel vessel which also serves as the cathode, graphite rods passing through an arched refractory roof serving as the anode. Magnesium floats on the surface of the electrolyte and is periodically removed.

Comparative data for the operational parameters of the Dow and IG-MEL processes are shown in Table 21.8 (Hanawalt, 1964; Emley, 1966). Since magnesium chloride itself is unsuitable as an electrolyte because of its

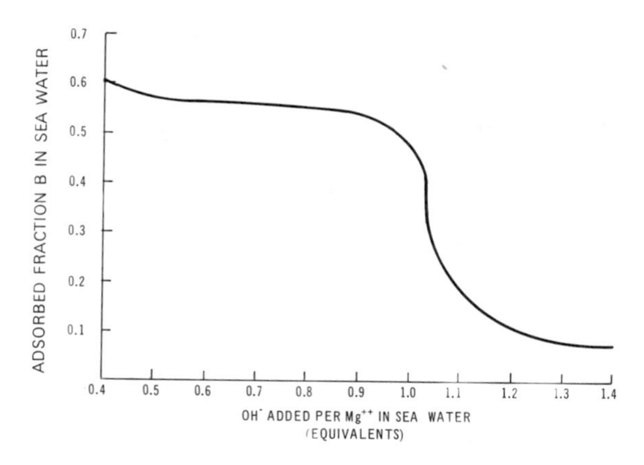

FIG. 21.19. Effect of OH^- concentration on boron adsorption by $Mg(OH)_2$ (Robinson *et al*, 1946).

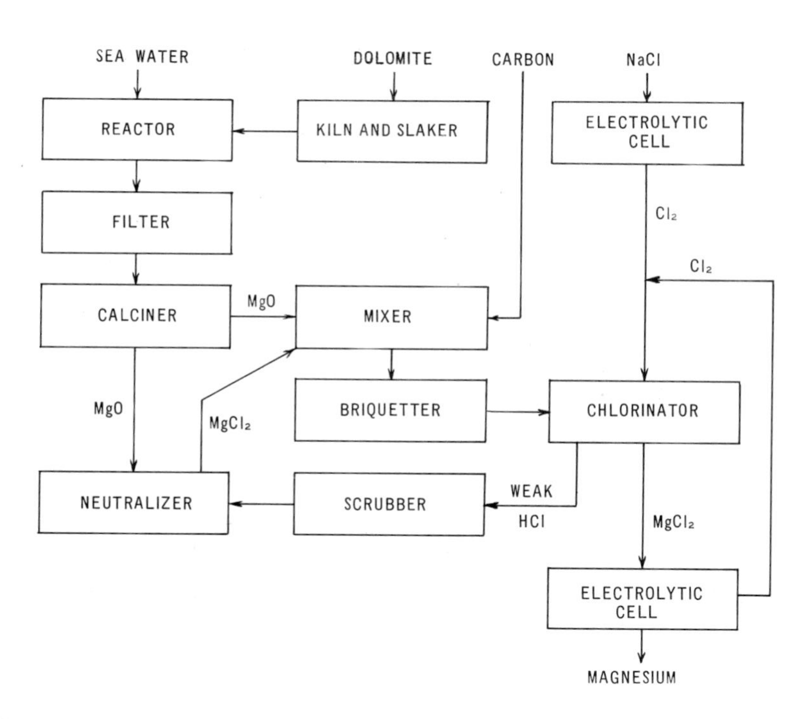

FIG. 21.20. Magnesium production from sea water, IG-MEL process flowsheet.

TABLE 21.8

Comparison of Dow and IG-MEL electrolytic processes for the production of magnesium (Hanawalt, 1964)

	Dow	IG-MEL
Cell feed	Hydrous	Anhydrous
Current, A	100000	38000
DC power (cells only) (kWh g^{-1} of Mg)	0·0194	0·0185
A.C. power (whole process) (kWh g^{-1} of Mg)	0·0209	0·0247
Operating temp., °C	700	750
Anode consumption (g C ton^{-1} of Mg)	100×10^3	15×10^3
Cathode efficiency, %	78	90
Cell feed analysis, wt. %		
$MgCl_2$	72	96·3
$CaCl_2$	0·5	1·4
NaCl	1·0	2·2
H_2O	25·0	—
MgO	1·5	0·1
Cell bath analysis, wt. %		
$MgCl_2$	20·0	12·0
$CaCl_2$	20·0	35·0
NaCl	57·0	28·0
KCl	2·0	24·0
CaF_2	1·0	1·0

low electrical conductivity and its unsuitable specific gravity other salts are added in order to improve the physical properties of the molten salt bath. The compositions of the molten salt mixtures used in the two processes are shown in Table 21.8.

Both the Dow and IG-MEL processes produce chlorine that can be recycled to earlier stages of the process. Alternatively, it may be sold as a by-product if a source of cheap HCl or low grade chlorine is available for neutralization or chlorination. Extra chlorine is required in both processes to replenish production losses.

Although electrolytic reduction of the chloride is the only process used for production of magnesium from sea water, it is also produced from terrestrial ores by the thermal reduction of magnesium oxide with silicon or ferro-silicon

$$4MgO + SiFe \xrightleftharpoons{1170°C} 2Mg + Mg_2SiO_4 + Fe.$$

The ΔH for the reaction of pure MgO is 65 kcal mol^{-1} at 1330°C and for dolomite it is 60 kcal mol^{-1} at 1180°C (Emley, 1966).

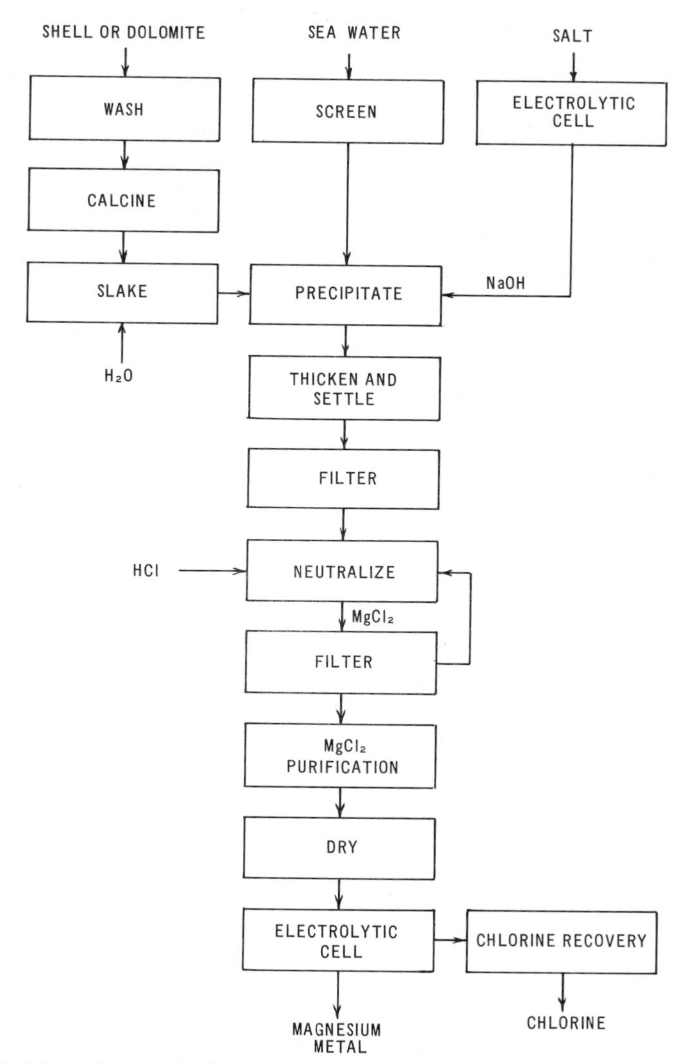

FIG. 21.21. Magnesium production from sea water, Dow process flowsheet.

Carbothermic reduction is carried out at 1800°C or more

$$MgO + C \underset{}{\overset{1800°C}{\rightleftharpoons}} Mg + CO$$

The reaction must be shock quenched to prevent recombination (Gross, 1967). Either of these commercially developed processes can be used with magnesium oxide derived from sea water, but cannot compete economically with the electrolytic methods.

21.3.3. BROMINE

Bromine was first discovered by Balard in 1825. Although it is widely dispersed in the lithosphere, its abundance is low ($< 0{\cdot}1$ ppm). However, it is much more strongly concentrated in the sea in which it is the ninth most abundant of the dissolved constituents, having an average concentration of ca. 67 ppm. For this reason sea water is a major commercial source of bromine.

Bromine is a reddish brown, liquid halogen with an unpleasant odour. It is used as a disinfectant, and in the production of fire-extinguisher and hydraulic fluids, fumigants and industrial solvents. However, about 75% of the bromine produced in the United States is used in ethylene dibromide production; this is used as a scavenging agent for lead in petroleum to which lead alkyls have been added as anti-knock compounds (Miller, 1965).

Although some bromide cocrystallizes when NaCl is precipitated from sea water, most of it remains in the sea water concentrate, for this reason coastal lagoons and salt marshes tend to have a correspondingly high concentration of bromine. For example, the Sivash Sea, a coastal embayment of the Sea of Azov, contains Br concentrations up to 200 ppm, and the brine of the Rann of Cutch, an 18000 km^2 seasonally flooded salt pan near the mouth of the Indus River contains 250 ppm Br. The Sebkh-el-Maleh is a natural saline lagoon in Tunisia, having bromine concentration of up to 2500 ppm (Jolles, 1966). All of these sea water concentrates have been commercially exploited for the production of bromine. Processes for the recovery of bromine from the ocean must compete economically against those based on its extraction from terrestrial brines which are almost always the remnants of relict seas.

Bromine always occurs in commercial deposits as bromide ion and it must be oxidized either electrolytically or chemically to the free halogen before recovery. In the earliest processes for the recovery of bromine from brines, manganese dioxide or sodium chlorate were used as oxidants. For a short period these processes were replaced by electrochemical oxidation. However, about 1877 chlorine began to be used for the oxidation of bromide and this is now general commercial practice. The standard potentials for the aqueous systems are (Latimer, 1952)

$$\tfrac{1}{2}Cl_2 + e \rightarrow Cl^- \qquad E^\circ = +1{\cdot}356$$

$$\tfrac{1}{2}Br_2 + e \rightarrow Br^- \qquad E^\circ = +1{\cdot}065$$

The heats of hydration of the halide ions are 88 and 80 kcal mol^{-1} for Cl^- and Br^- respectively. The corresponding ionic hydration entropies are -20 and -16 kcal mol^{-1} deg^{-1} respectively (Sharpe, 1967).

Two separate and distinct processes for the production of bromine were developed about the same time. The Dow process in which, after chlorination, air is used as the stripping agent, was developed by Dr. H. H. Dow in 1896. The steaming-out process in which the liberated bromine is stripped with steam, was improved and made workable by Konrad Kubierschky in Germany in 1906 (Stenger, 1964).

The Dow blowing out process is the only process currently used for recovering bromine from sea water (Moyer, 1969). A flow diagram of the process is shown in Fig. 21.22 (Shigley, 1951; McIlhenny, 1968). The basic

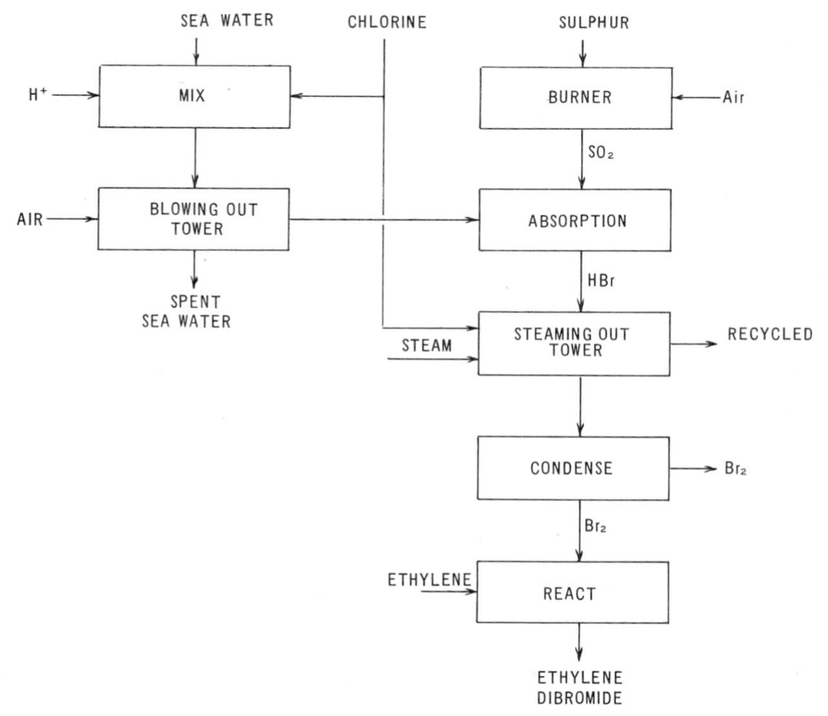

FIG. 21.22. Bromine production from sea water, Dow blowing-out process flowsheet.

chemistry of the process is simple, but there are no reliable published data on this complex system which comprises a solution in sea water of various acids, free and combined halogens and carbon dioxide (Fossett, 1971). The chlorination reactions in sea water include (Moyer, 1969; White, 1972):

$$Cl_2 + H_2O \rightleftharpoons 2HOCl + Cl^-$$
$$HOCl + Br^- \rightleftharpoons HOBr + Cl^-$$
$$HOBr + H^+ + Br^- \rightleftharpoons Br_2 + H_2O$$

The net result is the release of a mole of bromine for each mole of chlorine used. The rate constant for the first reaction is $5 \times 10^{14}\,l\,mol^{-1}\,s^{-1}$ showing the process to be practically instantaneous (White, 1972). The second reaction, for which the rate constant is $2{\cdot}95 \times 10^{-3}\,l\,mol^{-1}\,s^{-1}$, is likely to be rate controlling (Farkas, *et al.*, 1949).

The hydrolysis of bromine is very pH dependent, as shown in Fig. 21.23

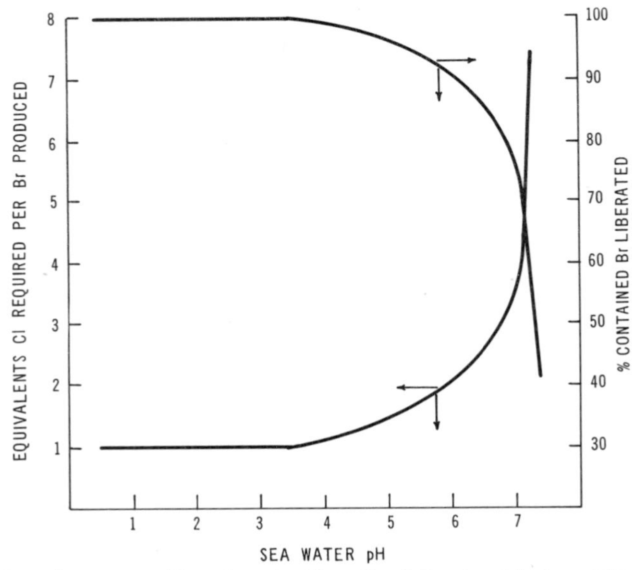

FIG. 21.23. Effect of sea water pH on bromine liberation following chlorine addition (Stewart. 1934). Upper curve shows the percent of the bromine present which is liberated. The lower curve shows the number of equivalents of chlorine required to liberate one equivalent of bromine as a function of pH.

(Stewart, 1934). Above pH 7, no bromine is liberated from sea water; at a pH of 3·5 or less, the liberation is practically complete. In the commercial process the acidification is partly accomplished with a mixed acid recirculated from the second bromine release step, the short-fall being made up by addition of extra acid. Although, theoretically, 0·44 units of chlorine are required to produce a unit weight of bromine, in practice about 15% excess chlorine is required when bromine is recovered from sea water because of the formation of the interhalogen compounds which interfere with the stripping of bromine by lowering its vapour pressure (Jolles, 1966; Moyer, 1969).

$$Cl_2 + Br^- \rightleftharpoons Cl^- + BrCl$$
$$Br_2 + Cl_2 \rightleftharpoons 2BrCl$$
$$BrCl + Cl \rightleftharpoons BrCl_2^-$$
$$Br_2 + Br \rightleftharpoons Br_3^-$$

Because of the low initial concentration of bromine in sea water (67 ppm) it is essential to have accurate control over the process if good recoveries are to be achieved. The rate of chlorine addition to the incoming water is controlled by measurement of the redox potential, which is kept at 0·88–0·97 V vs. S.C.E. The bromide concentration in the effluent varies between 5 and 15 mg l^{-1} depending upon the temperature and pumping rates (Stenger, 1964; Hart, 1947).

The liberated elemental bromine is stripped from the sea water in a large packed tower through which the sea water and air are passed countercurrentwise; because of the relatively low vapour pressure of the bromine over the sea water the volume of air required is generally large. This bromine laden air passes to an absorption system in which bromine stripping occurs, the residual air being vented. Two absorption systems, both still in commercial use, have been developed. In the earliest plant for the extraction of bromine from sea water (Kure Beach, North Carolina) the bromine in the air stream was absorbed by a solution of an alkali, such as sodium carbonate (Stewart, 1934).

$$3Na_2CO_3 + 3Br_2 \rightleftharpoons 5NaBr + NaBrO_3 + 3CO_2$$

The bromine concentration in the bromide–bromate liquor was concentrated about 800 times over that in the original sea water. The bromine was liberated once more by acidification (usually with H_2SO_4) and removed by steaming (Jolles, 1966).

$$5Br^- + BrO_3^- + 6H^+ \rightleftharpoons 3Br_2 + 3H_2O$$

This process was superseded in 1937 by an acid–gas process in which the bromine is allowed to react with sulphur dioxide (Heath, 1939)

$$Br_2 + SO_2 + 2H_2O \rightleftharpoons 2HBr + H_2SO_4,$$

the reaction partly taking place in the gas phase. The concentration of HBr in the absorbing solution reaches nearly 20% (Moyer, 1969).

The bromine is released from the solution with a second equivalent of chlorine. The solution is then steam heated almost to boiling and the bromine which distils is condensed, and either separated or reacted with ethylene to form ethylene dibromide. The acid remaining is recirculated to the initial sea water acidification.

The steaming out process is primarily used for brines having higher bromine concentrations than normal sea water, e.g. Dead Sea water, underground brines and sea salt bitterns. In this process (Fig. 21.24) the brines

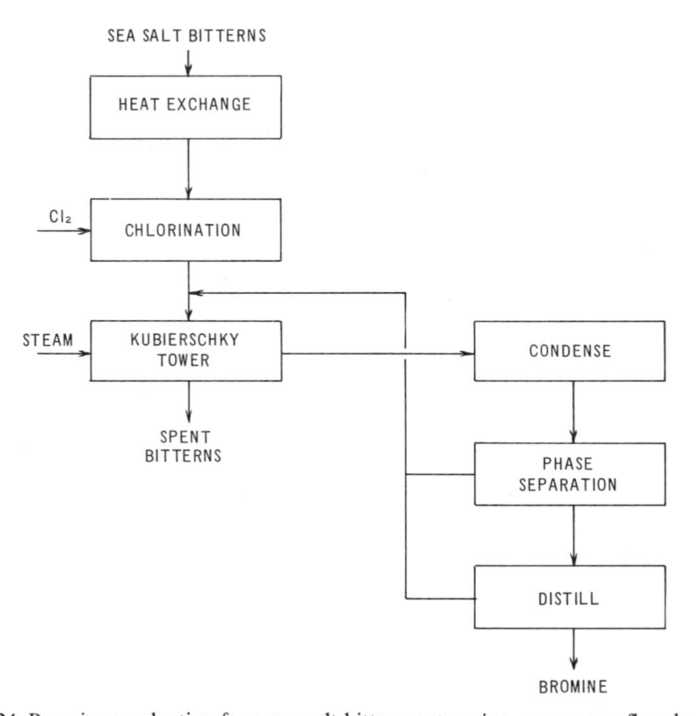

FIG. 21.24. Bromine production from sea salt bitterns; steaming-out process flowsheet.

are steam heated by cross exchanging to about 90°C and chlorinated in a packed tower. The halogen laden vapour is condensed and separated in a two-phase gravity separator. More than 95% of the bromine is recovered from the raw brine (Seaton, 1931; Stenger, 1964).

The earliest recovery of bromine from sea water took place in one of the more unusual episodes in marine chemical history. When the advent of leaded petroleum increased the demand for bromine, the Ethyl Corporation and DuPont developed a process for the fixing of bromine as tribromoaniline after chlorination (Stine, 1929):

$$3Br^- + 3Cl_2 + C_6H_5NH_2 \rightleftharpoons C_6H_2Br_3NH_2 + 3H^+ + 6Cl^-$$

Acidification of the sea water was necessary to depress the hydrolysis of the tribromoaniline. About 200 ppm of H_2SO_4 was required to reduce the hydrolysis to 11% at 15°C. At lower temperatures, less acid was required.

After a small pilot operation, a 4200 ton, 253 ft vessel was refitted as a floating chemical processing plant. Designed to produce 100000 lb of tribromoaniline, the ship could carry 500000 lb of H_2SO_4 and could process 7000 gallons of sea water per minute. Only a single voyage (off the coast of North Carolina) was made in which a process yield of 59% was attained. The process was not economically viable and was abandoned.

This was, almost certainly, the first shipboard chemical operation of any magnitude. It was not pursued because of economic rather than technical limitations.

21.3.4. POTASSIUM COMPOUNDS

Potassium is the seventh most abundant element in the earth's crust and is the sixth highest in concentration in sea water. Potassium salts are essential to plant growth and the element is a necessary component of a balanced agricultural fertilizer. Potassium compounds are widely produced from brines as well as from evaporite deposits.

Potassium compounds are major constituents not only of the concentrated brines remaining after salt has been crystallized from sea water by evaporation, but also of the ashes of many species of the larger marine algae. Because these sources are readily available, potassium salts of marine origin have over the years been recovered by a number of different processes. However, no potassium salts are produced directly from sea water in commercial quantities anywhere in the world at the present time.

From the 17th century until about 1845, when the LeBlanc soda ash process was developed, ashed seaweed (kelp) was a principal source of potash for soap and glass factories in Scotland and France. The species of algae which were used were mainly those of *Laminaria*, the ashes of which contain up to the equivalent of 33% K_2CO_3. A short revival of the Scottish seaweed industry occurred around 1860 when an industrial demand developed for iodine. There were by 1866, twenty manufacturers of iodine from ashed sea plants, most of whom also extracted potassium salts (Armstrong and Miall, 1946; Tressler and Lemon, 1951).

When the United States supply of potassium salts was interrupted during World War I, about 8% of the potassium requirements were met by the seaweed industry. The processes used were simple incinerations in rotary kilns followed by water leaching and fractional crystallization of potassium chloride (Turrentine and Shoaff, 1919).

The crystallization path of sea water at 25°C is shown in Fig. 21.15 (Hadzeriga, 1965; Jacobs, 1968). The first potassium-containing salt to precipitate is kainite ($4KCl \cdot 4MgSO_4 \cdot 11H_2O$), followed by carnallite ($KCl \cdot MgCl_2 \cdot 6H_2O$). About 0·02 units of potassium (as KCl) are available in the bitterns resulting from the crystallization of 1 unit of NaCl from sea water (Hildebrand, 1918).

About 1850, A. J. Balard, the discoverer of bromine, developed a process for the production of sodium sulphate, potassium chloride and magnesium chloride by fractional crystallization of the residual bitterns from French sea-salt manufacture. The process as later modified, is shown in Fig. 21.25. An impure carnallite was derived from the slurry produced by refrigeration of the mother liquor remaining from separation of the magnesium salts. This is typical of a number of methods used to recover potassium in California, Sardinia, Japan and India, all of which involve the precipitation of carnallite by cooling a magnesium–potassium–chloride solution.

A somewhat different approach was used by Niccoli (1926) in Eritrea and Italy. In his process, solar salt bitterns are evaporated in two steps to a density of 1·35 to 1.36. The resultant crystals consist of kainite, mixed with astrakhanite ($Na_2SO_4 \cdot MgSO_4 \cdot 4H_2O$), kieserite ($MgSO_4 \cdot H_2O$) and NaCl. They contain about 70% of the potassium originally present in the solution. The mixed salts are dissolved and schoenite ($K_2SO_4 \cdot MgSO_4 \cdot 6H_2O$) is formed by the addition of magnesium sulphate and evaporation or by selective leaching with cold (10–20°C) water. The potassium is recovered from schoenite after precipitating both magnesium and sulphate with $Ca(OH)_2$. The solution which contains only potassium and sulphate is filtered and evaporated; K_2SO_4 with a purity of 97–98% crystallizes. The overall yield is about 90%.

The Niccoli process has been modernized and modified and used in India for the production of schoenite on a small scale. In the Central Salt and Marine Chemicals process (Udwadia *et al.*, 1966) illustrated in Fig. 21.26, the second evaporation from a density of 1·312 to 1·336 forms the Niccoli salt. When the densities are properly controlled, the mixed salt contains 18–20% KCl, 30 to 35% $MgSO_4$, 6 to 8% $MgCl_2$ and 15 to 20% NaCl. Six to 7 units of mixed salt can be produced for each 100 units of salt. The overall conversion of the mixed salt to schoenite is 75 to 80%.

Several processes for the recovery of potassium from sea water and sea water brines have been studied as possible methods for the recovery of potassium from desalination brines (Salutsky *et al.*, 1964). A complex sulphate,

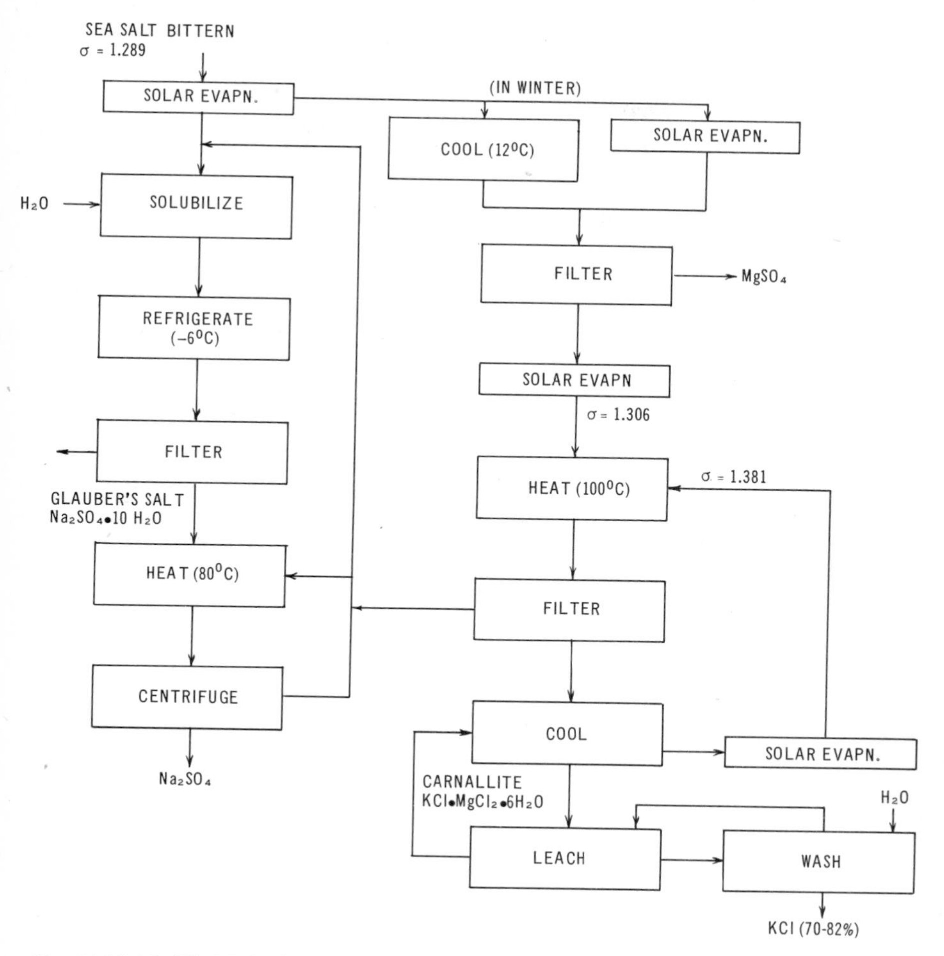

FIG. 21.25. Modified Balard process for recovering potassium salts from sea water by precipitation as carnallite.

georgyite ($5CaSO_4 \cdot K_2SO_4 \cdot H_2O$) can be separated by digestion of a brine containing 3 g l^{-1} or more expressed as K_2O with gypsum at 80°C. The complex salt can be extracted with fresh water and potassium sulphate recovered by crystallization:

$$2K^+ + SO_4^{2-} + 5CaSO_4 + H_2O \underset{20^\circ C}{\overset{80^\circ C}{\rightleftharpoons}} [K_2SO_4 \cdot 5CaSO_4 \cdot H_2O]$$

Potassium was also found to be efficiently separated from sea water or

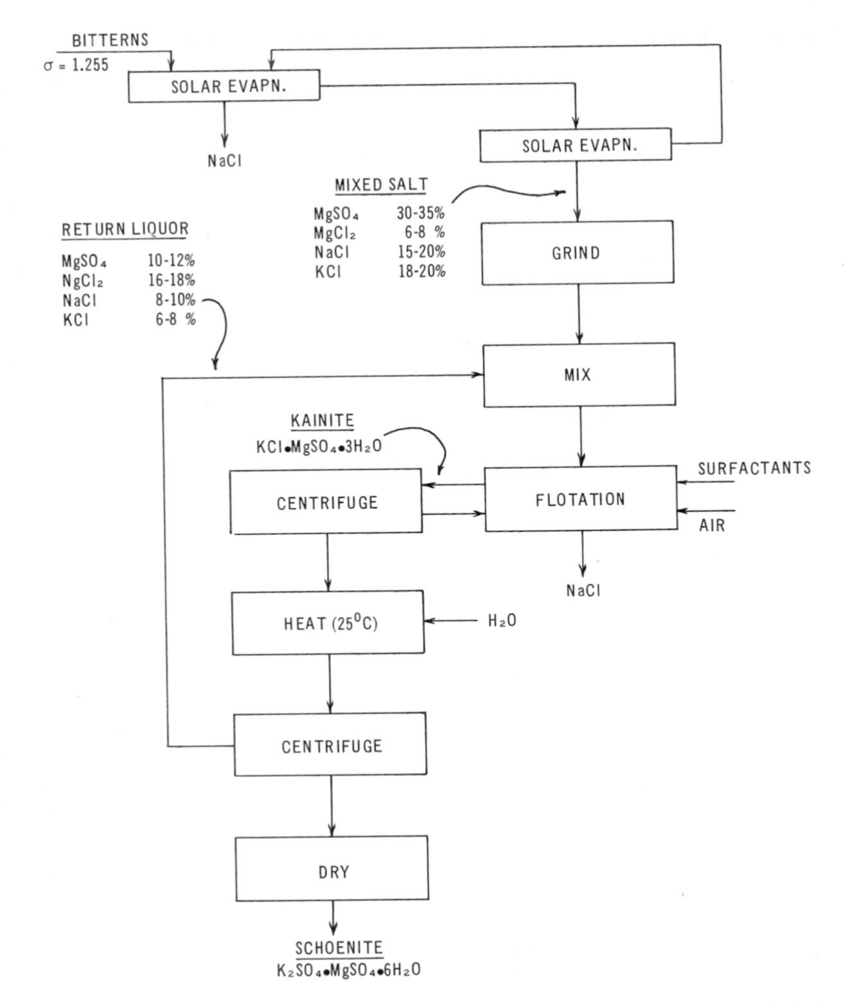

FIG. 21.26. Niccoli process for recovering potassium salts from sea water by precipitation as schöenite, as modified by the Central Salt and Marine Chemicals Research Institute (Udwadia *et al.*, 1966).

relatively dilute brines by precipitation as the insoluble double phosphate with magnesium. The recovery of potassium increases at high phosphate concentrations. A final pH of 9 to 10 is required for good yields and large excesses of the reagents are required to drive the reaction to practical completion:

$$Mg^{2+} + K^+ + Na_2HPO_4 + NaOH \rightleftharpoons MgKPO_4 + 3Na^+ + H_2O$$

Dipicrylamine (hexanitrodiphenylamine) forms a potassium complex that is comparatively insoluble in aqueous solutions and for which the solubility changes markedly with temperature (898 mg l^{-1} at 21°C and 1098 mg l^{-1} at 30°C; Butt, *et al.*, 1964). The magnesium, calcium and sodium complexes are much more soluble. A potassium recovery process using dipicrylamine as the separating agent from sea water has been tested on a pilot plant scale. The involved reaction is

$$H^+(DPA) + K^+ \rightleftharpoons H^+ + K^+(DPA)$$

where

(DPA) = NO_2 NO_2 NO_2 H N NO_2 NO_2 NO_2

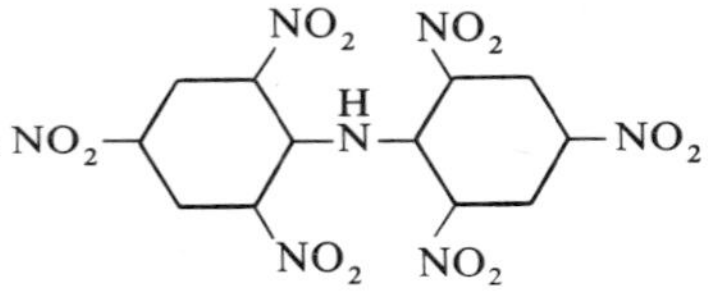

After precipitation and washing, the potassium amine salt is converted to the acid form and the potassium released into solution as the salt of the acid used. Recoveries of 97 to 98% of the potassium have been reported (Butt *et al.*, 1964). The insoluble acidic amine is converted to a salt by treatment with a base such as NaOH and recycled. After potassium removal, the sea water must be acidified to lower the solubility of the excess amine because of its high cost. A resin has been developed containing pentanitrodiphenylamine groups on a polyvinyl backbone, with a reported selectivity for K^+ ions over Na^+ (Skogseid, 1954).

In spite of the substantial research and pilot plant efforts, there has been no commercial production of potassium salts other than by fractional crystallization. Potassium salts are relatively inexpensive, and are widely available. Since the cost of transportation is an appreciable fraction of the total cost of production, the location of the plant is important (Jacobs, 1968).

21.3.5. OTHER MATERIALS

In addition to those materials presently being recovered, the ocean contains many other elements whose concentrations verge on levels of economic interest; some of these have been the subjects of investigations directed toward their commercial exploitation. As the concentrations of dissolved constituents become lower, the effects of the processes which control the entry of an element or its removal from the ocean dominate its distribution pattern and lead to considerable local concentration variations. Thus, for some of the minor sea water constituents, the concentration at a particular time and place in the ocean may deviate markedly from the average.

It has long been known that many metallic elements and some halogens can be accumulated to surprisingly large concentrations by marine organisms. Thus, enrichment of the transition metals can be as high as 100 to 100000 fold compared with the aqueous environment (Goldberg, 1965). Titanium is concentrated by a colonial ascidian (*Eudistoma ritteri*) over a million times, and vanadium is concentrated by a tunicate (*Phallisia mammillata*) from sea water to a similar extent (Bayer, 1964).

Such concentration factors would be useful for the concentration of elements from sea water if the extreme specificities of the natural complexing agents could be duplicated synthetically. It has been argued by Bayer (1964) that although the natural selectivity cannot be directly reproduced because of the requirement for a proper steric arrangement of a macromolecule, highly selective agents can be synthesized if proper attention is given to the type of bonding involved and to the choice of organic ligands.

21.3.5.1. *Gold*

The presence of dissolved gold in sea water has been known since 1872 and this fact has led to some of the more quixotic episodes in marine technology. More than fifty patents have been granted during the last hundred years on methods for recovering gold from sea water. Promoters, often fraudulent, have many times organized companies, sold stock, and raised money on the basis of proposed extraction schemes. The patent literature on gold recovery is extensive, and persistent. Most of the patented techniques depend on its sorption on various media, including charcoal, carbon, aluminium hydroxide, peat, coal, silica gel, pyrite, cellulose, fullers earth, sawdust, pumice, tars and similar materials (North, 1966). Capable and thorough scientific investigations have been undertaken· The result has always been the same. There has been no successful recovery of gold from sea water (Rosenbaum, 1969; Caldwell, 1938).

Fritz Haber, one of the most famous chemists of his time, conceived of paying the German war debt by recovering gold from sea water; the possibility of commercial recovery having first been seriously proposed by the eminent Swedish chemist Arrhenius. By 1920, a dozen men were at work under the direction of Johannes Jaenicke in a closed section of Haber's Institute. The laboratories were in an adjacent barracks, called Department M (for Meeresforschung—sea research) (Goran, 1967).

A recovery procedure was developed using synthetic sea water, in which both gold and silver were reduced to the metal by sodium polysulphide and removed in sulphur-coated sand filters. The analysis of gold in sea water was continually improved until it became one of the most sensitive analytical methods of wet chemistry. An experienced analyst was capable of determining 10^{-8} g l^{-1} of gold with an accuracy of $\pm$ 3% (Jaenicke, 1935).

After the development of the recovery process, Haber sought and obtained financial support. A specially equipped laboratory was installed in the liner *Hansa*. Two voyages were made in 1923 from Hamburg to New York; a third voyage was made on the *Wurtemburg* from Hamburg to Buenos Aires and a fourth in the North Sea. The results were all disappointing. Haber later estimated the concentration of gold to be 4 ng l^{-1}, about one-thousandth of the amount that had been expected. Haber himself said he had "given up my research for this dubious needle in a haystack" (Haber, 1927).

It has been reported (Bayer, 1964) that glyoxal bis-2-mercaptoanil is a specific chelating agent for silver, gold and mercury because of the favourable overlapping which it gives with the orbitals of metals with large ionic radii and the preferential coordination of sulphur ligands. Deep red to violet complexes are formed with gold (III), mercury (II) and silver at pH > 1.

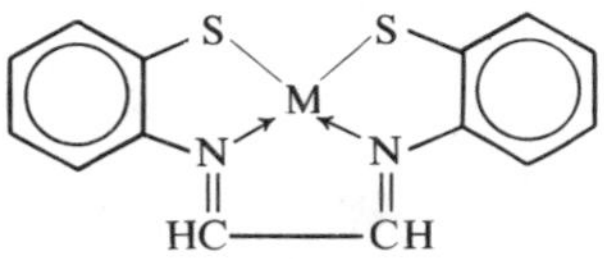

M = Au^{3+}, Hg^{+2}, Ag^{+}

Polycondensation of di- and triaminothiophenols with glyoxal gave macromolecules by means of which it was reported possible to extract gold from the Gulf of Naples sea water at a level of 1·4 μg l^{-1}.

21.3.5.2. *Uranium*

Uranium has a special interest because of its utility as a fuel for nuclear reactors, and as a component of nuclear weapons. Because it is relatively abundant in sea water (3·2 μg l^{-1}), it has received much technical attention, particularly from those nations without an assured, controllable supply.

Uranium, present in sea water as the stable carbonato complex $[UO_2(CO_3)_3{}^{4-}]$, is present at a greater concentration than are copper, nickel, vanadium or manganese. Methods of analysis for uranium in sea water are very well developed because of the nuclear implications. As with many other elements, the analytical procedures are suitable, after modification to be used as extraction procedures. Bayer (1960) has developed a macromolecular complexing agent based on glyoxal bis-2-hydroxyanil suitable for the selective sequestration of copper and uranium.

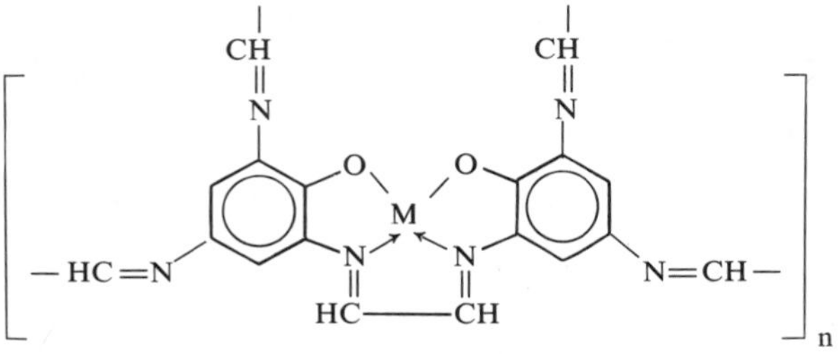

He found that recoveries of up to 100% could be obtained for copper and uranium by passing sea water containing 4·5–5 $\mu g\,l^{-1}$ copper and 0·5 $\mu g\,l^{-1}$ uranium through a column packed with the macromolecular complexing agent (Bayer, 1960, 1961, 1964).

Similar uptakes of uranium from sea water have been reported for other organic absorbents, polystyrene methylene phosphonic acid and 8-hydroxyquinoline. The best uptake was obtained with a resorcinol arsonic acid resin (Davies, *et al.* 1964).

The recovery of uranium from sea water by absorption on metal hydroxides has been carefully studied in the United Kingdom (Davies, *et al.*, 1964), Japan (Kanno, *et al.*, 1970) and India (Sani, 1966). Basic zinc carbonate, lead sulphide and lead pyrophosphate will extract uranium from sea water, but are all too soluble. Hydrated titanium oxide has good uptake characteristics and is stable in sea water. Tracer experiments with simulated sea water indicated that the hydroxide acted as a partially neutralized species $Ti_x(OHO)_y^-$ (Davies, *et al.*, 1964). In an actual test in flowing sea water, reasonable extraction efficiencies were obtained (62–93%) at flow rates of 0·55 to 0·79 $cm\,s^{-1}$. Removal of the uranium could be accomplished with Na_2CO_3 or $NaHCO_3$, but was best with $(NH_4)_2CO_3$ at higher temperatures (Kanno, *et al.*, 1970).

However, it is unlikely that recovery of uranium from sea water on a commercial scale will be economical until rich uranium ores become exhausted.

21.3.5.3. *Deuterium*

Heavy water is the oxide of deuterium. Its only commercial use is as a moderator for nuclear reactors because of its combination of a large scattering cross section, and a low absorption cross section for neutrons. The principal obstacle to its wider use is its relatively high cost compared with other moderators.

Surface waters in equatorial regions tend to be richer in the heavier isotopes because the vapour pressure of D_2O and HDO are lower than that of H_2O. To take advantage of the higher deuterium content of the Gulf stream, a plant producing 200 tons per annum was constructed at Glace Bay, Nova Scotia. The process to be used was the dual temperature exchange between hydrogen sulphide and water (Bebbington and Thayer, 1959) viz,

$$H_2O\,(liq) + HDS(g) \rightleftharpoons HDO\,(liq) + H_2S\,(g)$$

which has equilibrium constants of 2·18 and 1·83 at 30°C and 130°C respectively. The plant was designed to use five isotopic exchange stages. A fifth hot–cold tower was to take the enrichment to 99·8% D_2O. Many operating difficulties have been encountered and the plant is not yet in full operation.

List of Symbols

A	cross sectional area of heat transfer
A	active membrane area
a_n	coefficient of *n*th term of Fourier cosine series
A_{sp}	adjustable parameter (Table 21.6)
B_G	adjustable parameter (Table 21.6)
B	salt permeation constant
C_G	adjustable parameter (Table 21.6)
$\bar{c}$	bulk salt concentration of feed solution
C_a	heat capacity of air
C_b	heat capacity of brine
c_{do}	concentration of salts at dilution stream outlet
c_p	salt concentration of product
c_{pf}	measure of concentration polarization
C_p	specific heat at constant pressure
C_v	specific heat at constant volume
c_w	specific heat of sea water
c_f	salt concentration in feed at membrane interface
D	diffusion coefficient
E	evaporation rate
E	mass of water per unit area
E	voltage drop per cell pair
E_0	cell potential per cell pair
E^0	standard potential
F	correction factor
F	Faraday constant (96 500 coulombs)

f_w evaporation coefficient
h_i coefficient of heat transfer, convective
h_o coefficient of heat transfer, cover to atmosphere
h_w coefficient of heat transfer, solution to atmosphere
I ionic strength
I solar radiation flux
i current density
i coefficient of linear relationship of P_b with T_b
I_{lim} limiting current
I_n current
j number of heat rejection stages
j coefficient of linear relationship of P_b with T_b
J_1 membrane water flux
J_2 membrane salt flux
k ratio of specific heats (C_p/C_v)
K_A measure of cost of unit transfer area
K_C measure of cost of constant components
K_E measure of cost of unit energy
k_s coefficient of heat transfer between salt and soil
K_{sp} ionic solubility product
K_{sp}° ionic solubility product at $I = 0$
K_1 membrane water permeability
l function of evaporation rate and heat transfer rate, with time
m function of meteorological conditions and solar flux, with time
M_a mass of air circulating
M_w mass of water
n number of heat recovery stages
n number of cell pairs
N_b salt concentration of brine (normality)
P permeability
P_a partial pressure of water vapour in air
P_b partial pressure of brine
P_f pressure of feed
P_p pressure of product
P_1 pressure of gas, initial
Q quantity of heat transferred per unit time
R performance ratio (pounds per 1000 BTU)
R resistance per cell pair
r resistance of unit cross section
r compression ratio
S 1–1 limiting Debye–Hückel slope (modified for molal units)

s number of seconds in a day
T absolute temperature of process
T_d temperature of melting
T_{mt} transport number of ion in membrane
T_r effective temperature of surroundings as radiation receiver
T_{st} transport number of ion in solution
t time
T_a temperature of air (T_{a_c} near condensation; T_{a_b} near brine)
T_b temperature of brine
T_{bh} temperature of heated brine (maximum)
T_c temperature of cover
t_d temperature of distillate
t_s temperature of steam
t_g temperature of soil
t_{sw} temperature of sea water
U overall heat transfer coefficient
U_c overall heat transfer coefficient, condenser
U_e overall heat transfer coefficient, evaporator
U_1 overall heat transfer coefficient, freezing
U_2 overall heat transfer coefficient, melting
v specific volume
V_1 initial gas volume
w mass flow of water through membrane
w_b mass flow of brine
W_c work done by compressor
w_d mass flow of distillate
W_f work done by freezer
x mass ratio w_b/w_d
y coefficient of linear relationship of back radiation and T_b
z coefficient of linear relationship of back radiation and T_b
Z_b depth of brine
Z_{sa} depth of salt
γ amount of property being transferred
δ boundary layer thickness
ΔH change in enthalpy of product water
ΔT mean thermal driving force
Δt_b thermal difference of brine, maximum to reject
Δt_{fpd} freezing point depression
θ_n phase difference in nth term of Fourier cosine series
λ_f heat of fusion
λ_v heat of vaporization

η efficiency of transfer device
η salt rejection factor
η_b efficiency of compressor
η_c efficiency due to solution concentrations
η_e efficiency due to electrode reactions
η_i efficiency of current
η_m efficiency due to multiple manifold current paths
η_p efficiency of pump
η_R efficiency due to total stack resistance
η_s efficiency due to imperfect selectivity
η_w efficiency due to water transport
π osmotic pressure
π_f osmotic pressure at feed concentration
π_p osmotic pressure at product concentration
ρ resistance of unit area of transfer service
ρ density
σ Stefan–Boltzmann constant
Σ_c emissivity of cover
Σ_L total miscellaneous heat losses
Σ_r total energy loss by radiation
ϕ flux density

References

Anon. (1967). *Eng. Min. J.* 104.
Anon. (1966). *Eng. Min. J.* 94.
Armstrong, E. F. and Miall, L. M. (1946). "Raw Materials from the Sea". Chemical Publishing Co., New York.
Badger, W. L. and Assoc. (1959). Office of Saline Water Research and Development, Report No. 25, PB 161399, Suptd. of Documents, Washington, D.C.
Balard, A. J. (1826). *Ann. Chim.* (*Phys.*) **32**, 337.
Ball, C. J. P. (1956). *J. Inst. Metals*, **84**, 399.
Barduhn, A. J. and Hu, Yee-Chien (1960). Office of Saline Water Research and Development Report No. 44, PB 171031, Suptd. of Documents, Washington, D.C.
Bayer, E. and Fielder, H. (1960). *Angew. Chem.* **72**, 921.
Bayer, E. (1961). *Angew. Chem.* **93**, 659.
Bayer, E. (1961). German Patent 1102397, Sept. 28.
Bayer, E. (1964). *Angew. Chem. Int. Ed.* **3**, 325.
Bebbington, W. P. and Thayer, U. R. (1959). *Chem. Eng. Prog.* **55**, 70.
Bloch, M. R., Farkas, L. and Spiegler, K. S. (1951). *Ind. Eng. Chem.* **43**, 1544.
Boegelin, A. F. and Whaley, T. P. (1967). *In* "Encylopedia of Chemical Technology" (A. Standen, ed.), 2nd Ed., Vol. 12, pp. 724–731. Interscience, New York.
Borchert, H. (1965). *In* "Chemical Oceanography" (J. P. Riley and G. Skirrow, eds.), Vol. 2, 1st Ed. Academic Press, London.

Butt, J. B., Talmadge, J. A. and Savage, H. R. (1964). *Chem. Eng. Prog.* **60**, 50.
Caldwell, W. E. (1938). *J. Chem. Educ.* **15**, 507.
Christensen, J. J., McIlhenny, W. F., Muehlberg, P. E. and Smith, H. G. (1967). Office of Saline Water Research and Development Progress Report No. 245, Supt. of Documents, Washington, D.C.
Comstock, H. B. (1964). "Magnesium and Magnesium Compounds", Inf. Circular 8201. Bur. of Mines, U.S. Department of the Interior.
Davies, R. V., Kennedy, J. McIlroy, Spence, R. and Hill, K. M. (1964). *Nature, Lond.* **203**, 1110.
Davison, R. R., Jeffrey, L. M. Whitehouse, U. G. and Hood, D. W. (1958). Office of Saline Water Research and Development Report No. 22, PB 161396, Supt. of Documents, Washington, D.C.
Dow Chemical Co. (1965). Office of Saline Water Research and Development Report No. 139. Supt. of Documents, Washington, D.C.
Emley, E. F. (1966). "Principles of Magnesium Technology". Pergamon Press, London.
Farkas, L., Lewin, M. and Bloch, R. (1949). *J. Amer. Chem. Soc.* **71**, 1988.
Fossett, H. (1971). *Chem. Ind.* 1161.
Garrett, D. E. (1965). *In* "Second Symposium on Salt" (J. L. Rau, ed.), pp. 168–175. Northern Ohio Geological Society, Cleveland, USA.
Gillam, W. S. and McCoy, W. H. (1966). *In* "Principles of Desalination" (K. S. Spiegler, ed.), pp.1–20. Academic Press, New York.
Goldberg, E. D. (1965). *In* "Chemical Oceanography" (J. P. Riley and G. Skirrow, eds.), Vol. I, 1st Ed. Academic Press, London.
Goran, M. (1967). "The Story of Fritz Haber", pp. 91–98. University of Oklahoma Press, Norman, Oklahoma, U.S.A.
Gross, W. H. (1967). *In* "Encyclopedia of Chem. Technology" (A. Standen, ed.), 2nd Ed., Vol. 12, pp. 661–708. Interscience, New York.
Haber, F. (1927). *Angew. Chem.* **40**, 303.
Hadzeriga, P. (1965). *In* "Second Symposium on Salt" (J. L. Ran, ed.), pp. 204ff. Northern Ohio, Geological Society, Cleveland, U.S.A.
Hanawalt, J. D. (1964). *J. Metals*, **16**, 559.
Hart, P. (1947). *Chem. Eng.* **54**, 102.
Havighorst, C. R. and Swift, S. L. (1965). *Chem. Eng.* **72**, 84, 98, 150.
Heath, S. B. (1939). U.S. Pat. 2,143,223, Jan. 10.
Hildebrand, J. H. (1918). *J. Ind. Eng. Chem.* **10**, 84.
Hittman Associates (1970). Office of Saline Water Research and Development Progress Report No. 611, Supt. of Documents, Washington, D.C.
Hunter, R. M. (1944). *Trans. Electrochem. Soc.* **86**, 21.
Jacobs, J. J. (1968). *In* "Encyclopedia of Chemical Technology" (A. Standen, ed.), 2nd Ed., Vol. 16, pp. 369–406. Interscience, New York.
Jaenicke, J. (1935). *Naturwissenschaften*, **23**, 57.
Johnson, J. S., Dresner, L. and Krauss, K. A. (1966). *In* "Principles of Desalinatiion" (K. S. Spiegler, ed.), pp. 346–433. Academic Press, New York.
Jolles, Z. E. (1966). "Bromine and its Compounds", Ernest Benn Ltd., London.
Kaufman, D. W. (1960). "Sodium Chloride". Reinhold Publishing Co., New York.
Kaho, M. and Watanabe, T. (1970). *In* "Third Symposium on Salt", (J. L. Rau and L. F. Oelwig, eds.), Vol. 2. Northern Ohio Geological Soc., Cleveland, U.S.A., pp. 103–109.

Kaliforschungs-Anstalt (1938). German Patent 661,443.
Kanno, M. Yoshihiro, O. and Takashi, M. (1970). *Nippon Genshiryuku Gakkaishi*, **12**, 708.
Langelier, W. F., Caldwell, D. H., Lawrence, W. B. and Spaulding, C. H. (1950). *Ind. Eng. Chem.* **42**, 126.
Latimer, W. M. (1952). "The Oxidation States of the Elements and their Potentials in Aqueous Solutions", 2nd Ed. Prentice-Hall, New York.
Löf, G. O. G., Eibling, J. and Bloemer, G. V. J. (1961). *Amer. Inst. Chem. Eng.* 7, 641.
McCutchan, J. W. and Sieder, E. N. (1969). Office of Saline Water Research and Development Report No. 411, Supt. of Documents, Washington, D.C.
McIlhenny, W. F. and Ballard, D. A. (1963). *In* "Desalination and Ocean Technology" (S. Levine, ed.) Dover Publications, New York.
McIlhenny, W. F. (1966). *In* "Water Production using Nuclear Energy" (R. G. Post and R. L. Seale, eds.), pp. 187–209. University of Arizona Press, Tucson, Arizona, U.S.A.
McIlhenny, W. F. (1967). *In* "Encyclopedia of Chemical Technology" (A. Standen, ed.), 2nd Ed., Vol. 14, pp. 150–170. Interscience, New York.
McIlhenny, W. F. (1968). *In* "Proceedings, Fourth Forum on Geology of Industrial Minerals", Bureau of Economic Geology, University of Texas, Austin, U.S.A.
Marshall, W. L., Slusher, R. and Jones E. V. (1964). *J. Chem. Eng. Data*, **9**, 187.
Marshall, W. L. and Slusher, R. (1968). *J. Chem. Eng. Data*, **13**, 83.
Mastin, M. G. (1938). U.S. Patent 2,12,002.
Merten, U. (1965). *Proc. Int. Symp. Water Desalination 1st*, Washington, D.C., 1965 (Pub. 1967), Vol. 1, p. 275. U.S. Dept of Interior, Washington, D.C.
Meschter, E. (1961). *Rock Prod.* 112.
Miller, W. D. (1965). *In* "Mineral Facts and Problems", Bulletin 630. Bureau of Mines. U.S. Department of the Interior.
Moyer, M. P. (1969). *In* "Encyclopedia of Marine Resources" (F. E. Firth, ed). Van Nostrand, New York.
Murphy, G. W., Taber, R. C. and Steinhauser, H. H. (1956). Office of Saline Water Research and Development Progress Report No. 9, PB 161384, Supt. of Documents, Washington, DC.
M. W. Kellog Co. (1965). Office of Saline Water Research and Development Progress Report No. 158, Supt. of Documents, Washington, D.C.
Niccoli, E. (1926). British Patent 261,991.
Niccoli, E. (1926). British Patent 247,405.
North, O. S. (1966). *Eng. Min. J.* **167**, 195.
Office of Saline Water (1971). Saline Water Conversion Report 1970–1971, Supt. of Documents, Washington, D.C.
O'Shaughnessy, F. (1970). "Desalting Plants Inventory Report No. 3", Office of Saline Water, U.S. Department of the Interior. Washington, D.C.
Othmer, D. F. (1970). *In* "Encyclopedia of Chemical Technology" (A. Standen, ed.), 2nd Ed. Vol. 22, pp. 1–65. Interscience, New York.
Porter, J. W. (1967). ASME Pub. No. 67-UNT-5 American Society of Mechanical Engineers, New York.
Rave, W. J. (1953). *J. Electrochem. Soc.* **100**, 179C.
Roberts, C. R. (1960). "Magnesium and its Alloys". John Wiley, New York.
Robinson, H. A., Friedrich, R. E. and Spencer, R. S. (1946). U.S. Pat. 2,405, 055.

Rosenbaum, J. B., May, J. T. and Riley, J. M. (1969). ASME Pub. No. 69-AS-82. American Society of Mechanical Engineers, New York.
Salutsky, M. L., Dunseth, M. G. and Waters, O. B. (1964). Office of Saline Water Research and Development Report No. 91, Supt. of Documents, Washington, D.C.
Sani, A. R. (1966), *Nucleus (Lahore)*, **3**, 102.
Schambra, W. P. (1944). *Trans. Amer. Inst. Chem. Eng.* **41**, 35.
Seaton, M. Y. (1931). *Chem. Met. Eng.* **38**, 638.
Sharpe, A. G. (1967). *In* "Halogen Chemistry" (V. Gutmann, ed.). Academic Press, London.
Shaffer, L. H. and Mintz, M. S. (1966). *In* "Principles of Desalination" (K. S. Spiegler, ed.), pp. 200–287. Academic Press, New York.
Shigley, C. M. (1951). *J. Metals*, **3**, 25.
Shreve, N. J. (1967). "Chemical Process Industries", 3rd Ed. McGraw-Hill, New York.
Silver, R. S. (1966). *In* "Principles of Desalination" (K. S. Spiegler, ed.) pp. 77–115. Academic Press, New York.
Simpson, H. C. and Silver, R. S. (1963). *In* "Desalination and Ocean Technology" (S. N. Levin, ed.), pp. 62–88. Dover Publications, New York.
Skogseid, A. (1954). Norwegian Patent 83, 579.
Snyder, A. E. (1966). *In* "Principles of Desalination" (K. S. Spiegler, ed.), pp. 292–343, Academic Press, New York.
Spiegler, K. S. (1962). "Salt-Water Purification". John Wiley, New York.
Stenger, V. A. (1964). *In* "Encyclopedia of Chemical Technology" (A Standen, ed.), 2nd Ed., Vol. 3, pp. 758–766. Interscience, New York.
Stewart, F. H. (1963). Data of Geochemistry, 6th Ed., Chapter Y. *U.S. Geol. Sur. Prof. Pap.* **440** Y.
Stewart, L. C. (1934). *Ind. Eng. Chem.* **26**, 361.
Stine, C. M. A. (1929). *Ind. Eng. Chem.* **21**, 434.
Stoughton, R. W. and Lietzke, M. H. (1965). *J. Chem. Eng. Data* **10**, 254.
Strathman, H. (1968). Office of Saline Water Research and Development, Progress Report. No. 336, Supt. of Documents, Washington, D.C.
Strelez, C. L., Taiz, A. J. and Guljanitzki, B. S. (1953). "Metallurgie Des Magnesiums". Vebverlag Technik, Berlin.
Tallmadge, J. A., Butt, J. G. and Solomon, H. J. (1964). *Ind. Eng. Chem.* **56**, 44.
Texas A & M Res. Fndn. (1960). Office of Saline Water Research and Development Report No. 35, PB 161769, Supt. of Documents, Washington, D.C.
Tressler, D. K. and Lemon, J. M. (1951). "Marine Products of Commerce". Reinhold, New York.
Tsunoda, Y. (1965). *Proc. Int. Symp. Water Desalination 1st*, Washington, D.C., 1965 (Pub. 1967), Vol. 3, p. 225. U.S. Dept of Interior, Washington, D.C.
Turrentine, J. W. and Shoaff, P. S. (1919). *J. Ind. Eng. Chem.* **11**, 864.
Udwadia, N. N., Gadre, G. T., Kawa, R. M., Lele, V. N., Mehta, D. J. and Datar, D. S. (1966). *Res. Ind. (India)*, **11**, 71.
Usiglio, J. (1849). *Liebig. Ann. Chem.* **27**, 92, 179.
Ver Planck, W. E. (1958). "Salt in California", Bulletin 175. Division of Mines, State of California.
White, G. C. (1972). "Handbook of Chlorination", p. 708. Van Nostrand, New York.
Williams, L. R. (1965). *In* "Mineral Facts and Problems", Bulletin 630. Bureau of Mines, U.S. Department of the Interior.
Wiegandt, H. F. (1960). *Adv. Chem., Ser.* **27**, 82.

Yamane, R., Ichikawa, M., Mizutani, Y. and Onoue, Y. (1969). *I&EC, Process Des. and Develop.* **8**, 159.
Yeatts, L. B. and Marshall, W. L. (1972). *J. Chem. Eng. Data*, **17**, 163.

CHAPTER 22

Seaweeds in Industry

E. BOOTH

275 Haslingen Old Road, Rossendale, Lancashire BB4 8RR, England

22.1. MAINLY HISTORICAL

Burning seaweed to make crude sodium carbonate is certainly older than Christianity, and was recorded by Pliny the Elder who even distinguished the two common alkalis of this period when he noted that seaweed ash produced a hard soap, whereas wood ash gave a soft soap. This acute observation was not improved on until 1735, when Monceau showed that the alkali of seaweed ash contained soda and was chemically different from that of wood ash. The common name, soda ash, reflects the original way of making sodium carbonate.

Up to the time of the Napoleonic wars, this crude alkali satisfied the needs of a rural economy. The introduction, in 1791, of the Le Blanc process for the manufacture of sodium carbonate was extended to England by James Muspratt in 1822 and, with the repeal of the barilla tax in 1820, kelp burning became obsolete as a source of soda ash, even though the synthetic product cost £90 a ton. For the next fifty years, kelp remained the only commercial source of potassium salts and was used in the manufacture of potassium nitrate and as a fertilizer; it was also used in glass-making and in the manufacture of soap.

In 1813, Courtois isolated iodine from kelp and the medicinal uses of tincture of iodine developed quickly; the effect of iodine on goitre was demonstrated in 1818 by a Geneva physician, Dr. Coindet, and Buchanan used it as an antiseptic in 1828 but it did not come into wide scale use for this purpose until the American Civil War (1861–5). Kelp burning was still the only source of potash and iodine when Stanford (1862) wrote his classic paper on the "Economic Applications of Seaweed".

The fluctuations of the kelp industry in the early years of the 19th century have been described by Chapman (1950, 1970) and Newton (1951), who based their accounts mainly on the work of Stanford (1862, 1884a, 1887) and Hendrick (1916). The manufacture of iodine from kelp was initiated by Cournerie at Cherbourg in 1825, and did not reach Scotland until 1841. The production of kelp in Scotland declined from 8500 tons in the early 1790's (Sinclair, 1791) to 2565 tons in 1841; the value of kelp declined from £20 per ton in 1800 to less than £2 in 1831. Adulteration was commonplace and Sinclair records the ton of kelp was $22\frac{1}{2}$ cwt (1·14 ton).

Stanford's "char process" began in 1863 with the formation of the British Seaweed Co. Ltd. which was bought by the North British Chemical Co. Ltd. in 1876; this company became the British Chemical Co. Ltd. in 1891 and reorganized itself under the same name in 1913, and finally closed by voluntary liquidation in 1951. Chapman (1950) quoting from the discussion on Hendrick's paper (1916) records that W. G. O'Beirne, who was works-manager of the British Chemical Company, said that "... factories have been constructed in Tiree, North Uist and Clydebank, £40,000 has been sunk in them and all has been lost." Naturally, Chapman took this statement at its face value, but this claim of total loss has no foundation in fact. Indeed, the North British Chemical Co. was paying dividends of 40% by 1878. Examination of the records of these companies (Public Records Office 1863; Scottish Records Office, 1876, 1891, 1913) shows continuous success over the period 1863–1951, and each change in the name of the company was accompanied by substantial distributions to the shareholders. Only three years before O'Beirne's mischievous statement, the British Chemical Company reformed itself and the only change was that the shareholders received three shares in the new company for each share held in the old one. Ultimately, the works was damaged by enemy action in 1940, and in 1949 the company went into voluntary liquidation; eventually £76 897 was distributed—approximately £1.50 per £1 share.

The "char" factory on Tiree operated from 1864–1901 but the record of the factories at Loch Eport and Freagh, Co. Clare is obscure. Before the Tiree factory was completed, Stanford was building a new factory, the first in Clydebank, which used the traditional lixiviation process. Although it was cheaper to make iodine from the South American caliche deposits, a price agreement allowed the company to operate profitably until 1935, when the agreement was ended by G. Steedman, manager of the British Chemical Company. At this date kelp was scarce and was imported from Norway and in small amounts from Ireland. Öy (1949) records the export of 4800 tons of kelp in 1932, but says that both kelp burning and the production of iodine ceased in 1933. The production of iodine from kelp was revived in Japan during the war, and production in France continued until the 1950's. Kelp burning still lingers on in the Channel Islands (Toms, 1965) where 4 tons of wet seaweed is valued at £2.50 but, when burnt, is worth £4.00. Finally, reports suggest that the kelp industry is to be revived in Ghana as a means of making soda ash (Reynolds, 1972).

Outlines of the history of the American seaweed industry over the period 1910–1920 has been given by Chapman (1950) and Newton (1951), and a more complete account of this phase of the industry which now has only historical interest has been given by Scofield (1959). In the mid-1920's, after an abortive attempt to make algin in Orkney, F.C. Thornley (Thornley *et al.*, 1922) was attracted to California by the abundance of *Macrocystis*. The manufacture of algin started in 1927, and two years later the Kelco Company took control and became the first successful manufacturer of alginates. From this early start, Kelco developed both the product and its uses, and became the world's largest supplier of seaweed products.

22.2. Seaweed Harvesting

The collection of seaweed invariably meets with strong resistance from conservationists. This is based largely on prejudice and ignores the evidence available that storms and natural wastage remove more seaweed than all the efforts of the collectors.

When kelp burning was first introduced to Scotland, at Anstruther in 1694, Sinclair (1792) recorded ". . . one of the baillies protested against, as being prejudicial to the health of the inhabitants, and his own family; but at length they accepted the offer, on condition that the kelp should be burnt at the west end of the town, and only when the wind blew from the east." The industry extended to Orkney in 1730 and ". . . the opposition said it would drive away the fish, destroy corn and grass and might prevent the women having children." (Sinclair, 1793). Similarly, the first harvest of *Macrocystis* had hardly been landed on the coast of California when there was a complaint that the sardine fishing would be ruined (Scofield, 1959).

The situation has changed little over the years. The Fisheries Act of British Columbia gives the Minister power to prohibit harvesting if, in his opinion, it ". . . would tend to impair or destroy any bed or part thereof on which kelp or other plants grow, if harvesting kelp or other aquatic plants would tend to impair or destroy the supply of any food for fish and if harvesting of kelp or other aquatic plants would be detrimental to fish life." Similarly, in December, 1966 the State Legislature of Western Australia passed a Fisheries Act Amendment Bill which controls seaweed harvesting on the grounds that "... uncontrolled harvesting could do considerable damage to the habitat of many fishes . . ." These precautions ignore all the experience gained in other parts of the world and they are hardly necessary because the seaweed industry of British Columbia is small, and there is no such seaweed industry whatever in Western Australia at present.

These objections to seaweed harvesting cover a period of almost 300 years and have proved quite groundless. Undoubtedly the best example of continuous seaweed harvesting is the collection of *Macrocystis* off the coast of California; this has been cropped continuously since 1911 under the supervision of the California Fish and Game Commission. The study of these beds started in 1910, and their history was recorded by Scofield (1959), who worked for the Department for 34 years; Scofield has discussed all the objections to harvesting and has shown them to be completely without foundation. The harvest is considerable and the crop is fairly constant from year to year (Table 22.1).

Before seaweed can be harvested, it is necessary to establish with reasonable accuracy the quantity of seaweed which grows in the area, and to ensure that the growth is sufficiently dense to make collection worthwhile. It is also necessary to consider the rate of regrowth after harvesting, so that the stock may be conserved and, above all else, it is vital to persuade the harvesters to

TABLE 22.1

Annual harvest of Macrocystis *from off California*

Year	Harvest (short tons, wet weight)
1960	120 230
1961	129 256
1962	140 324
1963	121 032
1964	127 254
1965	135 129
1966	119 463
1967	131 495
1968	134 853
1969	131 239
1970	127 039
1971	155 560

Data: California Fish and Game Commission.

use recommended harvesting methods which will assist the regeneration of the plant.

Many survey methods have been tried in the past; these include not only sophisticated techniques, such as aerial photography (Spenser, 1947) and echo-sounding (Worthy and Walker, 1948) but also the assessment of plant growth in the intertidal zone (Walker, 1946, 1947a, 1948; Baardseth, 1970). Walker (1947b) has developed the idea of sampling along transects, and Grenager and Baardseth (1966) have described an accurate method for estimating the growth of the intertidal algae. Although MacFarlane (1966a) was not the first to use SCUBA diving techniques, her work on the survey and ecology of *Chondrus crispus* serves as a model for workers in this field. Aerial photography is probably only suitable in special instances, such as the delineation of large beds of *Macrocystis*, and echo-sounding has made little contribution to survey methods, which now tend to rely on extensive field-work backed by SCUBA diving.

Plant predators and disease are among the many factors which affect seaweed harvesting, and the industry must rely on the algologist for guidance with these problems. Typical of these difficulties are the "black rot" which affects *Macrocystis* when the water temperature exceeds 20°C, and the predation of the plant by sea urchins (North, 1964); this attack has been decreased by newly developed techniques of treating the urchins with lime (Leighton *et al.*, 1966).

Enlightened self-interest has, in general, prevented over-harvesting, and only a single instance of depletion of the stock by over-harvesting has been noted; this occurred in the Kattegat around 1964, when the stock of *Furcellaria* was seriously depleted (Lund and Christensen, 1969).

Often there are virtually no regulations governing the collection of seaweed. In the U.K. for instance, the Crown Estates control all harvesting below L.W.O.S.T. but, as yet, no sub-littoral seaweeds are harvested in the U.K. The intertidal zone may be privately owned, or it may come under the jurisdiction of the Crown Estates, but long-established rights of the local residents have been held paramount by the Scottish Land Court. In Orkney, the ancient Udal Law gives collecting rights to residents, and a similar situation is found in Ireland. In France, a series of Decrees dating from 1868 onwards virtually restricts the collection of littoral seaweeds to local residents, but cast seaweed can be freely collected. Sublittoral seaweed can be cut by anyone throughout the year, but any form of mechanical harvesting is prohibited. However, special permission to use a prototype harvester was granted to Nourylande and Maton in 1962, but this machine, which was essentially a suction pipe into which divers fed cut seaweed, proved uneconomic.

In Nova Scotia, the collection of *Chondrus crispus* is restricted to a definite season, usually May to October, with minor local variations, and mechanical devices are not allowed over the friable rock off the north coast of Prince Edward Island. The Sea Plants Harvesting Act of 1959 applies particularly to *Ascophyllum nodosum*, and it specifies that:

(1) a cutting instrument must not be used,
(2) the holdfast must not be intentionally damaged,
(3) at least 5 inches (12·7 cm) of the plant must be left intact,
(4) not more than 95% of any area may be harvested,
(5) harvested rockweed shall not be sold for agricultural use,
(6) a licence to harvest is required, and it shall cost at least $50, but not more than $500,
(7) a royalty of 10 cents per wet ton shall be charged,
(8) if the royalty exceeds the licence fee, the latter shall be deducted from the total payment,
(9) each licensee shall keep a record of the quantity harvested, and shall report annually.

In general, these regulations are in accord with long established practice in Scotland and Norway, but general practice in these countries allows cutting, and at least 10 inches (ca. 25 cm) of the plant would be left uncut. Experience gained in Scotland and Norway over many years shows that the recovery rate of *Ascophyllum nodosum* varies from place to place. Complete recovery may take 2–3 years, but in a few favourable locations, an annual harvest is possible; similar experience has been reported from Norway (E. Baardseth, personal communication). The regrowth of *A. nodosum* has been studied by Baardseth (1955) and Walker (1948) whose experiments were carried out on a pure stand of the seaweed on an uninhabited island in the Skerry of Work, Orkney, where the mean density of the seaweed beds was $13\,kg\,m^{-2}$, i.e. moderately good, but not exceptional. Four separate areas were cut to leave 2, 5, 8 or 11 inches of frond (5, 13, 20 or 28 cm), and the plots were re-examined after two years when the recovery was shown to be 40, 48, 83 and 100% respectively.

The collection of *Chondrus crispus* is usually restricted to a limited season; in Canada the season extends from June–December; in France it is defined by the spring tides, and extends from late May to early October. In Portugal the season extends from 15th May to the end of the year.

Harvesting may be carried out in five very different ways viz.:
(a) gathering cast seaweed,
(b) netting floating seaweeds,
(c) collecting littoral seaweeds,
(d) mechanically harvesting sub-littoral seaweeds,
(e) collecting by divers.

Each of these methods is important commercially, and it is only possible to give a few examples of each method. Winter casts of *Laminaria hyperborea* are usually divested of the frond and holdfast by wave action, and the quantity thrown on the beaches is largely governed by the direction of the wind, i.e. an off-shore wind is desirable, since an on-shore wind gives rougher seas, which tend to scour everything off the beach. The cast stipes are collected in France, Ireland, Norway and Scotland, and stacked on a base of stones to air-dry; few statistics are available and the harvest is somewhat erratic, but 2–3000 tons are collected annually in Orkney, and the total for Europe is probably about 10 000 tons. Similarly, autumn storms give large casts of *Furcellaria fastigiata* on the western shores of Prince Edward Island; these are collected and dried. The harvest, which amounts to about 15 000 tons of wet seaweed is collected and dried to yield about 5000 tons of the dry weed.

Although most of the algae are attached to rocks, a few free-floating types are well known, e.g. *Ascophyllum mackii*, *Gracilaria* spp. and, in the Kattegat there is an abundance of *Furcellaria fastigiata* (Hudson) Lamouroux *F. aegapropila* (Reinke) Svedelius, which is the main seaweed harvested by netting. To some extent, the attached sub-littoral form is dislodged by a dredge, and this is also gathered by the nets. This harvest started about 1940, and reached 15000 tons $year^{-1}$ (wet weight) by 1958 (Lund and Bjerre-Petersen, 1953, 1964). During the 1960's the harvest increased, but by 1964 there was a notable decline of the free-floating variety in the Tangen area (Lund and Christensen, 1969) and by 1967, the area produced practically no seaweed. This is one of the few recorded instances of the depletion of a seaweed stock by over-harvesting. Despite this local decline, the collection in the Kattegat increased during the 1960's, largely through an extension of the collecting area (Table 22.2).

TABLE 22.2
Annual harvest of F. Fastigiata (tons, wet)

Year	Harvest	Year	Harvest
1961	20 557	1965	23 647
1962	31 143	1966	25 000
1963	24 242	1967	27 500
1964	25 346		

The collection of *Ascophyllum nodosum* is the best example of the collection of a littoral seaweed. As the tide recedes, the harvesters are able to cut the fronds (pulling the seaweed from the rocks is discouraged) and throw them into nets which usually hold about a ton; the nets are tied when full. This work is continued until the tide turns and covers the seaweed; the netted seaweed floats, and can be towed to a suitable landing point, or loaded into lorries. This seaweed has a tendency to decompose if larger quantities are left in a heap, and it should be dried as soon as possible after landing. When immediate drying is impracticable, the netted seaweed can be moored and left in the sea, where it will remain in good condition for at least a week. There is no reliable data on the amount of seaweed collected by this method, except for Norway where the annual production of the dried meal has been approximately 15000 tons year^{-1} over the last decade. About the same quantity is produced in Ireland, and probably about 7000 tons is made in Scotland.

Sub-littoral seaweeds are harvested by a wide variety of methods. The Japanese, for example, have devised a variety of rakes, dredges and grapnels which appear to work well, but which require a considerable amount of labour. *Chondrus crispus* is gathered from depths of up to twelve feet by long-handled rakes or, in some localities by dredging, and as much as 700 kg per tide has been gathered by raking; the author has seen a man collect, during one tide, 380 kg, which was landed, weighed and paid for in a few hours. The harvest of this seaweed in Eastern Canada reflects the state of the carrageenan industry; it started in 1940, rose to 11 000 tons (wet weight) by 1952, and reached a record harvest of 48 000 tons in 1970, then declined slightly to 36 600 tons in 1971. There is also a considerable harvest of this alga on the Brittany coast, probably around 5000 tons year^{-1} (dry), probably 1000 tons is collected in Spain and Portugal, and smaller amounts of the other *Gigartinaceae* are commercially available, but the only reliable data is the production of 565 tons (dry) of *G. canaliculata* in Mexico (Gusmán del Proo personal communication).

Many attempts have been made to construct harvesters which will work in the vast forests of *Laminaria hyperborea* which grow in in-shore waters in the temperate zone. Jackson (1957) has describe the equipment designed at the Institute of Seaweed Research, and there is a recent account of a Norwegian harvester which appears to be successful where the sea-bed is suitable. This is essentially a dredge which cuts the seaweed, followed by a trawl net which collectes the plants (Svendsen, 1972). The only really successful mechanical harvesters work on the giant *Macrocystis* beds off the American coast; these have been in operation for almost 60 years, and their efficiency and seaworthiness has been developed to a high state by the Kelco Company. In addition to the 155 560 tons harvested in the U.S.A. during 1971, 65 000 tons were collected in Mexico (Gusmán del Proo, personal communication) but no details are available of the quantity harvested in Chile.

Diving is undoubtedly the oldest way of collecting the sub-littoral seaweeds, and women divers have provided the seaweed for the Japanese agar industry for many years. These expert swimmers dive with a basket which they fill with seaweed, and then surface for breath and unload their haul. At least 3500 people are still engaged in this simple method of gathering seaweed, but their efforts will doubtless be replaced by the diving suit and SCUBA diving, which are now used in Japan, where some 700 divers produce 269 tons of seaweed per annum (Okazaki, 1971). Diving for agar-bearing seaweeds is also practised in other parts of the world, but few figures are available, except those for the annual harvest of *Gelidium robustum* in Mexico, which amounted to 565 tons (dry) in 1971. This harvest seems to have suffered a decline since a crop of 1000 tons (dry) of *G. cartilagineum* (= *G. robustum*) was reported in 1967 (Gusmán del Proo, 1969). This type of harvesting is practical only with the more expensive seaweeds, e.g. *Gelidium* spp. which sells at about £500 per ton, compared with the brown seaweeds which are worth up to £100 per ton.

22.3. The Manufacture of Agar

22.3.1. THE STRUCTURE OF AGAR

Our knowledge of the composition of agar is largely based on the work of Araki (1966) who showed that the agar from *Gelidium amansii* could be separated into agarose and agaropectin by fractionation of the acetylated agar followed by hydrolysis of the two products with alcoholic alkali. Agarose was shown to be a neutral, linear polysaccharide formed from a chain of 1,3-linked β-D-galactopyranose units attached to 1,4-linked 3,6-anhydro-α-L-galactopyranose residues. The yield of agarobiose (**I**) can exceed 70%, which suggests that these are the only units in the agarose molecule. Enzymatic hydrolysis of agarose splits the chain at the alternative point to give neoagarobiose (**II**) which confirms the basic structure suggested by Araki.

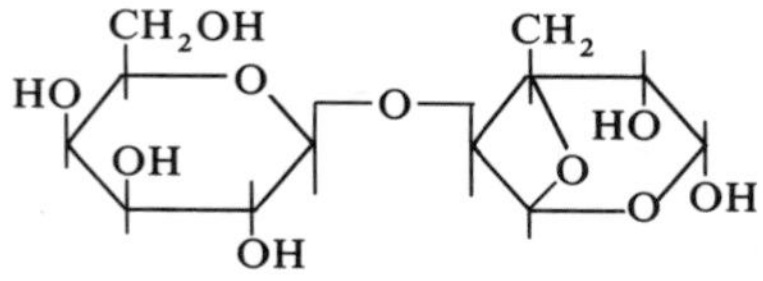

(**I**)

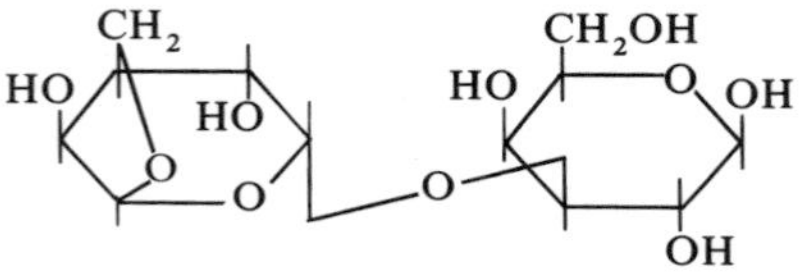

(**II**)

The agarose from different seaweeds show slight differences. These are attributable to differences in the proportions of 6-*O*-methyl-D-galactose. Only small amounts, ca. 1–3%, are found in the agarose from *Gelidium* spp., but that from *Gracilaria verrucosa* contains about 16% of the methyl-galactose, whereas *Ceramium boydenii* yields 20%. The presence of the 6-*O*-methyl-D-galactose modifies the structure of agarose in proportion to the amount present, and the chain must be considered to have the following basic structure (**III**) where x and y are determined by the percentage of D-galactose and 6-*O*-methyl-D-galactose respectively.

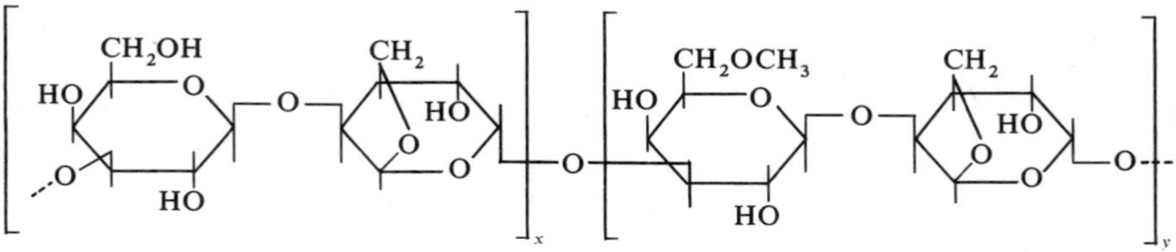

(**III**)

According to this scheme, the agarose from *G. amansii*, which yields 51%

D-galactose and 1·4% of the methyl-D-galactose, will have 50x units to 1y unit. At the other extreme, the agarose from *C. boydenii*, which gives 32% D-galactose and 21% of the 6-*O*-methyl derivative, suggests that $x = 3$ and $y = 2$.

The agaropectin fraction contains the same basic structure as agarose, but a considerable quantity of acidic residues are present, and the amount of these varies according to the alga used for the extraction of the agar. Sulphate is the main acidic residue, and it varies from 3–10%; D-glucuronic acid and pyruvic acid are also present but to a lesser extent. Little is known about the location of the sulphate and D-glucuronic acid residues, but, by the isolation of the dimethylacetal derivative (**IV**) Hirase (1957) showed that the pyruvic acid residue is attached to D-galactose.

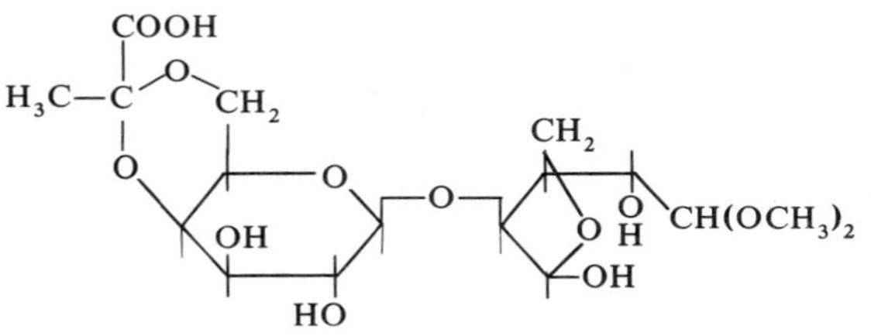

COOH
H_3C—C O CH_2
O O —O— CH_2
OH O OH $CH(OCH_3)_2$
HO —OH

(IV)

Much more work is necessary before any more definite statements can be made about the structure of agaropectin.

Araki's classical work on the structure of agar was largely based on the agar from *G. amansii* and was modified to account for the presence of 6-*O*-methyl-D-galactose. This concept of the presence of two polysaccharides in agar is an over-simplification, and the situation is further complicated by differences between the various algae, and even by seasonal variations in each alga. Nevertheless, this basic structural theory has only recently been challenged by the studies of Duckworth and Yaphe (1971), who suggested that agarose is a mixture of polysaccharides. These authors separated agar (Difco Bacto agar) into three fractions, subjected each fraction to enzymic hydrolysis and, on examination of the products, concluded that agar is basically a mixture of three products, viz. agarose, pyruvated agarose and a sulphated galactose. The reported composition of these three fractions is given in Table 22.3.

The degraded pyruvated agarose yielded a sulphate-free hexasaccharide containing one pyruvic acid residue, which suggests that the pyruvate groups are remote from the sulphate groups in the molecule. Further work on the

TABLE 22.3

Fraction	Neutral oligosaccharides (%)	Charged oligosaccharides (%)
Agarose	95	5
Pyruvated agarose	28	72
Sulphated galactose	18	82

structure of the sulphated polysaccharides must await the discovery of an unambiguous method of degrading the molecule.

22.3.2. THE MANUFACTURE OF AGAR

The most unusual feature of agar, which was called agar–agar until about twenty years ago, is that a frozen agar gel liberates water as it thaws (syneresis) and this cycle can be repeated to remove most of the water, which carries away the soluble impurities. About 350 years ago, an innkeeper in Osaka Prefecture (Japan) threw out some surplus seaweed jelly one frosty night and, the next day, he noticed water dripping from the thawing mass. He watched this phenomenon for several days and then redissolved the dried residue and found that he had reconstituted the original jelly. On the basis of this observation, he started to manufacture agar, which became known as "kanten", the Japanese word for the type of clear, night sky which presages a radiation frost.

Although the industry spread quite quickly, all the production was consumed in the Far East. Stanford (1884a) records that agar was first imported into France from China in 1856. The idea of using agar in nutrient media was suggested to Robert Koch about 1881 by Dr. Hesse, a bacteriologist whose wife had used it for some years to make fruit and vegetable jellies (Tseng, 1954). This development revolutionised bacteriology, and became the major outlet for agar.

The manufacture of agar by this method is absolutely dependent on the right climatic conditions, and is essentially a winter occupation for farmers in the mountainous regions of Japan. According to Okazaki (1971), some 400 manufacturing units are dispersed through the Prefectures of Gifu, Kyoto, Osaka and Hyogo. These small factories seldom use only one type of seaweed, and usually base their production on *Gelidium* spp. with the addition of *Gracilaria* and *Ceramium* spp. The seaweed (ca. 100 kg) is boiled with water (4,500 l) for about 4 hours in the presence of a small quantity of sulphuric acid (200 ml). It is then filtered and allowed to gel in wooden boxes (ca.

75 × 30 × 5 cm); about 300 boxes are needed to accommodate each preparation. (Filtration is always difficult with seaweed extracts, and the residue will doubtless be re-extracted and filtered to give a dilute liquor, which can be used to extract a fresh batch of seaweed.) The gel is removed from the boxes and cut into squares or sliced into strips, which are then exposed on bamboo mats for several days until the agar is sufficiently dehydrated by alternate freezing and thawing. During this period, the liquor which drains from the agar carries away most of the colour and impurities. Complete drying of the final product is very dependent on the weather, and is usually finished off in barns and other farm buildings.

Despite the fact that few bacteria are able to decompose agar, it is only natural that these organisms should thrive in the somewhat primitive conditions associated with this "cottage industry". When the weather is unseasonably warm, trouble is frequently experienced with bacterial degradation of the extract. The action of the agarase is inhibited by traces of hypochlorite, or by lowering the pH of the extract with acetic acid (Fujisawa and Sukegawa, 1956). Hydrogen peroxide enhances the antiseptic effect of hypochlorite (Fujisawa and Sukegawa, 1958).

This is essentially a crude method of extraction, without any scientific control of the ingredients, the process or the product. Nevertheless, it is economically important to the farmers concerned, and the total production by this method is said to exceed 1000 tons per annum (Okazaki, 1971). Assuming that the producer averages about a third of the European price of agar, the average income per unit will be about £2000.

With the exception of some agar made by the American Agar and Chemical Company, practically all the world's supply of agar was made by this crude method until 1939. There are few locations in the world with a climate suited to this method, and mechanical freezing was introduced, notably in Spain, and ultimately spread to Japan by about 1946. About 30 factories now operate in Japan, often in conjunction with another process which can use the refrigeration plant, such as fish freezing.

The main genera used in the extraction of agar are *Gelidium*, *Gracilaria*, *Acanthopeltis* and *Pterocladia*, but the first two species account for over 70% of the total production in Japan. The American industry relies largely on *Gelidium robustum*, the Spanish producers use both *Gelidium* and *Gracilaria* spp., *Pterocladia lucida* is used in New Zealand, and the Russian producers rely largely on *Ahnfeltia plicata* and *Phyllophora rubens*.

Basically, the process is essentially a refinement of the old "natural" process, but more sophisticated equipment is used and the gel is frozen with brine from a refrigeration plant. There is also the added advantage of laboratory control of the raw material, the process and the product.

The actual manufacturing process has been described in a long series of papers by Hayashi and Nagata (1967), and there is a recent account by Okazaki (1971). Basically, the process involves cleaning the seaweed by washing with water, followed by extraction under a pressure of 1–2 atmospheres for about two hours; during this period the liquor is circulated in the vessel and it is usually macerated at the same time. Claims have been made recently that the yield can be increased by the addition of small quantities of phosphates, usually sodium pyrophosphate, in the extraction stage (Matsuhashi, 1971); a similar effect is experienced in the extraction of carrageenan.

The suspension is then treated with a filter-aid and filtered while still hot, on either a plate and frame filter press or on a rotary vacuum filter. The clear solution is run into cans (with a capacity of about 200 litres), cooled with running water and allowed to stand overnight to gel. Following this, the gel is expelled from the container, cut into strips and cooled to about -2°C over 24 hours, and then further cooled to -15°C. The frozen gel is then sprayed with tepid water and the thawing mass is allowed to drain, when most of the water escapes, leaving a residue containing up to 10% agar. More water can be removed by pressing, either in a screw press or a hydraulic press, and the agar is finally dried in conventional circulating drying ovens. The only practical variation to this rather tedious process is spray-drying, which is said to give a satisfactory result with an inlet air temperature of 130°C, but the gel strength is adversely affected at 150°C (Matsuhashi, 1970).

In the manufacture of bacterial grades of agar, the seaweed is usually bleached prior to extraction. Ideally, the seaweed is sun-bleached as it is collected, but this operation is very dependent on weather conditions; a misty morning followed by a sunny afternoon gives ideal conditions for bleaching any of the red seaweeds. Alternatively, the seaweed can be bleached with sodium hypochlorite (ca. 0·1%) or hydrogen peroxide during the washing stage.

Gracilaria spp. give only weak gels when extracted by this method, but the gel strength can be greatly improved by treating the extract with alkali. These seaweeds first gained prominence during the war when the shortage of agar was acute (Wood, 1942). At this time, most of the work was carried out in America, Australia and British Columbia, but the process was really developed commercially in Japan by Ohta and Tanaka (1965). This work showed that the gel strength could be increased five-fold by heating the seaweed with sodium hydroxide or, preferably, with a mixture of sodium hydroxide (1·5%) and calcium chloride (0·05%) at 90°C for 5 hours. More elaborate treatments have been patented; one of these involves treatment at 20°C with sodium hydroxide (1%), followed by soaking first in lactic acid, then in a mixture of calcium, magnesium and sodium hydroxides. This is followed by a further treatment with lactic acid, when a good agar gel is finally extracted in 24% yield by boiling (Steinmetz *et al.*, 1969). Treatment with alkali under pressure, as in the manufacture of carrageenan, does not seem to be employed, but reduction with an alkali borohydride, a standard laboratory practice with carrageenan, has been used in the laboratory with agar (Duckworth *et al.*, 1971). Theoretically, the treatment with alkali converts galactose-6-sulphate residues to 3,6-anhydrogalactose; in practice this proceeds most readily with *Gracilaria* spp., less readily with *Eucheuma* spp. and least readily with *Chondrus crispus*.

For many years the Japanese attempted to maintain their near monopoly in the agar market and, for instance, the Korean production was bought and refined in Japan but, by 1960, the South Korean Government restricted these exports. By 1958, the productivity of the *Gelidium* beds in Japan started to decline; this was attributed to marine pollution (U.S. Dept. Interior, 1962). This caused the Japanese manufacturers to scour the world for supplies of seaweed and their imports rose from 3659 tons in 1958 to 6131 tons in the first six months of 1961. At the same time, these changed conditions caused a world shortage of agar, and the price rose steadily until about 1965. The main sources of supply developed by the Japanese were in Argentina, Chile and Portugal, and the manufacture of agar was also developed in these countries in the second half of the 1960's.

A secondary effect of the decline of the *Gelidium* beds was the use of fertilizers to improve their growth. The numerous patents granted in the last decade suggests that this practice is rewarding. A recent patent describes a mixture of urea, ammonium sulphate, ammonium chloride, calcium carbonate and sulphuric acid which is heated to 90°C and poured into polyethylene sacks; this mixture floats on water, and the sacks are pierced so that the fertilizer can slowly dissolve—presumably the sacks are moored in a fixed position (Kishimoto *et al.*, 1971).

22.3.3. THE USES OF AGAR

Agar is sold in various physical states (e.g. square, strip and powdered) and each type is marketed in three grades. In addition, there is a superior grade of powdered agar and, at the other end of the scale, there are two grades of scrap agar. The superior grade of powdered agar has a gel strength of 600g cm^{-2}, whereas No. 1 grade is standardized at 350, No. 2 at 250 and No. 3 at 150 g cm^{-2}; the crude protein rises from less than 0·5% in the best grade to not more than 3% in No. 3 grade. Similarly, the insoluble matter rises from less than 0·5% in the best grade to about 4% in the lowest grades, and the colour varies from near white to a buff colour.

The main outlet for agar is its well-known use in the plate culture of bacteria; this use demands a good quality product with little colour and the minimum of insoluble matter; it is also necessary for the agar to give a firm gel in 1–1·5% solutions.

Another large outlet is in canned food and canned petfoods but, over the last few years, carrageenan has largely replaced agar in the petfood market, and gelatin is also widely used in food canning. Nevertheless, agar is still widely used in this field; it withstands the sterilization process and is much more stable to acidic conditions than either carrageenan or gelatin. Agar is especially useful in canned foods sold in the tropics, since it sets at about 30°C and will not melt below 85°C.

The use of agar in compositions for making dental impressions is waning, at least in the United Kingdom where alginate compositions have gained favour in the last decade and now dominate this market. There is a small outlet for dental adhesives which are based on a mixture of agar and pectin (Beachner, 1968).

In the pharmaceutical field, there are several uses for agar. Perhaps the oldest is in emulsions with liquid paraffin for the treatment of constipation, but it is also used in lubricating jellies, emulsions, ointments and to some extent, as a tablet disintegrant, but in this use it has largely been replaced by alginates.

Agar is used in confectionery particularly in the Far East. In Europe, the competition from other seaweed extracts has seriously reduced the sale of agar in this market.

By the early 1960's agar was being used as a molecular sieve for the separation of substances of high molecular weight, but the presence of agaropectin was an objectionable feature. This led to the manufacture of pure agarose, which is also used in gel filtration and electrophoresis. The original method for the isolation of agarose by acetylation (Araki, 1937) is unsuitable for manufacturing purposes, and many alternative methods of purification have been proposed in the last few years. Tsuchya and Hong (1965) used both cetylpyridinium bromide (cf. Ghetie *et al.*, 1965) and polyethyleneglycol (cf. Polson, 1965) to prepare pure agarose, and Blethen (1966) introduced co-precipitation of agaropectin and carrageenan with quaternary ammonium salts to give a more readily filtered precipitate. Other methods of purifying agarose include:

1. enzymic removal of agaropectin (Morse, 1967),
2. purification on cation exchange resins (Zabin, 1969),
3. the use of sodium iodide or dimethylsulphoxide (Hayashi, 1970),
4. separation with acrinol (Fuse, 1971).

Derivatives of agarose with improved properties have been the subject of many patents (e.g. Ghetie *et al.*, 1968: Schell and Ghetie, 1972). Porath *et al.*, (1967) reacted agarose with cyanogen bromide and used the complex in column chromatography to couple with large protein molecules. They claim that they were able to isolate 350 mg of protein per gram of agarose, and state that this method is ideal for coupling solutes of high molecular weight to an insoluble base. It is difficult to estimate the demand for agarose as a laboratory reagent, but the voluminous literature, and the number of patents on agarose and its derivatives, suggest that there is considerable commercial interest in this field.

22.4. The Manufacture of Alginates

22.4.1. HISTORICAL

When Stanford (1881) discovered alginic acid, he was faced with two major problems which all innovators meet: how to manufacture and develop a market for the new product. These two problems proved particularly difficult to the early alginate producers, and they were not solved until the Kelco Company entered the field about fifty years after Stanford's original discovery.

In the first instance, Stanford had some difficulty in floating a company to manufacture algin, and his first effort, The British Algin Co. Ltd., never went into production, and it was agreed to form a new company under the aegis of the North British Chemical Company (Public Records Office, 1885). This second company, the Algin Co. Ltd., was formed in 1888 and went into voluntary liquidation two years later (Scottish Records Office, 1888). On this evidence, it has been widely assumed that Stanford failed to exploit his discovery of alginic acid, but this is not the whole story. In fact, the first company formed by Stanford was the Carbon Cement Co. Ltd., and this was very successful and ultimately, by 1929, the entire share capital was held by Turner and Newall Ltd., and the Carbon Cement Co. lost its identity in 1932 (Scottish Records Office, 1884).

In one of his early publications, Stanford (1884a) suggested several uses for alginic acid, e.g. as a textile finish, a mordant for dyestuffs, its use in foods, jujubes and lozenges, or in the manufacture of emulsions. He added two other potential outlets, the treatment of boiler feed-water and, mixed with the seaweed charcoal from the Tiree factory, as a lagging compound for steam boilers and pipes; the Carbon Cement Co. was formed to exploit these two outlets for alginates. Evidence from the records of these companies shows that the Carbon Cement Co. was selling several hundred gallons per week of crude extract at nine pence per gallon for the treatment of boiler feed-water. It is difficult nowadays to imagine unlagged boilers, and the lagging of steam boilers and pipes must rank as one of Stanford's (1882; 1884c) major innovations. Naturally, the demand was great, and the Glasgow shipbuilders provided a splendid outlet for the new product. Unfortunately, Stanford died in 1899, and the next attempt to manufacture alginates was made by Thomas Ingram of Liverpool.

Ingram's company was also called the British Algin Co. Ltd. (Public Records Office, 1900) and was formed to exploit a patent which was claimed to reduce the cost of manufacture from £65 to £25 per ton (Ingram, 1900). The main use suggested for the alginate was a binder in the manufacture of briquettes from anthracite dust (Ingram, 1901) and the Company prospectus clearly records that Ingram had made a high quality briquette (Anon., 1900). Production was started at Holywell, North Wales in 1902, but the equipment was sold by public auction in 1908. Although Ingram was unsuccessful in this venture, he continued experiments with alginates, and in the 1920's he was trying to float a company to manufacture alginate films (Longley, personal communication); Ingram's samples are in the possession of C. Aberdein of the International Seaweed Exchange.

The idea of making briquettes from anthracite dust and a crude alginate extract was next exploited by F. C. Thornley, who established a factory at Stromness, Orkney about 1923 (Thornley *et al.*, 1922). Only one cargo of anthracite dust was shipped from South Wales to Orkney, and the attempt to make the briquettes failed (Captain A. Irving, personal communication). After the collapse of this venture, Thornley went to San Diego where, by 1927, Thornley and Company produced some alginate for the American Can company, who used it in sealing cans. This manufacturing venture seems to have had no success, and the company changed its name to Kelp Products Corporation, which was also short-lived, since it was re-organized in 1929 as Kelco Company. From this rather insecure base, Kelco became the first viable alginate manufacturer, and has always been the world's largest producer of these products (Scofield, 1959).

Stanford (1899) was the first to recognize the potential use of alginate films, but he died at about the same time that his patent was granted. Ingram made a crude product in the 1920's and it was left to Bonniksen (1934) to develop the idea. Bonniksen, who was financed by the late Sir Thomas Merton, F.R.S., formed Cefoil Ltd. to exploit his invention, but the company made little headway over the next twenty years. By this time, the company had changed its name to Alginate Industries Ltd. and it entered a successful phase. It now ranks second to Kelco, and makes more alginate than the several producers in France and Norway.

In 1965, the production of alginate was started in Tasmania, where there are good beds of *Macrocystis*, but the Tasmanian company ceased production in 1974. The alga, however, is still harvested for sale on the world market. There are three manufacturers in Japan, but their total annual production amounts to only 1600 tons (Okazaki, 1971). The production in the rest of the world is negligible.

22.4.2. THE STRUCTURE OF ALGINIC ACID

The difficulty in obtaining a pure sample of alginic acid hindered all attempts to elucidate its structure until Nelson and Cretcher (1929, 1930) were able to demonstrate the presence of D-mannuronic acid and to show that the carbon dioxide released on heating was quantitatively equivalent to a uronic acid structure. The work of Hirst *et al.* (1939) suggested the molecule consisted largely of β-1,4-linked D-mannuronic acid residues (**V**) and this was confirmed by Lucas and Stewart (1940).

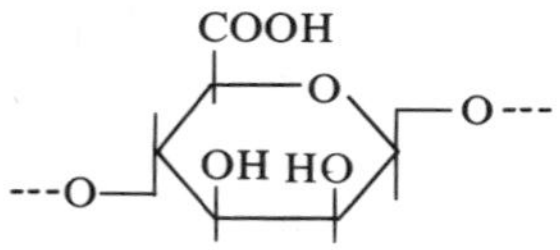

(**V**)

This basic structure remained unchallenged for many years until Fischer and Dörfel (1955) showed the presence of L-guluronic acid in the molecule. These workers examined samples from seventeen brown algae, and showed that the ratio of D-mannuronic acid to L-guluronic acid (M/G) was variable. This variation was confirmed by work at the Norwegian Institute of Seaweed Research (Haug and Larsen, 1962), and further research showed that the M/G ratio varied from 0·45 in *Ectocarpus* spp. to 2·35 in a specimen of *Laminaria hyperborea* (Haug, 1964). This work was purely analytical, i.e. paper chromatography, and, although it was reproducible, there was some doubt whether the L-guluronic was an artifact. This doubt was removed when Hirst *et al* (1964) isolated 4-O-β-D-mannosyl-L-gulose. This was the first direct proof that both acids are present in the same molecule, and it also evident that the two acids are linked one to the other.

The 1,4-linkage of both acids was firmly established by Hirst and Rees (1965) and the Norwegian workers went on to show that the molecule contains blocks of D-mannuronic acid residues, blocks of L-guluronic acid residues as well as a fraction in which the two acids alternate (Haug *et al.*, 1967).

The molecule is linear and the commercial samples have a molecular weight of about 200 000, which corresponds to a degree of polymerisation of almost 1000 (Cook and Smith, 1954).

Until 1964, alginic acid was considered unique to the brown seaweeds, but Linke and Jones (1964) discovered the acid in culture of a *Pseudomonas* isolated from human sputum. This was followed by the isolation of a partially acetylated alginic acid from the cultures of *Azotobacter vinelandii*; this alginic acid had a high M/G ratio (Gorin and Spenser, 1966). Work on cultures of *A. vinelandii* has revealed the presence of a C-5 epimarase which, when incubated in the presence of calcium ions, can convert D-mannuronic acid residues in the polymer chain to L-guluronic acid residues (Haug, 1971). It seems probable that a similar epimarase exists in the brown seaweeds and that this could be responsible for the variation in the M/G ratios. It is interesting to note that a bacterial culture from *Beijerinckia indica* gives an exocellular polysaccharide, which contains only L-guluronic acid (Haug and Larsen, 1970).

Both glucuronic acid and xylose have been reported in alginic acid (Massoni and Duprez, 1960), but it seems most probable that these were impurities from a contaminating polysaccharide.

22.4.3. THE PRODUCTION OF ALGINATES

There is only a limited amount of information available on the methods of making sodium alginate, and what is available is in patents taken out by Kelco and the Algin Corporation of America in the 1930's; Maass (1959) has carried out an excellent survey of the literature on alginates. Alginate Industries do not appear to have patented any aspect of their process, although there is a paper by Bonniksen (1951) which is remarkably uninformative.

The first meaningful patent (Thornley and Walsh, 1931) describes the preparation of calcium alginate, and two subsequent Kelco patents (Clark and Green, 1936; Green, 1936) refine the process but describe only the preparation of alginic acid. Any mention of a method of converting the acid to the commercially important sodium salt is studiously avoided but, in all other respects, these patents give a fair description of the process.

The patent applications by le Gloahec and Herter (1936) were first made in Britain and France in 1934. Le Gloahec then emigrated to America, and the American patent was assigned to the Algin Corporation of America; this appears to the be the earliest mention of this company which, presumably, was formed to exploit these patents. There is little significant difference between the English versions of these patents and they appear to be very detailed, but they obscure more than they reveal; a commonplace trouble with patents.

A method for the extraction of alginate from *Ascophyllum nodosum* has been described by Myklestad (1967).

Basically, all these processes involved five stages. The seaweed is first washed with water and disintegrated; the alginate is then extracted with about three times the theoretical requirement of sodium carbonate, and the solution is filtered from the residual seaweed. The calcium salt is then precipitated by running the clear alkaline extract into a concentrated solution of calcium chloride, which gives a fibrous precipitate of calcium alginate. This is then converted to alginic acid by treatment with dilute hydrochloric acid, and the excess acid and any residual calcium chloride is removed by washing with water.

The main difficulty is the retention of water by the calcium salt and the free acid. Some of this water can be removed mechanically, e.g. by a screw press, although this operation is very dependent on the form in which the calcium alginate is precipitated. Some seaweeds, e.g. *Laminaria digitata*, give a long-fibred precipitate which is relatively easy to handle, but the *Fucoids*, for instance, tend to give short-fibred precipitates which are very difficult to process. Also, the initial treatment of the seaweed, such as drying and storing, can degrade the alginate, and this adds to the difficulty of processing. No information is available on the methods of converting alginic acid to the sodium salt. Another difficulty which is not always realized is the large amount of water used in the process, which can amount to 1 million gallons per ton of product.

There is no available data on the world production of alginates, but Percival and McDowell (1967) consider that production exceeds 10 000 tons per annum, and the quantity of seaweeds known to be used by industry suggests the annual production is approaching 20 000 tons.

There is one reliable description of a laboratory method for the preparation of sodium alginate (Rose, 1951). In this method, the seaweed (*Laminaria digitata*; 160 g, dry weight) is minced, mixed with water (2 litres) and calcium hydroxide (5 g), and heated to 60°C for 30 minutes when it is filtered. The residue is stirred with 2 litres of cold 0·2 N sulphuric acid for 30 minutes, filtered and washed free from acid. The pulped seaweed is then stirred with 1 l of a 3% solution of sodium carbonate for 2 hours at 50°C, when it is diluted with 2 l of water, stirred a further 3 hours and allowed to settle overnight. The liquor is then centrifuged and added to 1 litre of 25% calcium chloride solution and the precipitated calcium alginate is strained, and washed three times with 1 litre aliquots of 0·5 N hydrochloric acid to convert the calcium salt to the acid. This is then washed to remove excess acid and any residual calcium chloride. The wet alginic acid is then suspended in 4·5 litres of water and titrated with 110 ml of 2N sodium hydroxide using phenolphthalein as indicator and the viscous solution is precipitated by adding it slowly, with adequate stirring, to 7 litres of alcohol. The sodium alginate is then strained on muslin, washed with alcohol and ether and dried in vacuo; the yield is about 35 g (22%).

22.4.4. THE USES OF ALGINATES

Although sodium alginate is the commonest commercial alginate, the ammonium, calcium and potassium salts are also available, and the free acid and its propyleneglycol ester are also widely used. Alginates are often used to modify the physical properties of water by forming gels or emulsions, acting as a stabilizer, or restricting the evaporation of water from exposed surfaces, or in giving a film which resists the penetration of ink into paper. These properties can be modified by the addition of metallic salts, especially calcium ions, phosphates, and by the acidity of the mixture. The readiness with which the properties can be varied has encouraged the manufacturers to produce blended products for special applications, and well over one hundred different modifications are available on the British market. The first modern use of alginate was in food, where it is on the GRAS list (Generally Recognised As Safe), but there is an increasing outlet for other uses and the development of these markets will be listed historically.

About 1908, the Liverpool Borax Co. Ltd. started to manufacture a crude alginate extract for use in ceramics but, while this use of alginates persists, the company never isolated a pure alginate and they discontinued production in 1966. The next stage, in which pure alginates were produced, started with the formation of the Kelco Company in 1929. In the early Kelco patents (see, e.g. Clark, 1936) there is special mention of a gradual improvement in the ability of the product to suspend cocoa in chocolate milk drinks, and this outlet was undoubtedly important in the first production stages. The use of alginates as an ice-cream stabilizer followed, and the extension into the general food market was a natural consequence. Alginates are now used in a very wide range of foods such as desserts, milk shakes, puddings, dairy products, bakers' sundries, icings, meringues, cake mixes, salad dressings, sherbets, cream cheeses etc. There is also an outlet in frozen foods, convenience and slimming foods. The list is endless, and new applications appear almost daily.

The next landmark in the development of the outlet for alginates was the introduction of propyleneglycol alginate (Steiner, 1947); this patent gave Kelco a monopoly with a new product which is stable under acidic conditions and was soon widely used to suspend the pulp in fruit drinks, with a notable improvement in their appearance. Another new use for this product arrived with the introduction of "keg" beer; the addition of 50–100 ppm stabilizes the froth and gives this type of beer its distinctive appearance. The rapid expansion in the manufacture of this type of beer created a substantial market for this product. It is interesting to recall that a Mr. E. M. Holmes suggested "... perhaps the algae might be employed for producing a head on beer ..." in the discussion on an early paper by Stanford (1884a).

Another idea pioneered by Kelco is the use of alginates as a surface size in papermaking (Vallandigham, 1949). The original patent claims that the addition of 0·2–2·0% of alginate to the starch normally used as a surface size on paper, reduces the penetration of ink into the paper. This new idea met with mixed success, but it was used in a Lancashire paper-mill before 1953, to coat the paper used for the magazine "Picture Post"; this mill used 5% sodium alginate in the starch size, and the printing quality of a relatively cheap paper was remarkably good. Improvements in the speed of off-set litho-printing in the 1960's created a demand for the firmer surface which alginates impart to the paper, and increased this outlet considerably (Thomin, 1972).

In 1956, I.C.I. introduced the first of their Procion dyestuffs; these have many desirable characteristics and were unique in reacting with the hydroxyl groups in cellulose to give outstanding fastness to washing. This type of dyestuff also reacts with the hydroxyl groups in the gums normally used in textile printing, but they do not react with alginates. This happy chance created a new demand for alginates. For the first few years the number of these new dyestuffs was relatively small, but the range and the number of manufacturers increased, and the demand for alginates grew proportionately. The textile printers also find alginate thickeners are much more convenient to prepare than the traditional gums, and this use of alginate, which is already large, is expected to increase considerably.

The British industry was interested initially in the production of alginate films and, while films seem to have little application in the United Kingdom. they have been extensively patented in Germany for use as sausage casings (Woolf and Co., 1966). With the advent of war, the Ministry of Supply directed attention to the production of alginate fibres, as a jute substitute, for use in camouflage netting, and several factories were established for this purpose. This venture was not particularly successful, but considerable experience was gained in fibre production (Speakman and Chamberlain, 1944). After the war it was hoped that alginate yarns would be widely used in the production of light-weight woollen goods, and also in simulating "drawn-thread" effects; in each case, the alginate would be woven into the fabric and then dissolved in the finishing process. For various reasons, neither of these ideas gained much popularity. On the other hand, alginate yarn is widely used in the manufacture of socks which are knitted in a continuous tube, with shaping at the heel and toe; a few rows of alginate yarn are included between the toe of one sock and the welt of the next, and the individual socks are separated by dissolving the alginate during the finishing process. A similar principle is used in knitting scarves; again a continuous process where alginate yarn is introduced between each scarf and then dissolved in the finishing process. Alginate yarn is also used in weaving elasticated materials; here the yarn is doubled with the elastomer, woven and then removed by washing. Another growing use for alginate yarn is in sewing disposable hospital laundry bags; these are usually made of two polyethylene sheets sewn together with alginate thread which dissolves in the alkaline washing liquid and releases the contents of the bag.

Medical dressings made from woven alginate yarn have been available for a long time, but they have become increasingly popular in the last few years. These are woven as calcium alginate and then modified to a mixture of sodium and calcium alginates, which gives a gel in contact with the wound and has good haemostatic qualities (Miller and Caldwell, 1971). This type of fabric is available as first-aid dressings and it is particularly useful for burn dressing. Alginate wool is also available and is widely used in dentistry for its haemostatic properties.

Amongst the miscellaneous uses of alginates are their use as a lubricant in extruding welding rods, and in creaming latex and latex emulsion paints. It is also used in making dental impressions where the product "Kromopan" appears to have completely replaced mixtures based on agar products in the United Kingdom. New uses for alginate are constantly being sought, and the most dramatic of these is the recent patent claims for an alginate tobacco substitute (Hedge, 1972). Several well-known companies are interested in this new development, and the patents are based on a paper-like film made from calcium alginate in sheets weighing 50–250 g m^{-2}, which are coloured and flavoured to resemble tobacco (Prouse *et al*., 1972).

22.5. CARRAGEENAN

22.5.1. INTRODUCTION

Carrageenan, with various spelling modifications, is the name used for a type of sulphated polysaccharide extracted from a number of red seaweeds. *Chondrus crispus* and *Gigartina stellata*, often known individually or collectively as Irish Moss, are principally used, but other algal species give extracts with somewhat similar properties, and *Eucheuma*, *Hypnea* and, to a lesser extent, *Iridea* and *Polyoides* are all commercially important.

Landsborough (1857) pointed out that the chief supply came from Carrageen in Ireland, and he said the market price had been as high as 2s 6d ($12\frac{1}{2}$p) per lb but "The fashion, however, has gone out and the price has fallen . . ." Simmons (1862) dates the earliest recorded use of carrageenan to 1842, when Zimmermann introduced the use of Irish Moss into Bavarian breweries, but he added ". . . I believe it had been used previously both in England and Ireland for this purpose." This use of Irish Moss is still practised in the U.K. today. By 1857 Simmons had recorded the use of carrageenan in sizing silk, and this use evidently developed rapidly, since Thomson (1879) describes the method of extraction and its use in ". . . Oxford shirtings, to produce a more mellow feel and to make the fabric more pliable and less likely to curl at the edges when dry."

By the end of the 19th century, the use of carrageenan as a fining agent in brewing, and as a textile size led to its manufacture in England by two companies; Blandola Ltd. specialized in the textile field, and Savilles (1902) Ltd. made products for the brewer and, some time later, Manchester Finings Ltd. entered the field. These three companies still operate in their respective fields.

The industry spread to the United States, where several small companies were operating in the 1930's. The amalgamation of Seaplant Chemical Corporation with the Algin Corporation of America to form Marine Colloids Inc. in 1959, really marks the turning point in the development of the market for carrageenan as a speciality product. In the next nine years, the total harvest of Irish Moss on the Atlantic coast of America rose three-fold, from 12 200 tons to 37 000 tons (wet weight) per annum.

In Europe, the industry developed in France, and the sale of carrageenan, under the name of "Coreine", for the treatment of constipation, colitis and diarrhoea was established in 1911. Later, the main manufacturer was Satia, who are now the second largest makers in the world. About 1960, the Copenhagen Pectin Company entered the field and quickly became prominent manufacturers. Meanwhile, some of the smaller producers in England and Norway went out of business, but the industry spread to Spain about 1968. Apart from two small firms in Japan, which together produce only 270 tons per annum (Okazaki, 1971) and a small production in Brazil from *Hypnea musciformis*, there is little known about the production in the rest of the world. The three main manufacturers are understandably reticent about their production data, but a knowledge of the harvest of suitable seaweeds in various parts of the world, suggests a total annual production of about 10 000 tons. The U.K. consumption by food manufacturers for human use has been estimated at 120 tons per annum.

22.5.2. SOURCES OF SEAWEED

In addition to the data already quoted, there is a fair amount of *Chondrus* produced in other parts of the world. Only a small amount is harvested in Ireland, probably less than 300 tons per annum, and the quality is variable. The production in Maine in 1967 amounted to 1400 tons (wet weight) and the harvest of Moss in Newfoundland started in 1968, but no figures are available. The quality of Canadian Moss has improved in recent years, through the introduction of mechanical driers; only one drier was operating in 1965, but eleven were reported in 1970 (Ffrench, 1970). The dried Moss usually contains less than 20% moisture, and is remarkably free from other seaweeds. The main objectionable contaminant is mussels, and it appears that more sprat of *Mytilis edulis* settles on cut than on raked Moss (MacFarlane, 1966).

Relatively small amounts of other members of the *Gigartinaceae* are collected in various parts of the world, e.g. *G. pistilata*, *G. acicularis* and *G. radula*, as well as *G. skottsbergii*; a reasonable estimate of the total annual harvest of this group of seaweeds is 1500 tons.

Eucheuma spp. are in great demand for the production of carrageenan; all the members of this group grow in tropical waters, and they were first exported from Singapore and Zanzibar. A considerable quantity is exported through Singapore, but both its origin and quantity are unknown; it is usually sold as *E. spinosum*, but the nomenclature of this algal group is in a very confused state. The exports from Zanzibar amount to 700 tons per annum (Jassund, personal communication); this is sold as dry seaweed, but it usually contains 45–48% moisture. The most important species, since it yields high gelling grades of carrageenan, is *E. striatum*—(= *E. cottonii*), usually known as "thick Zanzibar", and the commercial product usually contains the thick, fleshy cakes of *E. platycladum*, which also gives a high gelling extract. *E. spinosum* (= *E. serra*), generally known as "thin Zanzibar" is also available, but extracts of this alga do not gel and the demand is slack, mainly because better quality material can be obtained from Singapore.

Eucheuma spp. are widely harvested in the Pacific, but details are available only for the Philippines. According to Michanek (1971) the collection declined from 1100 tons in 1966, to 400 tons in 1969, but Velasquez (personal communication) states that both *E.cottonii* and *E. spinosum* are cultured in the Philippines, and the exports in 1971 amounted to 500–600 tons.* Some *Eucheuma* has been imported into the U.K. from Japan; it is believed this material originated in China, and was a cultured product.

About 1966, it was realized that the industry was likely to be restricted by the availability of the Moss, as long as it was dependent on the collection

* Doty and Alvarez (1974) gave the 1973 production as 1200 tons and expect this to increase five-fold in 1974.

of "wild" plants, and serious attempts were started to apply scientific principles to the cultivation of seaweeds (Neish, 1968; Allen *et al.*, 1971). Chen and McLachlan (1972) have completed a cultural study of the life cycle of *Chondrus crispus*, which will assist further work in this field. Several of the edible algae have been cultured in Japan, and other parts of the Pacific, for many years, and the possibility of growing *Gracilaria* spp. from transplants was demonstrated in Florida by Williams (1945), and similar trials were carried out in Australia and British Columbia during the war. The commercial cultivation of *Eucheuma* spp. in the Pacific is a notable step towards the treatment of seaweeds as an agricultural crop. It should be added that these seaweeds are moderately expensive (£300–£350 per ton c.i.f. Liverpool) and the demand is increasing, because they readily yield a high gelling grade of carrageenan.

The other seaweeds used for the extraction of carrageenan are not, as yet, particularly important. The unusually good gel which small quantities of the extract from *Hypnea musciformis* give with milk, will probably create a demand for this seaweed. About 500 tons of this alga have been processed in the U.K.; this material is believed to have been collected in West Africa. Michanek (1971) reported the collection and processing of an unspecified quantity of this alga in Brazil, and de Goracuchi *et al.* (1969) have claimed that *H. musciformis* is abundant in the Gulf of Trieste, and shows a seasonal variation in yield.

22.5.3. THE STRUCTURE OF CARRAGEENAN

Studies on the structure of carrageenan were started in Edinburgh during the war and, a decade later, the separation of λ- and κ-carrageenans was accomplished by fractional precipitation with potassium chloride (Smith *et al.*, 1954). In the following year, O'Neill (1955) isolated derivatives of 3,6-anhydro-D-galactose from the hydrolysis products of carrageenan; incidentally, this is thought to be the only naturally occurring sugar anhydride. A few years later Rees started his classical study of the carrageenans which led to a fairly complete understanding of the family of related polysaccharides in carrageenan, and their behaviour in solution and in gels (Rees, 1969, 1972).

Basically, carrageenan consists of a chain of alternating molecules of sulphated 1,3- and 1,4-linked D-galactose units (**VI**); the sulphate content varies somewhat, and is usually about 25% in κ-carrageenan, and some 10% higher in λ-carrageenan (Black *et al.*, 1965).

This basic structure, with the following three variations, covers the structure of the various carrageenans:

(a) the degree of sulphation and the location of the sulphate groups may vary,

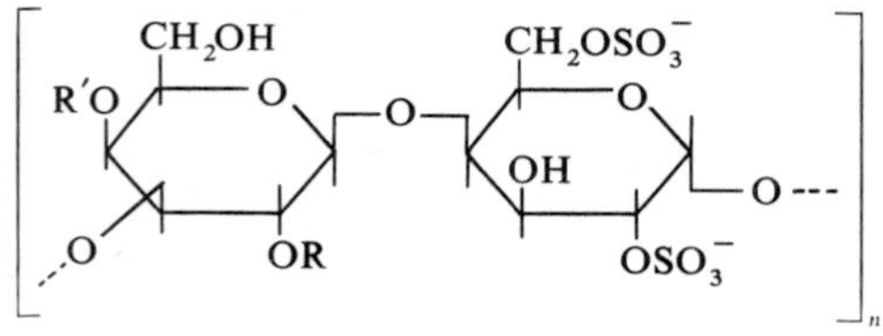

R′ is usually H but can be SO_3^-. R = H or SO_3^-.

(VI)

(b) the 6-sulphate may be converted, by enzymes within the plant or by the chemical methods, to the 3,6-anhydro group and
(c) superimposed on these variations is what Rees called the "masked repeating structure", which means that the basic residues are not identical throughout the entire molecule, but may differ in sulphate content. On this basis, the constitution of the main carrageenans may be outlined as follows (**VII**):

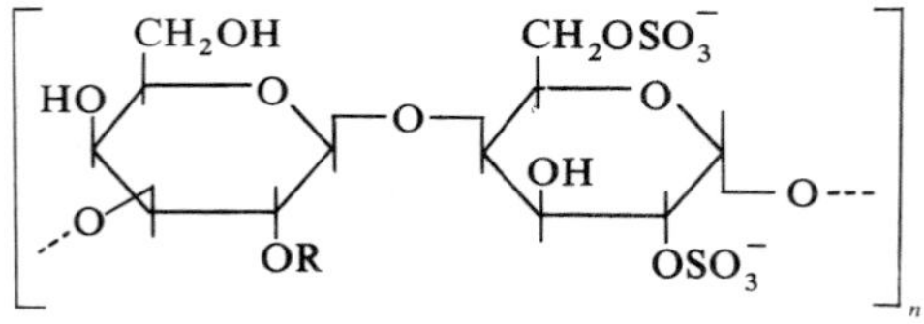

R varies from unit to unit but is usually SO_3^- or, less frequently, H.

(VII) λ-carrageenan

Ring closure to the 3,6-anhydro-D-galactose in the 1,4-linked residue gives κ-carrageenan (**VIII**); this desulphation is not complete throughout the chain, and some of the 1,4-linked residues remain intact; the 4-position in the 1,3-linked unit is usually sulphated.

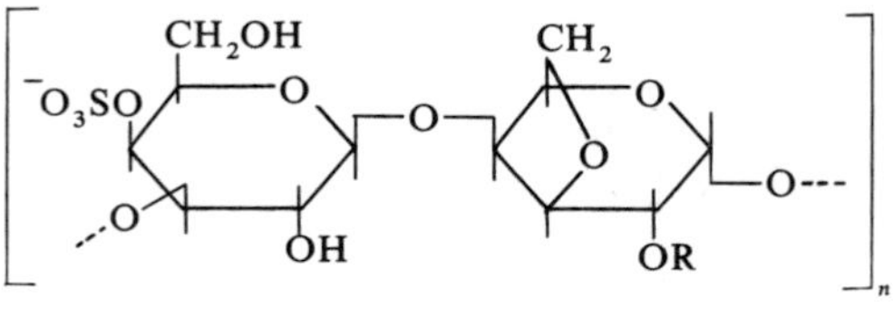

R = H or SO_3^-

(VIII) κ-carrageenan

The alga, *Eucheuma spinosum*, yields *i*-carrageenan (**IX**) which is structurally similar to κ-carrageenan, but about 10% of the D-galactose-2,6-disulphate in the 1,4-linked unit remains unchanged, and the 2-position is usually sulphated.

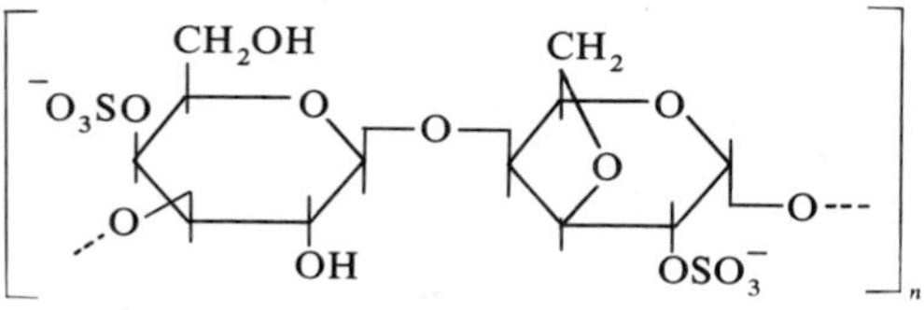

(**IX**) ι-carrageenan

A fourth modification, μ-carrageenan, occurs in *Chondrus crispus*, but it has not yet been separated from λ-carrageenan. Treatment of these two components with alkaline borohydride, followed by precipitation with potassium chloride, gives a product which is similar to κ-carrageenan. From this and other evidence, Anderson *et al.* (1968) deduced the formula shown in (**X**) for μ-carrageenan where the replacement of D-galactose-6-sulphate by 3,6-anhydro-D-galactose is not a dominant feature; it is variable, dependent upon the source of the seaweed, and there is always a small proportion of 2-sulphate on the 1,4-linked residues.

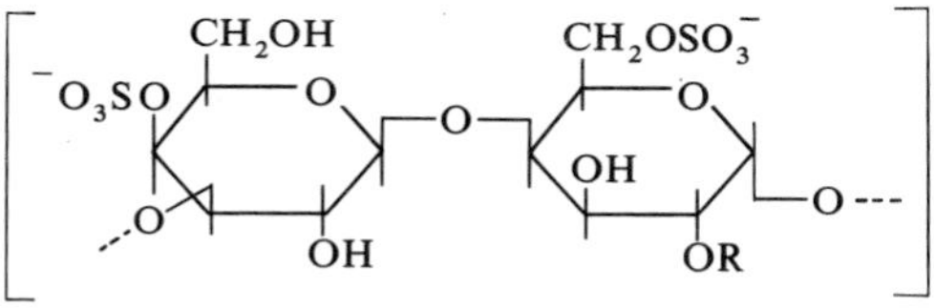

R is predominantly H but occasionally SO_3^-

(**X**) μ-carrageenan

Estimates of the molecular weight of carrageenan vary from about $3{\cdot}5 \times 10^5$ (Johnson and McCandless, 1968) up to 7×10^5 (Smith *et al.*, 1954). The degraded carrageenan used in the treatment of ulcers (see p. 250) has a molecular weight of about 30 000 (Anderson, 1967).

While both λ- and κ-carrageenan are usually present in the alga in varying amounts, it has been shown that only λ-carrageenan is found in *Gigartina atropurpurea*, and that *G. chauvinii* contains only κ-carrageenan (Stancioff and Stanley, 1969).*

22.5.4. THE MANUFACTURE OF CARRAGEENAN

When the raw material arrives at the factory, it contains sand, shells, stones and other foreign matter which must be removed by washing before the seaweed can be processed further. This washing stage is usually carried out in rotary-drum washers, or the seaweed may be washed in tanks with a

* Since this account was written, Rees and his collaborators (Rees *et al.*, 1973) have published a series of papers which cover many details of the fine structure of carrageenan. See also Chen *et al.* (1973).

perforated bottom. A surprising amount of sand is removed in the process, and sand-traps in the main drain are essential.

Prior to extraction, the washed Moss may be soaked in a solution of sodium chloride, calcium chloride or potassium chloride. The effect of various metallic ions on the gel strength of carrageenan has been studied in detail by Zabik and Aldrich (1965, 1967, 1968). It is common practice to bleach at this stage, when sodium hypochlorite is usually used at a concentration of about 0·1 %; any excess is removed by washing with water, and sodium bisulphite is used to remove any residual hypochlorite. In general, ion exchange with sodium gives a viscous product with low gel strength, calcium treatment gives an elastic gel, and treatment with potassium produces firm gels. This effect is well illustrated with *Eucheuma cottonii* which, after soaking in sodium chloride (10 %) and extracting in the presence of a little sodium carbonate gives a viscous non-gelling product; without the pre-treatment with sodium chloride, a strong gelling product of relatively low viscosity is obtained from this alga.

In the actual extraction stage, the seaweed is stirred with water for about two hours, at a temperature just below 100°C to avoid frothing; the actual quantities vary a little with different seaweeds, but about 30 kg to 1000 litres is usual. A diatomaceous filter-aid is then mixed into the solution and the hot liquor is filtered as quickly as possible; relatively large amounts of filter-aid are necessary, and it is vital to keep the solution hot. At this stage, the clear solution contains about 1 % dissolved solids, and it must be concentrated in a vacuum evaporator to 3–4 % solids; the practical limit is around 3 % with gelling grades, but 4 % can be slightly exceeded with the other types of extract.

Initially, carrageenan was isolated by drum-drying, but Bougarde (1898) mentioned the superior product produced by precipitation with alcohol, and Pfister (1941) patented the use of *iso*-propanol as a precipitant for carrageenan. With the exception of a small quantity of drum-dried carrageenan, precipitation with *iso*-propanol is the usual method adopted for its isolation. It has been shown (Wills *et al.*, 1969) that *iso*-propanol shows no sign of toxicity when ingested daily by man. Okazaki (1971) mentions precipitation with methanol, but this should be avoided since methanol is toxic to humans.

The hot concentrate is then mixed with an equal volume of *iso*-propanol (up to 15 % water can be tolerated) and the carrageenan precipitates in a fibrous form. After the temperature of the mixture has been allowed to fall below 30°C, the carrageenan is separated using a centrifuge; totally enclosed "pusher-type" machines are usually used. At this stage the wet cake may contain 60 % aqueous *iso*-propanol, and about half of this can be removed by passing the fibrous mat through a conical screw press. The press cake is then disintegrated mechanically and dried at about 80°C; the air escaping from the drier is passed through an activated carbon absorption system to collect the *iso*-propanol, which is recovered periodically by steaming the charcoal.

Recovery of the *iso*-propanol from the aqueous liquor is readily accomplished by distillation, when an azeotrope containing 14–15% water distils at 80–81°C. The still-bottoms present an effluent problem, since they contain traces of *iso*-propanol and a considerable quantity of dissolved solids with a high B.O.D.; the total dissolved solids approximates to 9–10% of the weight of carrageenan produced.

There are many variants on this basic process, the most important being the production of high-gel strength grades by hydrolysis with dilute potassium hydroxide. This method was introduced during the war (Marshall *et al.*, 1949) and a somewhat similar method, which used extraction in the presence of lime, followed by hydrolysis with potassium hydroxide, was patented by Byrne and Powling (1946). This method produced an excellent substitute for bacteriological agar from Irish Moss, and it is still the basis for the manufacture of high-gelling grades of carrageenan. Today, the best products are made from the so-called "thick Zanzibar" seaweed, by treating the filtered extract (9,000 litres) with potassium hydroxide (25·4 kg) at a pressure of 2 atmospheres for 2 hours. The excess alkali is then neutralized to about pH 7·5 with dilute hydrochloric acid, when the liquor can be concentrated and precipitated in the usual fashion. The basis of this reaction is the formation of 3,6-anhydrogalactose rings from galactose-6-sulphate residues in the molecules.

It is common commercial practice to add small quantities of soluble phosphates (up to 1 kg/10000 litres) in the extraction stage, since this treatment appears to increase the yield; this practice has been adopted recently in the extraction of agar. In some instances, the extraction liquor is made alkaline, say to pH 8, with lime or soda ash but, for most extractions, the liquor is only slightly on the alkaline side, say pH 7·2. Acidic conditions must be avoided throughout the process, because the carrageenan degrades rapidly in hot aqueous solutions.

The patent literature on carrageenan is voluminous, and only a few of the more important patents can be quoted. The first suggestion that carrageenan was a mixture of two polysaccharides is in a patent (Krim-ko Corpn., 1949) which used extraction at 82°C to manufacture a gel-forming carrageenan followed by extraction under pressure to produce a fraction which reacts with cold milk. This was followed by another patent (Stancioff, 1963) which aimed at the selective extraction of λ- and κ-carrageenan. Products with improved cold milk reactivity are claimed when the carrageenan is slightly degraded with hydrogen peroxide during the extraction stage (Gordon and Jones, 1966).

The use of a degraded carrageenan in the treatment of peptic ulcers was first described by Anderson and Watt (1959), and the subsequent patent (Anderson and Hargreaves, 1960) disclosed that the carrageenan was degraded with hydrochloric acid in aqueous acetone. More recently Colquhoun and Dewar (1968) have described the use of sodium hypochlorite for producing degraded carrageenan.

Poor solubility in cold water is one of the main disadvantages of carrageenan, and it is claimed (Gordon *et al.*, 1966) that this can be overcome by soaking the Moss in a solution containing 2% sodium chloride and 2% potassium chloride, followed by extraction under mildly alkaline conditions. It is said that this product can be readily made into a 1% dispersion when stirred into water at room temperature; almost any carrageenan would reach this standard. The claim (Société d'Auby, 1970) for a mixture of cassava starch and carrageenan co-precipitated from hot aqueous solution by alcohol, is unusual and the product is said to be rapidly soluble in cold water. A similar claim was made long ago for mixtures with alginate (Gloahec, 1950).

Without doubt, the major factor in the manufacture of carrageenan is a control of the raw material, but even the major producers, who have drying facilities at the collection sites, have to cope with large seasonal variations in the composition of the seaweed. This variation was studied in detail by Black *et al.* (1965), and the analysis of a series of Moss samples collected at the end of August, 1970 (Booth, unpublished results) shows the wide range in yield and properties which the manufacturer encounters. (Table 22.4).

TABLE 23.4

Properties and Yield of Carrageenan

Source	Yield (%)	Gel strength (F.I.R.A.)
Ireland	33	27
Halifax Co. N.S.	54	46
Ketch Habour, N.S.	58	50
Miminegash, P.E.I.	43	104

Similar wide variations in yield and physical properties are experienced with other seaweeds, e.g. four successive deliveries of *Hypnea musciformis* gave yields of 14, 20, 30 and 21%, and the gel strength (F.I.R.A.) of a 0·3% solution of the extract in milk gave values of 28, 58, 94 and 72 respectively. With variations of this magnitude in the raw material, the manufacture of carrageenan resembles a game of chance, and it is possible to produce uniform products only by blending to an agreed standard.

22.5.5. THE USES OF CARRAGEENAN

The first industrial use of carrageenan was in fining beer and, until about 1960, six factories in England were engaged in the production of extracts for the brewing industry, which used one pint per barrel of beer. The change to the so-called keg beers obviates the need for fining at this stage, and only two factories make these extracts today. Similarly, the use of carrageenan as a textile size has declined, particularly over the last twenty-five years, and cheaper substitutes are largely used today. Carrageenan is still used for a few special products, notably in printing very deep shades of vat dyestuffs on the more expensive types of curtain fabrics; here, the advantages of carrageenan are that it is practically unaffected by metallic ions, alkali or heat, and it is readily washed from the fabric during the finishing process. Similarly, the use of carrageenan in wallpaper printing has almost ceased, but mica disperses unusually well in carrageenan pastes, and this unique property provides an outlet when mica finishes are in fashion.

Another traditional use for carrageenan is in cough medicines, and this use still lingers on; for instance one branded product, which is popular in the United Kingdom, uses about one ton per annum. In fact, there may be some substance in this traditional use of carrageenan, since a recent patent claims that it extends the activity of commonly used alkaloids such as codeine and ethylmorphine (Weber and Mony, 1967).

Until the 1950's, the industry was very small, and the uses of carrageenan were only developed to their present level with the formation of Marine Colloids Inc. in 1959. The two major outlets for carrageenan were in the manufacture of ice cream and chocolate milk drinks. Carrageenan, often mixed with locust bean gum and sometimes with the further addition of sodium alginate, is widely used as an ice cream stabilizer, to produce a smooth texture and to retard the growth of ice crystals and the separation of fat globules. The mechanics of the system has been studied by Govin and Leeder (1971). The market for chocolate milk drinks is extensive in North America; these drinks are based on a suspension of cocoa (about 1 %) in milk with added sugar; carrageenan is unrivalled as a suspending agent for cocoa and, therefore, enjoys a substantial market. Some very fine powders (ca. 300 mesh) are now available which can be used in cold processes.

The food market provides an outlet for carrageenan which has increased considerably with the demand for convenience foods and slimming diets. In general, carrageenan has the unique property of complexing with milk casein (Hansen, 1970) and other proteins, and it is often used at very low concentrations, usually less than 0·05 %. In food manufacture, carrageenan competes with all the natural gums, synthetic cellulose derivatives and the bacterial gum, Keltrol; since cost is of prime importance, carrageenan is only used when it gives some effect which cannot be obtained with cheaper materials. Among the many food uses of carrageenan are milk-based puddings and pie fillings, dried milk products for reconstitution, skimmed milk products modified so that they do not appear "watery", milk substitutes for use in vending machines, sauces, whipping cream, meringues, custard powders and a host of other items.

The claims that large quantities of carrageenan are used in paints can be ignored; this use seemed full of promise a decade ago, but it never became established. Nevertheless, the industry still appears hopeful of reviving this market since a recent patent claims that carrageenan is useful in stabilizing titanium dioxide pigment suspensions (Stancioff and Witt, 1972).

There is some use of carrageenan in cosmetics, especially in shampoos and skin lotions. Advertisers, at least in Britain, are unduly loth to mention the word "seaweed", but skin lotions seem to be the exception to this generalization; unfortunately, experience suggests this outlet is small. Toothpaste is another outlet for carrageenan, which is used in one of the popular brands; to a large extent it has been replaced by cellulose derivatives, although it is still used for products sent to tropical climates, since it gives a toothpaste which is thermostable.

In the mid-1960's there was a shortage of agar, which followed the decline in the productivity of *Gelidium* beds in Japan; the price of agar rose from £1·32 to £2·75 per kg and this shortage, coupled with the sharp increase in price, opened a new market for carrageenan. Until then the makers of canned dog foods used a large tonnage of agar and suddenly they had to find a substitute. This demand was met with high-gelling grades of carrageenan, made from alkali-modified *Eucheuma* extracts. The supply of *Eucheuma* from Zanzibar was insufficient to meet the demands of this new market, and supplies were sought in the Philippines and in China. Some of the demand was also met by alkali-modified extracts of *Chondrus*. Most of the products sold to this market are blends of carrageenan extracted from *Chondrus* and *Eucheuma* spp., but some has been prepared from *Hypnea musciformis*. In practice, a mixture of the carrageenan with locust bean gum and potassium chloride is used, which is claimed to:

(a) help prevent the separation of water and fats,
(b) give a uniform product,
(c) have an attractive taste,
(d) ease removal from the can.

This outlet soon proved to be the largest market in Britain for carrageenan, with a demand of about 500 tons per annum, whereas the trade puts the human food market for it at about 100 tons.

Among the new uses suggested for carrageenan are:

(a) to give antistatic coatings to photographic emulsions. This has been patented in many countries by Ilford Ltd. (Wood, 1972),

(b) in deodorizing preparations (Airwick Industries Ltd., 1971),

(c) in tablets, to promote rapid disintegration (Vince and Lambelin, 1971),

(d) in protein-enriched meat products (Lipner, 1972).

These four outlets are selected not for their intrinsic merit, but to illustrate the diverse outlets for carrageenan.

The growth of this industry in the last decade has been remarkable, and all the indications suggest that only the availability of seaweed will limit the continued extension of the demand for carrageenan.

22.5.6. THE TOXICOLOGY OF CARRAGEENAN

Natural products are generally regarded as more suitable food additives than synthetic substitutes, but even products with a long history as human and animal food are subjected to exhaustive trials before their use is officially sanctioned. Such trials usually administer high dose rates which often create severe stress and malfunction of one or more organs in the test animal. The choice of animal is often governed by cost, rather than by the similarity of its reactions to those which might be anticipated in humans. Following tests at high dose rates, it is desirable to determine the dose level which gives "no response". Adverse results with a natural product are extremely unusual but, over the past four years, the results of trials with carrageenan have been so startling and contradictory that the topic merits special mention.

Nilson and Wagner (1959) showed that if rats and mice were fed 1, 5 or 15% carrageenan in place of maize meal, or 25% instead of mixed cereals in the diet, the life-span of the animals was not affected, nor were there signs of malignancy or post-mortem effects. Some of the rats on the diet containing 25% carrageenan showed some signs of cirrhosis of the liver, and this group attained only 90% of the weight of the controls. Subsequently the Food and Drug Administration (1961) approved the use of carrageenan and its salts; later (F.D.A., 1966) the use of polysorbate-80 was approved as a roll-release agent for use with drum-dried carrageenan. The World Health Organisation (1970) put the "acceptable daily intake" (ADI) at 500 mg kg^{-1}, i.e. about 35 g day^{-1} for the "average" man. A subsequent report (W.H.O., 1971) said "New evidence suggests a need to re-evaluate carrageenan ..." and the ADI was reduced to 50 mg kg^{-1}. In the United Kingdom, carrageenan is classed with a number of other substances in Group A, which "... the available evidence suggests are acceptable for use in food." (Food Additives and Contaminants Committee, 1970).

Lethal effects have been reported only after intravenous injection of carrageenan; this was first reported with dogs (Houck *et al.*, 1957) and subsequently with rabbits and guinea pigs (Morard *et al.*, 1964). Anderson and Duncan (1965) found the lowest lethal dose to rabbits over a 24 hour period was 1–5 mg kg^{-1} for λ-carrageenan, and 3–15 mg kg^{-1} for κ-carrageenan (both ex *Chondrus crispus*). On the other hand, Johnson and McCandless (1968) found that repeated intravenous doses of 15 mg kg^{-1} were well tolerated by rabbits. Intra-peritoneal injections (300 mg kg^{-1}) are tolerated by the guinea pig (Schwarz and Leskowitz, 1969) and sub-cutaneous infiltration (2000 mg kg^{-1}) is tolerated by the rats (Selye, 1965).

Carrageenan does not appear to be absorbed by the gut of the rat (Hawkins and Yaphe, 1965; Dewar and Maddy, 1970) and the dog can tolerate large doses in the drinking water (0·5%) for thirty days (Houck *et al.*, 1960). Lambelin *et al.* (1966) reported that carrageenan protected the intestinal membrane against experimentally induced enteritis in rats.

Since Irish Moss has a long history as an edible seaweed, and carrageenan itself has been used in medicine since 1911 with no adverse reports, either clinical or experimental, Marcus and Watt (1969) caused some consternation when they reported adverse effects in guinea pigs, rabbits, mice and rats. They claimed changes in the mucosa of the caecum, colon or rectum, which caused lesions similar to those found in human ulcerative colitis when carrageenan (1%) or degraded carrageenan (5%) was added to the drinking water over periods of 30–365 days. In one form or another, this paper was read at official meetings of four learned Societies in a period of seven months, causing widespread concern. This series of incidents resembled a "chemical witch-hunt" and probably caused the W.H.O. to reduce the ADI in 1971. It must be added that some doubt exists about the carrageenan used by Marcus and Watt, which was obtained from a merchant. Enquiries at the time of the original publication suggested that the sample had been manufactured by Longleys (Compstall) Ltd. some years earlier, and as a toothpaste grade containing ca 9% Calgon.

The Marcus and Watt articles brought forth a spate of papers, mainly in *The Lancet*. Bonfils (1970) pointed out that twelve years clinical experience had been gained with the degraded product, as well as toxicological and pharmacological experimental work, from which he concluded that degraded carrageenan was free from side effects, and offered no toxicity risk to the human colon. Maillet *et al.* (1970) found no effect on rats and mice. However, when using twice the dose used by Marcus and Watts, they did find mucosal erosion but no ulceration in guinea pigs. These erosions rarely extended to the colon and the lesions were very different from those found in human ulcerative colitis. Sharratt *et al.* (1971) pointed out that the observed lesions were probably an osmotic effect, since similar results can be produced by the addition of sodium sulphate to the drinking water. Using iron-labelled carrageenan, they were able to show that no part of the gut of ferrets, mice, rats or squirrel monkeys adsorbed any carrageenan; guinea pigs and rabbits, however, were affected and some ulcerative colitis was detected. An exhaustive study with a high molecular weight carrageenan from *Chondrus crispus*, using Rhesus monkeys, guinea pigs, rats and gerbils (Abraham *et al.*, 1972) showed no abnormalities in any of the animals. In quite separate trials on rats fed 1–3% carrageenan in the diet, a lowering of plasma cholesterol was recorded (Ito and Tsuchiya, 1972) and no adverse result was noted.

Recent work on the calcium metabolism of rats (Hong *et al.*, 1972a, 1972b) suggests that a carrageenan from *Hypnea japonica*, after reduction with sodium borohydride, increased calcium uptake in the bone, and decreased serum calcium. Unfortunately, neither *H. japonica* nor reduction with sodium borohydride are used commercially, and it is doubtful if this result is applicable to commercial carrageenan.

There is also medical evidence to show that carrageenan therapy is beneficial in the treatment of human ulcerative colitis from the use of Coreine (Laboratoires Daniel Brunet) which has been used in France for over 60 years for the treatment of constipation, colitis and diarrhoea (Friedel, 1914; Mathieu and Alivisatos, 1916; Alivisatos, 1920). Two medicinal grades of carrageenan, Coreine and Ebimar or C 16 are sold in France, and the manufacturers have disclosed annual sales approaching 100 tons. They estimate the dose rate of a maximum of 50 000 patients per month at 5–10 grams per day, and have not encountered any untoward effect from this treatment.

The general picture, as seen at this time, is that an unusual result with a sample of carrageenan of dubious origin, initiated widespread research which has shown that some samples of carrageenan can produce colonic ulceration, particularly in the caecum of two herbivores, guinea pigs and rabbits. Other experimental animals, ranging from mouse to baboon, do not appear to react to carrageenan, nor is this reaction applicable to all herbivores since the hamster is unaffected; limited tests indicate that humans are also resistant to carrageenan (Grasso *et al.*, 1973).

The commercial acceptability of carrageenan depended on research into this ulceration problem and several important papers were published while this account was in the press. It has now been shown (Fabian *et al.*, 1973) that a degraded carrageenan prepared from *Eucheuma spinosum* causes ulceration in the colon of rats. In another paper (Benitz *et al.*, 1973), a degraded product from *E. spinosum* caused ulceration in Rhesus monkeys, but an undegraded preparation from *Chondrus crispus* proved innoxious.

The present situation is adequately summarized (W.H.O., 1974) by the statement that "... a particular preparation of degraded carrageenan (used as a therapeutic agent and not as a food additive) is ulcerogenic in certain cases. There is no definite evidence that any of the various forms of undegraded carrageenan are capable of eliciting pathological changes in experimental animals or man." On the basis of this new evidence, the ADI has now been raised to 75 mg kg^{-1} body weight.

22.6. Furcellaran

The manufacture of furcellaran is confined to two firms in Denmark, where it was first produced in 1943 as a substitute for agar. It was initially known as Danish Agar, but the name furcellaran is more commonly used today. The supply of seaweed for this industry has already been outlined, and the harvest has increased considerably since the Canadian sources came into production a few years ago

Structurally, furcellaran resembles κ-carrageenan, but the molecule has not been studied so intensively. Yaphe (1959) showed that 56% of the molecule is exactly like κ-carrageenan; the ester sulphate content is lower (ca. 20%) and the molecule probably has a branched structure. Rees (1961) showed that only a small proportion of the ester sulphate residues can be converted to 3,6-anhydrogalactose residues.

Despite the chemical similarity to κ-carrageenan, the commercial method of isolation is very different. After washing with water to remove sand, shells etc., the seaweed is steeped in dilute alkali for about a week and is then dried. The seaweed is then extracted with boiling water, filtered and the filtrate evaporated to ca. 3–4% solids, as with carrageenan. The concentrate is then sprayed into a cold solution of potassium chloride (1·5%) when the furcellaran gels in the form of threads which are pressed between rollers to remove surplus liquor and then frozen for 24 hours. On thawing, furcellaran behaves like agar, and most of the water drains away. The wet fibre is re-treated with potassium chloride, frozen, thawed and the surplus moisture expelled to give a fibrous mass containing about 85% moisture. The fibrous mat is dried and ground to give a light, buff-coloured, free-flowing powder. This method of manufacture produces all the commercial material as the potassium salt, and it usually contains 8–15% of potassium chloride which enhances the gelling characteristics of the product. Production is about 1200 tons per annum, and it sells at up to £2.20 per kg, which puts the annual value of this product at approximately £2 million.

The food industry provides an outlet for about 90% of the production, and the rest is used by the pharmaceutical industry as an emulsifier and as a tablet disintegrant. Like carrageenan, it is frequently mixed with locust bean gum, which increases the elasticity of the gel; sugar is commonly added to furcellaran, since it increases both the clarity and the strength of the gel. As would be expected, the properties of furcellaran are midway between those of agar and carrageenan, and it is used in all the food applications of these older products. It is especially useful for making "instant" puddings when boiled with milk; these milk products are possibly better than those obtained with other commercial seaweed products.

22.7. Miscellaneous Seaweed Products

Although claims have been made for the production of mannitol and laminaran from the brown seaweeds, there is no evidence that either compound is produced from seaweed. Similarly, there are suggestions that Irish Moss is used in leather finishes, but the amount involved is trivial, e.g. it is used by one company in Lancashire, who buy a few tons at irregular intervals. A company in the British Midlands is known to make a small amount of erythritol from seaweed, but details of the process have never been disclosed.

The only substantial production of a seaweed extract, other than those already discussed, is the manufacture of funori in the Far East, principally in Japan. This industry is based on *Gloiopeltis* spp., principally *G. furcata*, *G. tenax* and *G. complanata*, which grow in the inter-tidal zone. These seaweeds are air-dried after collection and the extract is prepared by the user; it is mainly used as a textile size, especially on silk goods, but it finds some application as a size for paper and as a ceramic glaze. According to Chapman (1970) the production declined from 1000 tons in 1900, to 714 tons in 1936, and it has since fallen to 500 tons (Okazaki, 1971).

22.8. Edible Seaweeds

There is abundant evidence to show that seaweed was eaten in London in the middle of the last century, but its use as a food is practically unheard of today; exceptions are the use of laverweed in South Wales, and Irish Moss which is available at Health Food shops. Dulse could be obtained in parts of Scotland, e.g. Aberdeen market in the 1950's, but it had disappeared completely by 1970. There is evidently some sale for these products in Ireland, where one company, Ulster Vitamins Ltd., specializes in the sale of Irish Moss and Dulse. It seems a pity that seaweed is not more widely used, because most of the brown seaweeds have a pleasant taste and Dulse, when freshly collected and boiled, is exceptionally good. When the Shackleton Expedition was stranded on Elephant Island, Hussey (1949) records: "Seaweed was also used to provide a meal. It was tasty, although it did not agree with everyone. Unfortunately it was an appetizer, a serious thing as there was often no other course to follow it up with."

In South Wales, laverweed is considered a great delicacy and is usually fried with bacon and eggs. Even here, there was a decline in the collection from 162 tons in 1960 to 127 tons in 1965; 40% of the latter figure was collected in Wales, 21% in Cumberland and 39% came from Scotland (Preston and Jeffries, 1967). At this time, the collectors were paid £47 per ton nett for the seaweed, and collected 1–2 cwt per tide; they also collected cockles while waiting for the tide to ebb, and after the flood tide had covered the seaweed. Two men at Dunbar collected 75 tons in 1965.

There is a considerable quantity of Dulse collected in the Bay of Fundy, especially in New Brunswick where it is prominently displayed in drug stores, and is apparently widely chewed. The harvest of air-dried Dulse in 1960 amounted to 45000 kg, and this figure has not varied much over a 40-year period (MacFarlane, 1964).

The general picture is that the consumption of seaweed is very localized in the western world, but it is extensively used in tropical South and South East Asia (Zaneveld, 1959) and reaches a peak in Japan, where stocks of dried seaweed are available in even the remotest village.

Tamiya (1959) outlined the methods by which seaweeds are cultivated in Japan, and claimed that 56000 families were engaged in the industry, which produced about 3500 tons of edible seaweeds per annum. He outlined some crude methods of propagation which had been practised for at least 250 years, but Kang (1972) claims that *Porphyra* has been cultivated in Korea for about 1600 years. There is also some doubt about the quantity harvested, since Okazaki (1971) puts the production of kombu (*Laminaria* spp.) at 4000 tons and both wakama (*Undaria* spp.) and nori (*Porphyra* spp.) at 11000 tons, but he does not mention aonori (*Monostroma* spp.) which is said to be second to nori in its importance as a cultivated seaweed (Ohno, 1972). Okazaki also omits any mention of *Enteromorpha* and *Ulva* spp. which are known to be used on a large scale in Japan. Other quantitive statements give the annual production of nori as 17200 tons (Baker, 1965), and the total value of the annual production has been estimated at $140 million (Imada *et al.*, 1972). Despite the conflict between the various estimates of the annual production, it is obvious that edible seaweeds are very widely used in Japan. In addition to their own production, the Japanese import large quantities of nori from Korea.

In the early stages, the culture of seaweed must have evolved by trial and error, and systematic culture methods are based on the classic work of Drew (1949, 1953) which was honoured in Japan by the erection of a statue in her memory on Sumiyoshi Hill, Kumamoto Prefecture in 1963. The method of cultivation has been described by Baker (1965) and, more recently by Imada *et al.* (1972). After collection, the seaweed is pulped with water, made into sheets by straining the slurry on bamboo gauze, air-dried and cut into sheets 20 × 16 cm; these sheets are often cut to smaller sizes.

It is usually believed that the main species under cultivation is *Porphyra tenera* Kjellmann, but it now appears that this species has been largely replaced by *P. yezoensis* Ueda in Korea and Tokyo Bay (Kang, 1972). Disease has invaded the culture beds in the last decade and *P. tenera* has proved more susceptible than *P. yezoensis*. This disease has been described as cancerous and has been blamed on the pollution of the water with 2-chloroanthraquinone (Ishio *et al.*, 1972a), but it has also been ascribed to nitrogenous aromatic compounds in the mud of the River Ohmuta (Ishio *et al.*, 1972b) and Shirotori (1972) has reported the presence of the potent carcinogen, 3,4-benzopyrene, in Japanese nori.

Wakame (*Undaria pinnatifida*, *U. undarioides* and *U. peterseniana*) is cultivated on crude rope rafts all round the coast of Japan except in the extreme north. The growth of this seaweed is encouraged by the removal of other seaweeds, and by adding rocks to the beach. It is usual to air-dry it, but the better qualities are air-dried, washed with water and re-dried.

Kombu is a general term given to *Laminaria* spp. which is collected with a variety of grapnels in Northern Japan; the collection is made during the summer months when conditions for air-drying are good. Before drying, the seaweed may be cut into strips or sliced into thin shavings and it is sometimes soaked in acetic acid or even dyed; the haptera are also used, usually pickled. One type is infused like tea, and tastes like chicken soup. At least a dozen different types of kombu are available and a fair quantity is exported.

22.9. Seaweed Meal

Seaweed was highly regarded as a manure during the nineteenth century, and obviously it must be dried before it can be transported economically; a patent for a dried seaweed manure was granted in the middle of the century (Gardissal, 1856). The first recorded production of a dried seaweed manure was that from a factory at Boothby, Maine, which sold its product to tobacco growers in Connecticut (Anon., 1902). The first drying factory in Scotland, Neptune Mills, was opened in Oban in 1914 and was destroyed by fire in

1925; the factory was not rebuilt, but the company continued in business until it was absorbed by Scottish Agricultural Industries Ltd.

Seaweed meal has been produced in California by Philip R. Park Inc. since 1928; this firm claims to have three driers, so production must be on a fairly large scale. The product is sold for use in animal feeding stuffs. A small venture was started by J. M. Walsh in 1933; this company produces kelp tablets for human consumption and was still operating in 1968. Kelp Organic Products have also operated in this field since 1950.

Although a little seaweed meal was produced between the wars, the industry was on a small scale until the late 1949's when factories were established in Ireland, Norway and Scotland; all the production is based on *Ascophyllum nodosum*. In the early stages, practically all the output was sold as cattle food to farmers and compounders in Britain by Seaweed Agricultural Ltd., but the Norwegian producers, whose annual output has been about 15000 tons over the last decade, widened the market and they now export about 4000 tons to the United Kingdom, 3000 tons to Finland, almost 500 tons to Sweden and practically 1000 tons to the U.S.A. Most of the factories in Norway produce seaweed meal as a side-line to fish drying, which is their main work. The factories in Ireland and Scotland only process seaweed. Grass and fish meal driers are commonly used to dry seaweed.

By 1960, about a dozen factories operated in Ireland, but only half are in production today, and at least a third of their output is used in the production of alginates. All the Scottish production is used by Alginate Industries Ltd., who also import from Norway, and some of the Norwegian meal is converted to alginate by Protan and Fagertun A/S. The basic reason for the change in outlet for seaweed meal is price; as long as the meal was available at £25–30 per ton, it could be used in animal feeding stuffs, but now that it exceeds £80 per ton it can only be included in the rations of prize cattle and racehorses. The rate at which the demand for alginates is expanding suggests that practically all the better quality meal will be used in the production of alginate in the foreseeable future.

There is but little information available on the production of seaweed meal in the rest of the world, but some is produced in British Columbia, Chile, Poland and South Africa.

The uses of seaweed meal in agriculture have been discussed in some detail by Stephenson (1973), the trials evaluating its benefits to livestock have been reviewed by Jensen (1972), and Booth (1966) has reviewed its use as a fertilizer.

22.10. Liquid Seaweed Extracts

A patent for the manufacture of a liquid seaweed extract was granted over sixty years ago (Penkals, 1912), but the market was not ready to receive this type of product and liquid seaweed fertilizers made no progress until about 1950, when another patent was under consideration (Milton, 1962). This patent was exploited by Maxicrop Ltd. a few months after the application date, and their product was launched in 1950 and was soon followed by "Bio Extract" which is now made by Pan Britannica Industries Ltd. In the early stages, "Maxicrop" was sold only to commercial growers, and "Bio" was sold through the retail trade, but this distinction no longer exists. Also they were both purely seaweed extracts, but the range has been extended to include extracts fortified by the addition of nitrogen, phosphorus and potassium salts.

The sale of these products gathered momentum in the late 1950's when foliar feeding became an orthodox method of plant nutrition. Several similar products came on the market in the 1960's, e.g. "Marinure", "SM-3" and "Trident" brands were made in the United Kingdom and "Algifert" was produced in Norway. The total annual sale is now about 5×10^6 litres.

These products are made by heating the seaweed under pressure with a slight excess of sodium carbonate; after heating for several hours, practically all the seaweed dissolves. The liquor is then filtered, the solid content is adjusted to 8% and formalin is added to prevent bacterial attack. A large export market has been developed for these extracts, which are drum-dried to give a readily soluble powder.

To a large extent, these extracts are used to supply trace elements, and it is possible to chelate added metals during the process of manufacture to produce special products (Milton, 1962). The manufacture and properties of these seaweed fertilizers have been reviewed by Booth (1969), and Stephenson (1973) has given an extensive account of his commercial experience with these products.

22.11. Maërl

The use of coral deposits in agriculture was probably implied by Pliny when he write: "The people of Gaul have discovered another method of nourishing the soil with a substance called marga." These deposits, which usually consist mainly of *Lithothamnion calcareum*, are not common, but the occasional coral beach can be found around the British coast, e.g. there are two such beaches near Dunvegan, Isle of Skye, another south of Heads of Ayr, a raised coral beach exists on the west side of Cumbrae and there is an extensive one near Ballyconneely on the northern shore of Galway Bay. By far the largest and best known deposit of maërl is found a little to the west of St. Malo in France, where it is harvested by a grab in 1 ton lots from depths of up to 30 m, and landed at Le Legue. This huge deposit has been worked for centuries and the annual harvest rose to 200 000 tons in the 1960's (Stephenson, 1973); the source of this vast supply of maërl is unknown but, since the individual plant is relatively small, it is evident that it is deposited by some unusual tidal effect which probably garners the fragments of coral from a very large area.

The maërl is allowed to drain and is then dried and ground; several factories process maërl, the best known being Usine Timac at St. Servan. They give the following analysis for their product:
$CaCO_3$, 80–85 %; $MgCO_3$, 10–15 %; Fe, 0·48 %, with traces of Zn, B, Cu, Mn, I, S, and P.

Until about 1964, there was little demand for maërl outside Brittany, where it sold at the equivalent of £6 per ton, and was used at 1·5–1·9 ton per hectare every third year (P. Hadfield, personal communication). About this time, maërl was used in Guernsey instead of lime, since it could be imported at £4.50 per ton (F. Hubert, personal communication). In the second half of the 1960's, the price of seaweed meal began to increase and three companies started to import maërl into the United Kingdom as a cheap substitute for seaweed meal; it sells at about £15 per ton.

There is practically no scientific evidence for the value of maërl in agriculture, but there is no shortage of extravagant claims for the product. Taylor (1968) records that experience in France suggests that treating pasture with maërl conveys some immunity to foot-and-mouth disease in cattle; a similar claim has been recorded for seaweed meal by Chapman (1970). On a more mundane level, tests show very little gain from the use of maërl (Breole, 1967), but Augier (1972) has shown that β-indole-3-acetic acid and four other hormones with auxinic or gibberellic properties are present in the strains of *Lithothamnion calcareum* used in agriculture.

References

Abraham, R., Golberg, L. and Coulston, F. (1972). *Exp. Med. Pathol.* **17**, 77.
Airwick Industries Ltd. (1971). B.P. 1 241 914.
Alivisatos, A. S. (1920). *Arch. Mal. Appar. Dig.* November.
Allen, J. H., Neish, A. C., Robson, D. R. and Shacklock, P. F. (1971). Atlantic Regional Laboratory, Halifax, Nova Scotia. Tech. Rep. No. 15.
Anderson, N. S., Dolan, J. C. S. and Rees, D. A. (1968). *J. Chem. Soc.* 596.
Anderson, W. (1967). *Can. J. Pharm. Res.* **2**, 81.
Anderson, W. and Duncan, J. G. G. (1965). *J. Pharm. Pharmacol.* **17**, 647.
Anderson, W. and Hargreaves, J. G. G. (1960). B.P. 840 623.
Anderson, W. and Watt, J. (1959). *J. Pharm. Parmacol.* **11**, 318
Anon. (1900). *Liverpool Journal of Commerce*, 4th August.

Anon. (1902). *Rep. U.S. Comm. Fish.* p. 278.
Araki, C. (1937). *J. Chem. Soc. Jap.* **48**, 1338.
Araki, C. (1966) In Proceedings 5th International Seaweed Symposium (1965). (E. G. Young and J. L. McLachlan, eds.), p. 3. Pergamon Press, Oxford.
Augier, H. (1972) *C.R. Acad. Sci. (Paris), Ser. D*, **274**, 1810.
Baardseth, E. (1955). "Regrowth of *Ascophyllum nodosum* After Cutting." Institute for Industrial Research and Standards, Dublin.
Baardseth, E. (1970). F.A.O. Fisheries Synopsis No. 38, Rev. 1. F.A.O., Rome.
Baker, H. W. (1965). *Brit. Phycol. Bull.* **2**, 497.
Beachner, C. E. (1968). U.S.P. 3 410 704.
Beattie, I. A., Blakemore, W. R., Dewar, E. T. and Warwick, M. H. (1970). *Food Cosmet. Toxicol.* **8**, 257.
Benitz, K. F., Golberg, L. and Coulston, F. (1973). *Food Cosmet. Toxicol.*, **11**, 565.
Black, W. A. P., Blakemore, W. R., Colquhoun, J. A. and Dewar, E. T. (1965). *J. Sci. Food Agr.* **16**, 573.
Blethen, J. (1966). U.S.P. 3 281 409.
Bonfils, S. (1970). *Lancet*, ii, 414.
Bonniksen, C. W. (1934). B.P. 415 497.
Bonniksen, C. W. (1951). *Proc. Chem. Eng. Group Soc. Chem. Ind., London*, **33**, 11.
Booth, E. (1966). *In* "Proceedings 5th International Seaweed Symposium (1965)", (E. G. Young and J. L. McLachlan, eds.) p, 349. Pergamon Press, Oxford.
Booth, E. (1969). *In* "Proceedings 6th International Seaweed Symposium (1968)", Santiago, Spain (R. Margalef, ed.), p. 655. Direccion General de Pesca Marítima, Madrid.
Bougarde, G. (1898). U.S.P. 112 535.
Breole, M. (1972). Réunion de la Section Americaine de l'Association Français pour l'Etude du Sol, 29th May.
Byrne, J. F. and Powling, P. S. (1946). B.P. 577 533.
Chapman, V. J. (1950). *In* "Seaweeds and Their Uses", p. 61. Methuen and Co., London.
Chapman, V. J. (1970). *In* "Seaweeds and Their Uses", 2nd Ed., Methuen and Co., London.
Chen, L. C.-M. and McLachlan, J. L. (1972). *Can. J. Bot.* **50**, 1055.
Chen, L. C-M., McLachlan, J., Neish, A. C. and Shacklock, P. F. (1973). *J. Mar. Biol. Ass., U.K.* **53**, 11.
Clark, D. E. and Green, H. C. (1936). U.S.P. 2 036 992.
Clow, A. and N. L. *In* "The Chemical Revolution", *p.* 74. Blatchworth Press, London.
Colquhoun, J. A. and Dewar, E. T. (1968). U.S.P. 3 378 541.
Cook, W. H. and Smith, B. D. (1954). *Can. J. Biochem. Physiol.* **32**, 227.
Courtois, B. (1813). *Ann. Chim.* **88**, (i), 304.
Dewar, E. T. and Maddy, M. L. *J. Pharm. Pharmacol.* **22**, 791.
Doty, M. S. and Alvarez, V. B. (1974). *In* "Proceedings 8th International Seaweed Symposium", Bangor; (in press).
Drew, K. M. (1949). *Nature, Lond.* **164**, 748.
Drew, K. M. (1953). *J. Linn. Soc.* **55**, 84.
Duckworth, M., Hong, K. C. and Yaphe, W. (1971). *Carbohyd. Res.* **18**, 1.
Duckworth, M. and Yaphe, W. (1971). *Carbohyd. Res.* **16**, 189, 435.
Fabian, R. J., Abraham, R., Coultson, F. and Golberg, L. (1973). *Gastroenterology*, **65**, 265.
F.D.A. (1961). *Fed. Register*, **26**, 9411.

F.D.A. (1966). *Fed. Register*, **31**, 3116.
Ffrench, R. A. (1970). In "The Irish Moss Industry" p. 27. Department of Fisheries and Forestry, Ottawa.
Fischer, F. G. and Dörfel, H. (1955). *Hoppe-Seyler's Z. Physiol. Chem.* **301**, 186.
Food Additives and Contaminants Committee (1970). Report on the Review of the Emulsifiers and Stabilisers in Food Regulations, 1962. H.M.S.O., London.
Friedel, G. (1914). *Arch. Mal. Appar. Dig.* July.
Fugisawa, H. and Sukegawa, T. (1956). *Bull. Jap. Soc. Sci. Fish.* **22**, 30.
Fugisawa, H. and Sukegama, T. (1958). *Nippon Kaishi Suisangaku Kaishi*, **24**, 342.
Fuse, T. (1971). *Agric. Biol. Chem.* **35**, 799.
Gardissal, C. D. (1856). B.P. 2003.
Ghetie, V. F., Motet, D. and Dragomireren, C. (1965). *Studo Cercet. Biochim.* **8**, 337.
Ghetie, V. F., Motet, D. and Schell, D. H. (1968). F.P. 1 527 883.
Gloahec, V. C. E. le (1950). U.S.P. 2 513 416.
Gloahec, V. C. E. le and Herter, J. R. (1936). B.P. 405,358; U.S.P. 2 128 551.
Goracuchi, C. de, Lokar, L. C., Bruni, G. and Davango, S. (1969). *Trieste. Bull. Soc. Adriat. Sci.* **57**, 1.
Gordon, A. L. and Jones, J. J. (1966). U.S.P. 3 236 833.
Gordon, A. L., Jones, J. J. and Pike, L. F. (1966). U.S.P. 3 280 102.
Gorin, P. A. J. and Spenser, J. F. T. (1966). *Can. J. Chem.* **44**, 993.
Govin, R. and Leeder, J. G. (1971). *J. Food Sci.* **36**, 718.
Grasso, P., Sharratt, M., Carpanini, F. M. B. and Gangolli, S. D. (1973). *Fed. Cosmet. Toxicol.* **11**, 555.
Green, H. C. (1936). U.S.P. 2 036 934.
Grenager, B. and Baardseth, E. (1966). *In* "Proceedings 5th International Seaweed Symposium (1965), Halifax Nova Scotia (E. G. Young and J. L. McLachlan, eds.), p. 129. Pergamon Press, Oxford.
Gusmán del Proo, S. A. (1969). *In* "*Proceedings 6th International Seaweed Symposium* (1968)", Santiago, Spain (R. Margalef, ed.), p. 179. Direccion General de Pesca Marítima, Madrid.
Hansen, P. M. T. (1970). *Macromolecules*, **3**, 269.
Haug, A. (1964). "Composition and Properties of Alginates". Rep. No. 30. Norwegian Institute of Seaweed Research, Trondheim.
Haug, A. (1971). *Carbohyd. Res.* **17**, 297.
Haug, A. and Larsen, B. (1962). *Acta Chem. Scand.* **16**, 1908.
Haug, A. and Larsen, B. (1970). *Acta Chem. Scand.* **24**, 855.
Haug, A., Larsen, B. and Smidsrod, O. (1967). *Acta Chem. Scand.* **21**, 691.
Hawkins, W. W. and Yaphe, W. (1965). *Can. J. Biochem.* **43**, 479.
Hayashi, K. (1970). *Nippon Shokuhin Gakkaishi*, **17**, 575.
Hayashi, K. and Nagata, Y. (1967). *Nippon Shokuhin Gakkaishi*, **14**, 450.
Hedge, R. W. (1972). G.P. 2 150 388.
Hendrick, J. (1916). *J. Soc. Chem. Ind.* **35**, 565.
Hirano, T., Kikuchi, T. and Okada, I. (1956). *J. Tokyo Univ. Fish.* **50**, 87.
Hirase, S. (1957). *Bull. Chem. Soc. Jap.* **30**, 75.
Hirst, E. L., Jones, J. K. N. and Jones, W. O. (1939). *J. Chem. Soc.* p. 1880.
Hirst, E. L., Percival, E. and Wold, J. K. (1964). *J. Chem. Soc.* p. 1493.
Hirst, E. L. and Rees, D. A. (1965). *J. Chem. Soc.* p. 1182.
Hong, K. C., Cruess, R. L. and Skoryna, S. C. (1972a). *Can. J. Physiol. Pharmacol.* **50**, 784.

Hong, K. C., Cruess, R. L. and Skoryna, S. C. (1972b). *In* "Proceedings 7th International Seaweed Symposium (1971)", Sapporo, Japan (K. Nisigawa, ed.), p. 566. University of Tokyo Press, Tokyo.

Houck, J. C., Bhayana, J. and Lee, T. (1960). *Gastroenterology*, **39**, 196.

Houck, J. C., Morris, R. K. and Lazaro, E. J. (1957). *Proc. Soc. Exp. Biol. Med.* **96**, 528.

Hussey, L. D. A. (1949). *In* "South with Shackleton" p. 151. Sampson Low, London.

Imada, O., Saito, Y. and Teramoto, K. (1972). *In* "Proceedings 7th International Seaweed Symposium (1971)", Sapporo, Japan (K. Nisigawa, ed.), p. 358. University of Tokyo Press, Tokyo.

Ingram, T. (1900). B.P. 22 590.

Ingram, T. (1901), B.P. 1773.

Ishio, S., Yano, T. and Nakagawa, H. (1972a). *In* "Proceedings 7th International Seaweed Symposium 1971", Sapporo, Japan (K. Nisigawa, ed.), p. 373. University of Tokyo Press, Tokyo.

Ishio, S., Kawabe, K. and Tomiyama, T. (1972b). *Nippon Suisan Gakkaishi*, **28**, 17.

Ito, K. and Tsuchiya, Y. (1972). *In* "Proceedings 7th International Seaweed Symposium (1971)", Sapporo, Japan (K. Nisigawa, ed.), p. 558. University of Tokyo Press, Tokyo.

Jackson, P. J. (1957). *Engineer*, **203**, 400, 439.

Jensen, A. (1972). *In* "Proceedings 7th International Seaweed Symposium (1971)", Sapporo, Japan (K. Nisigawa, ed.), p. 7. University of Tokyo Press, Tokyo.

Johnson, K. H. and McCandless, E. L. (1968). *J. Immunol.* **101**, 556.

Kang, J. W. (1972). *In* "Proceedings 7th International Seaweed Symposium (1971)", Sapporo, Japan (K. Nisigawa, ed.) p. 108. University of Tokyo Press, Tokyo.

Kishimoto, K., Hirose, M. and Nozu, M. (1971). J.P. 71 18,563.

Krim-Ko Corporation (1949). B.P. 627 212.

Lambelin, G., Mees, G., Foster, M. and Parmentier, R. (1966). *Gastroenterologia, Basel*, **106**, 13.

Landsborough, D. (1857). *In* "A popular History of British Seaweeds". p. 221. Lovell Reeve, London.

Leighton, D. L., Jones, L. G. and North, W. J. (1966). *In* "Proceedings 5th International Seaweed Symposium (1965)", Halifax, Nova Scotia (E. G. Young and J. L. McLachlan, eds.), p. 141. Pergamon Press, Oxford.

Linker, A. and Jones, R. S. (1964). *Nature, Lond.* **204**, 187.

Lipner, S. (1972). U.S.P. 3 644 128.

Lucas, H. J. and Stewart, W. T. (1940). *J. Amer. Chem. Soc.* **62**, 1792.

Lund, S. and Bjerre-Petersen, E. (1953). *In* "Proceedings 1st International Seaweed Symposium (1952)", Edinburgh, p. 85. Institute of Seaweed Research, Edinburgh.

Lund, S. and Bjerre-Petersen, E. (1964). *In* Proceedings 4th International Seaweed Symposium (1961)", Biarritz (A. D. deVirville and J. Feldmann, eds.), p. 410. Pergamon Press, Oxford.

Lund, S. and Christensen, J. (1969). *In* "Proceedings 6th International Seaweed Symposium (1968)", Santiago, Spain (R. Margalef, ed.), p. 699. Direccion General de Pesca Marítima, Madrid.

Maass, H. (1959). "Alginsäure und Alginate." Strassenbau Chimie & Technik Verlagsgesellschaft m.b.H., Heidelberg.

MacFarlane, C. I. (1964). *In* "Proceedings 4th International Seaweed Symposium (1961)", Biarritz (A. D. DeVirville and J. Feldmann, eds.), p. 414. Pergamon Press, Oxford.

MacFarlane, C. I. (1966). *In* "Proceedings 5th International Seaweed Symposium

(1965)", Halifax, Nova Scotia (E. G. Young and J. L. McLachlan, eds.), p. 169. Pergamon Press, Oxford.
MacFarlane, C. I. (1966a). "A Report on Some Aspects of the Seaweed Industry in the Maritime Provinces of Canada". Department of Fisheries and Forestry, Ottawa.
MacFarlane, C. I. (1968). *In* "Chondrus crispus Stackhouse, A Synopsis". Nova Scotia Research Foundation, Halifax, Nova Scotia.
Maillet, M., Bonfils, S. and Lister, R. E. (1970). *Lancet*, ii, 414.
Marcus, R. and Watt, J. (1969). *Lancet*, ii, 489.
Marshall, S. M., Newton, L. and Orr, A. P. (1949). "A Study of Certain British Seaweeds". H.M.S.O., London.
Massoni, R. and Duprez, G. (1960). *Chim. Ind.* **83**, 79.
Mathieu, A. and Alivisatos, A. S. (1916). *Arch. Mal. Appar. Dig.* February.
Matsuhashi, T. (1970). *Nippon Shokuhin Gakkaishi*, **17**, 1.
Matsuhashi, T. (1971). *Nippon Suisan Gakkaishi*, **14**, 450.
Michanek, G. (1971). "A Preliminary Appraisal of World Seaweed Resources". F.A.O., Rome.
Miller, J. H. H. and Caldwell, R. (1971). B.P. 1 231 506.
Milton, R. F. (1959). B.P. 664 989.
Milton, P. F. (1962). B.P. 902 563.
Morard, J. C., Fry, A., Abadie, A. and Robert, A. (1964). *Nature, Lond.* **202**, 401.
Morse, P. (1967). U.S.P. 3 362 884.
Myklestad, S. (1967). Norg. P. **111**, 426.
Neish, A. C. (1968). *Science Forum*, **1**, 15.
Nelson, W. L. and Cretcher, L. H. (1929). *J. Amer. Chem. Soc.* **51**, 1914.
Nelson, W. L. and Cretcher, L. H. (1930). *J. Amer. Chem. Soc.* **52**, 2130.
Newton, L. (1951). "Seaweed Utilisation". Sampson Low, London.
Nilson, H. W. and Wagner, J. A. (1959). *Food Res.* **24**, 235.
North, W. J. (1964). *In* "Proceedings 4th International Seaweed Symposium (1961)", Biarritz (A. D. De Virville and J. Feldman, eds.), p. 248. Pergamon Press, Oxford.
Ohno, M. (1972). *In* "Proceedings 7th International Seaweed Symposium (1971)", Sapporo, Japan (K. Nisigawa, ed.), p. 405. University of Tokyo Press, Tokyo.
Ohta, F. and Tanaka, T. (1965). *Mem. Fac. Fish. Kagoshima Univ.* **13**, 38.
Okazaki, A. (1971). "Seaweeds and Their Uses in Japan". Tokai University Press, Tokyo.
O'Neill, A. N. (1955). *J. Amer. Chem. Soc.* **77**, 2837.
Öy, E. (1949). *U.N.S.C.C.U.R. Proc., Wildlife Fish Resources*, p. 177.
Penkals, L. (1912). B.P. 27 257.
Percival, E. and McDowell, R. H. (1967). "Chemistry and Enzymology of Marine Algal Polysaccharides". Academic Press, London.
Pfister, A. (1941). U.S.P. 2 231 283.
Polson, A. (1965). B.P. 1 006 259.
Porath, J., Axen, R. and Ernbeck, E. (1967). *Nature, Lond.* **215**, 1491.
Preston, A. and Jeffries, D. F. (1967). *Health Phys.* **13**, 477.
Prouse, R. E., West, A. F., King, D. A. and Poulson, R. (1972). G.P. 2 134 515.
Public Records Office (1863). File No. BT31/752/298c.
Public Records Office (1885). File No. BT31/3575/21916.
Public Records Office (1900). File No. BT31/9242/68621.
Rees, D. A. (1961). *J. Chem. Soc.* p. 5168.

Rees, D. A. (1969). *Advan. Carbohyd. Chem.* **24**, 267
Rees, D. A. (1972). *Biochem. J.* **126**, 257.
Rees, D. A., Anderson, N. S., Dolan, T. C. S., Lawson, C. J., Penman, A., Stancioff, D. J. and Stanley, N. S. (1973). *J. Chem. Soc., Perkin Trans.* **1**, 2173, 2177, 2182, 2188, 2191.
Reynolds, G. F. (1972). *Chem. Brit.* **8**, 536.
Rose, R. C. (1951). *Can. J. Technol.* **9**, 19.
Schell, H. D. and Ghetie, V. F. (1972). U.S.P. 3 651 041.
Schwartz, H. J. and Leskowitz, S. (1969). *J. Immunol.* **103**, 87.
Scofield, W. L. (1959). *Calif. Fish Game*, **45**, 135.
Scottish Registry Office (1876). File No. 700.
Scottish Registry Office (1884). File No. 1,408.
Scottish Registry Office (1888). File No. 1,698.
Scottish Registry Office (1891). File No. 2,157.
Scottish Registry Office (1913). File No. 8,518.
Selye, H. (1965). *Science, N.Y.* **149**, 201.
Sharratt, M., Grasso, P., Carpanini, F. and Gangolli, S. D. (1971). *Lancet*, i, 192.
Shirotori, T. (1972). *Tokyo Kasei Daigaku Kiyo*, No. 12, 47.
Simmons, P. L. (1857). *J. Roy. Soc. Arts*, **5**, 362.
Simmons, P. L. (1862). "Waste Products and Undeveloped Substances". Hardwicke, London.
Sinclair, J. (1791). *In* "Statistical Account of Scotland" Vol. 3, p. 78. W. Creech, Edinburgh.
Sinclair, J. (1793). *In* "Statistical Account of Scotland" Vol. 7, p. 539. W. Creech, Edinburgh.
Société d'Auby (1970). B.P. 1 194 682.
Smith, B. D., Cook, W. H. and Neal, J. L. (1954). *Arch. Biochem. Biophys.* **53**, 192.
Speakman, J. B. and Chamberlain, N. N. (1944). *J. Soc. Dyers Colour.* **60**, 264.
Spenser, D. A. (1947). *J. Roy. Soc. Arts*, **95**, 675.
Stancioff, D. J. (1963). U.S.P. 3 176 003.
Stancioff, D. J. and Stanley, N. F. (1969). *In* Proceedings 6th International Seaweed Symposium (1968)", Santiago, Spain (R. Margalef, ed.), p. 595. Direccion General de Pesca Marítima, Madrid.
Stancioff, D. J. and Witt, H. J. (1972). U.S.P. 3 663 284.
Stanford, E. C. C. (1862). *J. Roy. Soc. Arts*, **10**, 185.
Stanford, E. C. C. (1881). B.P. 142.
Stanford, E. C. C. (1882). B.P. 2132.
Stanford, E. C. C. (1884a). *J. Roy. Soc. Arts*, **32**, 717.
Stanford, E. C. C. (1884b). B.P. 13 312.
Stanford, E. C. C. (1884c). *J. Soc. Chem. Ind.* **13**, 297.
Stanford, E. C. C. (1887). *Chem. News, Lond.* **35**, 172.
Stanford, E. C. C. (1899). B.P. 18 075.
Steiner, A. B. (1947). U.S.P. 2 426 125.
Steinmetz, C. P., Gunter, V. and Hinterwalde, R. (1969). B.P. 1 154 137.
Stephenson, W. A. (1973). "Seaweed in Agriculture and Horticulture" 2nd. Ed. Educational Productions, Wakefield, Yorkshire.
Svendsen, P. (1972). *Fisken Hav.* (2), 34.
Tamiya, H. (1959). *Proc. Symp. Algol.* New Delhi, 1960, p. 379.
Taylor, C. (1968). *Christchurch Star*, 15th June.

Thomin, W. H. (1972). *Pap. Technol.* **13**, 120
Thomson, W. (1879). *In* "The Sizing of Cotton Goods". 2nd Ed., p. 110. J. Heywood, Manchester.
Thornley, F. C., Tapping, F. F. and Reynard, O. (1922). B.P. 211 174.
Thornley, F. C. and Walsh, M. J. (1931). U.S.P. 1 814 981.
Toms, C. (1965). *In* "Country Fair" August pp. 67–8.
Tseng, C. K. (1954). *Sc. Mon.* **58**, 24.
Tsuchiya, Y. and Hong, K. C. (1965). *Tohoku J. Agr. Res.* **16**, 141.
U.S. Department Interior (1962). Market News Leaflet No. 66. "Japan's Agar Agar Industry."
Vallandigham, V. V. (1949). U.S.P. 2 477 912.
Vince, A. and Lambelin, G. (1971). Belg. P. 758 264.
Vincent, C. W. (1876). "Chemistry Applied to the Arts and Manufactures." W. Mackenzie, Glasgow.
Walker, F. T. (1946). Scottish Seaweed Research Association, Report No. 22.
Walker, F. T. (1947a). *Proc. Linn. Soc. Lond.* Session 159, Part 2, 90.
Walker, F. T. (1947b). *J. Ecol.* **35**, 166.
Walker, F. T. (1948). Scotish Seaweed Research Association. Report No. 108.
Weber, A. and Mony, C. (1967). Fr. P. 5227.
W.H.O. (1970). *World Health Organ. Tech. Rep. Ser.* No. 445, Annex 5.
W.H.O. (1971). *World Health Organ. Tech. Rep. Ser.* No. 462, Annex 5.
W.H.O. (1974). *World Health Organ., Tech. Rep. Ser.* No. 539.
Wills, J. H., Jameson, E. M., and Coulston, F. (1969). *Toxicol. Appl. Pharmacol.* **15**, 560.
Williams, R. H. (1945). *Proc. Florida Acad. Sci.* **8**, 161.
Wood, E. J. F. (1942). *J. Counc. Sci. Ind. Res.* (*Aust.*) **15**, 1.
Wood, H. H. (1972). U.S.P. 3 655 384.
Woolf and Co. (1966). G.P. 1 213 211.
Worthy, E. J. and Walker, F. T. (1948). Scottish Seaweed Research Association. Report No. 106.
Yaphe, W. (1959). *Can. J. Bot.* **37**, 751.
Zabik, M. E. and Aldrich, P. J. (1965). *J. Food Sci.* **30**, 795.
Zabik, M. E. and Aldrich, P. J. (1967). *J. Food Sci.* **32**, 91.
Zabik, M. E. and Aldrich, P. J. (1968). *J. Food Sci.* **33**, 371.
Zabin, B. A. (1969). U.S.P. 3 423 396.
Zaneveld, J. S. (1959). *Econ. Bot.* **13**, 89.

Chapter 23

Marine Drugs: Chemical and Pharmacological Aspects

HEBER W. YOUNGKEN, JR.

and

YUZURU SHIMIZU

College of Pharmacy, University of Rhode Island, Kingston, Rhode Island, U.S.A.

23.1 INTRODUCTION

Biomedical aspects of the oceans are commonly called "Drugs from the Sea" and in recent years numerous articles have been published on the subject.* Marine organisms and sea water have been sources of medicines and pharmaceuticals since ancient times, and over the ages all sorts of claims have been made for the use of various kinds of marine plant and animal extracts and constituents as medicinal agents. In comparison with the large number of useful chemotherapeutic products which have come either directly or indirectly from terrestrial organisms, very few originate from marine sources and probably not more than a dozen or so of these are in widespread use

* For a review of the potential of the oceans to provide biomedical materials, see Youngken (1969). The organic chemistry of marine natural products has been reviewed by Scheuer (1973) and by Faulkner and Andersen (1974).

in therapy today. However, the upsurge over the past decade of interest in the oceans as a potential source of new drugs, has stimulated a flurry of activity at the research bench and in the clinics. Some of this stems from a desire to find out whether there is a scientific basis for folklore medicine. The development of modern chemical, pharmacological and engineering technology has facilitated the investigation and exploitation of hitherto "untapped" drug resources of the oceans. However, despite these developments, the full potentialities of the oceans as a source of natural products and pharmaceuticals have not been appreciated either in the academic sphere or in industry. There are many reasons for this; these include the problems of collecting, harvesting and isolating active constituents from marine life, their clinical evaluation, processing, and the lack of adequate financial support.

Extensive reviews on the subject of marine pharmaceuticals and marine pharmacology have been published by Der Marderosian (1969) and Baslow (1969). These provide basic information on the physical, chemical, biological and pharmacological characteristic of marine life. A consideration of the wide spectrum of chemical metabolites, i.e. organic acids, carbohydrates, proteins, amino acids, sterols, lipids, enzymes and the many other compounds, abundant in sea water and living marine organisms, underlines the importance of searching the oceans for medicinals. As will be noted in this account, biologically active compounds such as antibiotics, antitumour, anticoagulant, antiviral, antiulcer, haemolytic, analgesic, antilipemic and cardioinhibitory agents, stimulants, depressants and various fungicides and insecticides as well as other agents from marine sources have been described. Some pharmaceutical adjuvants and stabilizers, e.g. alginates employed as aids in dosage formulations and several highly poisonous marine toxins, are also described.

The present chapter emphasizes the chemistry related to the pharmacological and pharmaceutical applications of certain well tested marine drugs and a number of compounds being currently investigated.

23.2 Carbohydrates and Derivatives

Marine organisms are an abundant source of carbohydrates, including certain polysaccharides which are unique in their abundance and variety. Only a few have medicinal applications at present. These are found in cell wall structures and as intercellular mucilage in certain marine organisms and occur as galactose polymers, complex uronic linkages, sulphated polysaccharides, glucosamines and other forms. Classical drugs of this group include (i) agar, a dried mucilaginous substance extracted from red algae of the species *Gelidium, Gracilaria, Hypnea, Campyraephora, Pterocladia* (ii) carrageenan from *Chondrus, Gigartina,* and *Eucheuma* species (iii) *alginic acid* and *alginates* from these and certain brown algae i.e. *Laminaria, Fucus* and *Macrocystis.* The polysaccharide laminarin is also extracted from these brown algae (see Chapter 22).

A variety of pharmacological properties have been attributed to algal polysaccharides. Agar* is used in pharmaceuticals as a mild laxative and as a suspending and emulsifying agent in the formulations of ointment bases and cosmetics. Its chemical composition varies according to the starting material used. Typically, it consists of two components, agarose and agaropectin. Agarose is a polymer of 1,3-β-D-galactose, 1,4-linked 3,6-anhydro-α-L-galactose. Agaropectin is, essentially, a 1,3-β-D-galactose polymer which contains varying amounts of sulphate, pyruvate, methylated sugars and uronic acids in the complex molecule. The agarose and agaropectin contents of agars from different sources are shown in Table 23.1.

TABLE 23.1
Comparison of agarose and agaropectin contents of agars (*Tagawa, 1968*)

Agar source	Agarose fraction (%)	Agaropectin fraction (%)
Gelidium amansii (Korea)	58·0–63·0	20·0–25·0
Gracilaria verrucosa	91·5–93·8	3·0–3·4
Gracilaria spp.	85–87·5	3·6–5·0

Agarose can be extracted from the algae by highly polar organic solvents, e.g. dimethyl sulphoxide, whereas agaropectin is practically insoluble in these solvents. The gelling powers of agar arise mostly from the neutral polysaccharide agarose; the gelling power of the acidic polysaccharide, agaropectin, is considered to be weak, and for this reason the ratio of agarose to agaropectin in the starting material greatly affects the properties of the finished product.

The alginates are of some interest in chemistry and medicine because of their action as anticoagulants. They are used in absorbable haemostatic materials for the control of surface bleeding and have special application in radioisotopic intoxication. Alginic acid (Der Marderosian, 1969) continues to be of pharmaceutical value as a tablet-disintegrating agent (3–10% alginic

* Agar, alginic acid and alginates have extensive use as components of tissue culture media, as adhesives, stabilizers and as emulsifiers in food products and in the textile industry as sizing agents (see Chapter 22).

acid in tablets disintegrates tablets faster than does a 15% starch binding agent) and it is also reported of use in cases of intoxication with radio-strontium (Hesp and Ramsbottom, 1965; Sutton, 1967; Waldron-Edwards, 1968).

Probably the most important use of these marine polysaccharides, particularly the alginates, is in the food and confectionery industries where they are employed as additives and binders. Alginic acid is a complicated long chain series of 1,4-linked D-mannuronic and L-guluronic acid linkages viz.

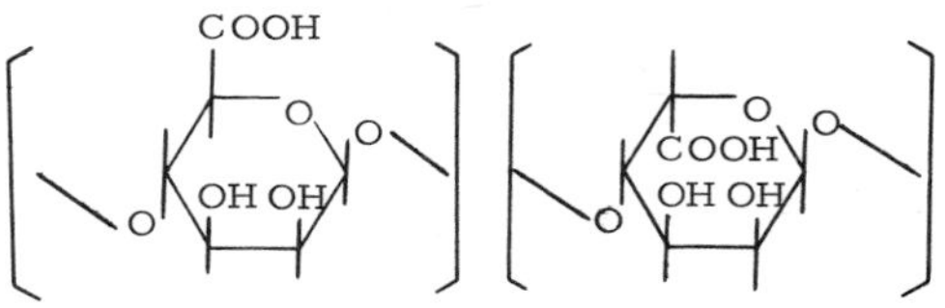

COOH O O OH OH O | O COOH O OH OH O

IVXD-mannuronic acid L-guluronic acid

(**I**) Partial Structure of Alginic acid (Hirst and Rees, 1965)

It is generally considered that the cell wall structures of marine algae contain material rich in L-guluronic acid linkages; the intercellular alginic acid is primarily a D-mannuronic acid polymer (Preston, 1962).

The ability of alginic acid to form salts with metal ions, especially alkali earth and heavy metals, is the basis for its use as a protective and therapeutic agent for those exposed to intoxication by radioisotopes. Hesp and Ramsbottom (1965) have reported the removal of strontium-85 and strontium-90 from the intestine following dosage with alginic acid and its salts and this has been confirmed by Sutton (1967) and Waldron-Edwards (1968). Tanaka (1969) has examined the possible use of other polyuronic acids and their derivatives with some success. He also suggested their possible therapeutic use following the ingestion of toxic ions such as ferric and ferrous ions. The probable chemical interaction of alginate with strontium in the body can be expressed in the following abbreviated chemical scheme (Tanaka, 1969).

Sr^{2+}/diet
↓
Sr^{2+} + 2Na (Alg.) ⇌ $2Na^{+}$ + $Sr(Alg)_2$
↓ insoluble
Sr^{2+}/blood excreted
↓
Sr^{2+}/bone

Tanaka has also reported a haemostatic action of alginate preparations, but this has, so far, had limited application.

The source of carrageenan, of current interest in medicine, is carrageen, a dried, bleached powder isolated from the Irish moss (*Chondrus crispus*) and from other red algal species, *Gigartina*, *Hypnea*, and *Furcellaria* (Yaphe, 1959). It consists of two components, κ-carrageenan (degraded carrageenan or Ebimar) which can be gelatinized by the addition of KCl solution, and λ-carrageenan (also degraded carrageenan) which remains in solution. The chemical structures of the carrageenans are very complicated and for κ-carrageenan the basic sugar moiety is 4-β-D-galactosyl-3,6-anhydro-D-galactose (carrabiose). This can polymerize in various ways with sulphate esters at the C-4 position of the galactose residue (O'Neill, 1955).

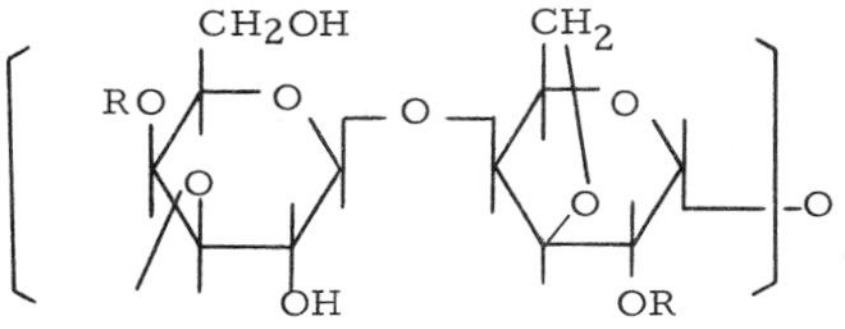

R = H or $-SO_3^-$

(**II**) Partial structure of κ-carrageenan
(Some of the linkage is 1,4 and sulphate at C-6)

Carrageenan is reported to have antiviral activity for certain influenza viruses. Other natural and synthetic sulphated polysaccharides have similar activity on viruses as reported by Takemoto and Spicer (1965) and Kathan (1965).

Anderson and Duncan (1965) have reported that carrageenan and other naturally occurring sulphated polysaccharides exhibit anticoagulant activity. This activity may be explained by the similarity of the chemical structure to that of heparin (**III**), an anticoagulant factor found in human and other mammalian tissues.

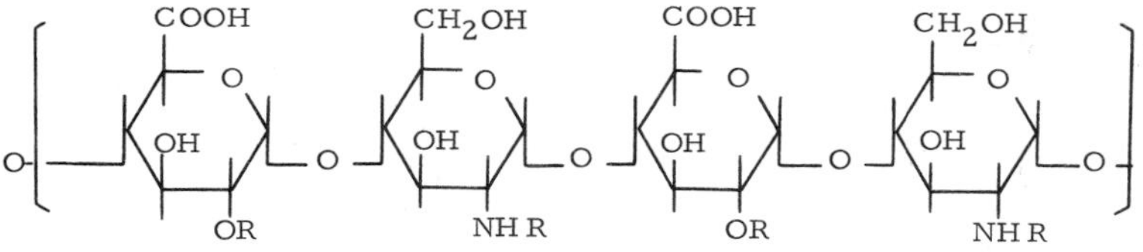

R = H or SO_3^-

(**III**) Heparin

Baslow (1969) has concluded that degraded marine polysaccharides such as carrageenan are potentially useful antiulcer agents which can be used with other medications, and that this action is the result of its stimulation of a local "protective" mechanism in the intestinal tract.

The highly sulphated laminarins which are prepared from laminarin, a 1,3-glucan, found in many species of brown algae of the *Laminaria* family, are also reported to have an anticoagulant activity comparable to that of heparin (Dewar, 1956).

Some antilipemic action (reduction of blood cholesterol levels) is reported for these sulphated polysaccharides. However, some degree of sulphation, seems to be essential for either antilipemic or anticoagulant activity since, as the number of sulphate groups in the molecule increases, so does the anticoagulant activity (Besterman and Evans, 1957). Laminarine, which contains few sulphate groups in its molecule, possesses antilipemic activity without anticoagulant effects.

Polysaccharides from animal sources. Marine animals are good sources of chondroitin sulphates; these medicinally active mucopolysaccharides and polymers of N-acetyl-D-galactosamine sulphate and glucuronic acids.

The soft bones of sharks and others of the fish orders *Squaliformes* and *Rajiformes* contain large amounts of a characteristic chondroitin sulphate D. (Seno, 1961; Tsuchiya and Suzuki, 1962).

The exoskeletons of lobsters, shrimp and crab are the major sources of the polysaccharide, chitin (**IV**). This polysaccharide is a polymer of 1-4 linked *N*-acetyl-D-glucosamines having a molecular weight of 4500–6000 (Foster and Webber, 1960). It occurs in two forms, α- and β-, which can be distinguished by their X-ray diffraction patterns. Chitin can be deacetylated to produce the product commercially known as Kylan® which is used as a cationic protective colloid to impart shrink resistance to wool (Merck Index, 1968). It is also the starting material for the preparation of an important amino sugar, glucosamine (Organic Syntheses, 1955) an important chemical reagent used in drug synthesis.

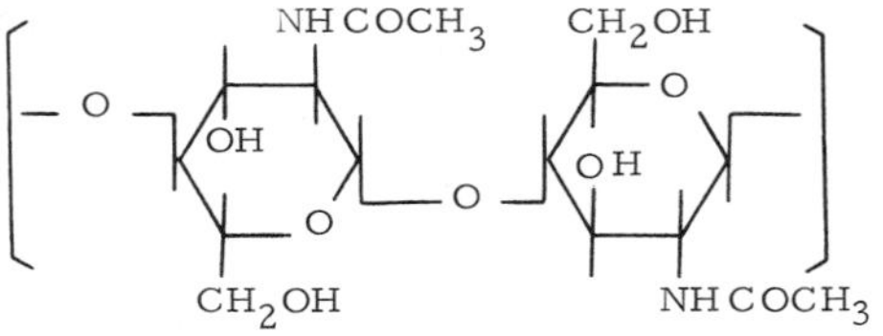

(**IV**) Chitin

One of the recent applications of chitin in medical science is its use as a selective binding agent with certain enzymes. Katz *et al.* (1972) separated human lysozyme in pure form using deaminated chitin which was obtained from the backbone (pens) of the squid (*Loligo vulgaris*).

23.3 Aliphatic Acids and Derivatives

Fresh water and marine algae contain free fatty acids of various degrees of unsaturation. The antimicrobial activity reported for extracts from marine species of *Laminaria*, *Sargassum*, *Porphyra* and *Codium* is probably due to their fatty acid components.

Pratt and his coworkers (1951) reported the presence of acrylic acid as an antibacterial agent. Sieburth (1961) recognized an antibacterial substance in the intestines of penguin and traced it back to acrylic acid derived from algae used in the food chain of this Antarctic bird. Acrylic acid is a potent antibacterial agent, and its toxicity is reported to be minimal (Sieburth, 1960, 1961). Its use is limited, however, to the preservation of non-food materials.

The biosynthesis of acrylic acid is probably the result of a retro-Michael type decomposition of dimethyl-β-propiothetin, a compound which exists in the cells of many algae (Cantoni and Anderson, 1956). In the reaction,

$$(H_3C)_2\overset{+}{S}-CH_2CH_2COOH \rightarrow (H_3C)_2S + CH_2=CHCOOH$$

dimethyl-β-propiothetin → dimethyl sulphide + acrylic acid

odoriferous dimethyl sulphide is formed (Challenger and Simpson, 1948). This can cause some annoyance and hazard to health when it occurs naturally on the beach during the decomposition of algae washed ashore. Straight chain C8-C16 fatty acids have also been reported as antibacterial components of algae (Proctor, 1957). For more details of algal aliphatic acids, comprehensive reviews such as that by Katayama (1962) should be consulted. In general, there is a high degree of polyunsaturation in algal fatty acids (Table 23.2).

Polyunsaturated fatty acids are also characteristic of lipids found in oils of marine animals, for example, linoleic acid, γ-linoleic acid, homo-γ-linolenic acid, arachidonic acid etc. in fish oils. These are hypocholesterolemic (antilipemic) agents, and lower the blood level of cholesterol. They become more effective pharmacologically as the extent of unsaturation increases (Peifer, *et al.*, 1960). Arachidonic, eicosatrienoic, and eicosapentaenoic acids

TABLE 23.2
The numbers of carbon atoms and double bonds of fatty acids of some marine algae (Wagner and Friedlich, 1964).

	C_{16}	C_{18}	C_{20}	C_{22}
Ulva lactuca	1, 2, 3	1, 2, 3, 4	4,5	
Codium spp.	1, 2, 3, 4	1, 2, 3, 4	4, 5	5
Lessonia fusca	1, 2	1, 2, 3, 4	4, 5	4
Porphyra columbina	1, 2	1, 2	3, 4, 5	

are essential fatty acids which are used in the biosynthesis of prostaglandin—a new drug of considerable interest in medicine (see below).

$$CH_3(CH_2)_4(CH{=}CHCH_2)_4(CH_2)_2{-}COOH$$

Arachidonic acid

$$CH_3(CH_2)_4(CH{=}CHCH_2)_3(CH_2)_5COOH$$

Eicosatrienoic acid

$$CH_3CH_2(CH{=}CHCH_2)_5(CH_2)_2COOH$$

Eicosapentaenoic acid

The prostaglandins are a large family of hydroxy-fatty acid compounds found in animal tissues. They possess interesting biochemical actions in tissue function at the cellular level. Recently, varieties of the Gorgonian coral, *Plexaura homomalla*, have been found to contain significant amounts of these compounds and can be used as a natural source (Weinheimer and Spraggins, 1969; Schneider *et al.*, 1972; Hinman, 1972). The discovery of these highly active compounds in animal tissues by Kurzrock and Lieb dates back to 1930 but it was not until some twenty years later that their existence was confirmed by Bergstrom and Sjovall (1952). These workers elucidated the chemical structure in 1960; subsequently Corey *et al* .(1968) and Schneider *et al.* (1968) achieved the total synthesis of several prostaglandins including their analogues.

Chemically, the prostaglandins (PG) in (V) are C_{20} unsaturated oxygenated fatty acids which are biosynthesized from C_{20} straight-chain fatty acids such as arachidonic and eicosatrienoic acid. They are found in tissues of many animals including mammals and amphibia.
They are extremely potent compounds and exhibit a variety of biological activities.* Their medical applications are now at the stage of extensive

* For an interesting summary of the progress of research with the prostaglandins, see "Research in Prostaglandins," Worcester Foundation, Shrewsbury, Massachusetts, U.S.A. 01535 (1972) and *Chemical and Engineering News*, October 16, 1972.

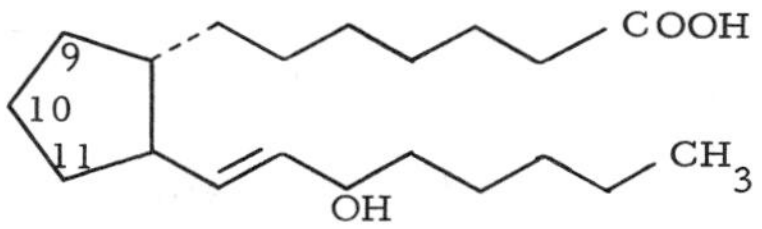

PGA series 9: = O, Δ^{10}
PGE series 9: = O
PGF series 9: = OH

(**V**) Mammalian prostaglandins

clinical trials in such areas as reproductive biology, i.e. fertility control, labour inductions and abortion; also in certain renal pathology and the treatment of intestinal ulcers.

Unfortunately, their synthesis in mammalian tissues of land animals is too limited to provide sufficient quantities from these sources for medicinal use. Hence, it has been more practical to synthesize it. Recently, the Gorgonian corals have been found to contain prostaglandins in amounts suitable for chemical isolation. In 1969, Weinheimer and Spraggins isolated a prostaglandin, 15 epi (R)-PGA_2, its acetate and methyl ester, in high yields (totalling 1·5% of the dry weight) from soft coral harvested off the Florida coast.

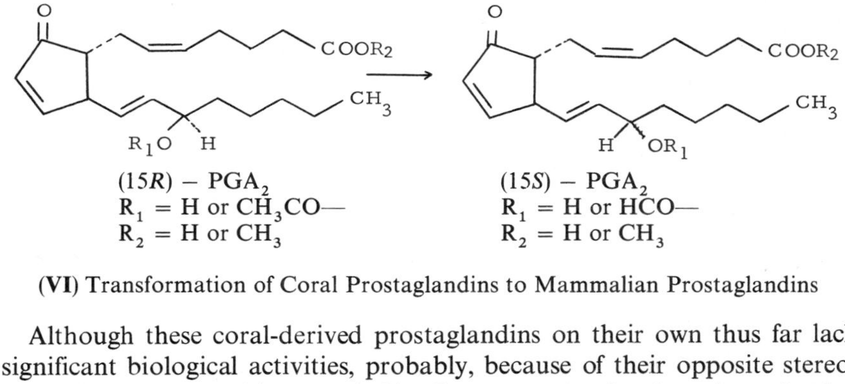

(15*R*) – PGA_2
R_1 = H or CH_3CO—
R_2 = H or CH_3

(15*S*) – PGA_2
R_1 = H or HCO—
R_2 = H or CH_3

(**VI**) Transformation of Coral Prostaglandins to Mammalian Prostaglandins

Although these coral-derived prostaglandins on their own thus far lack significant biological activities, probably, because of their opposite stereoisomerism at C-15, this remarkable discovery stimulated an investigation of the possibility of synthesizing bio-active prostaglandin from those present in marine animals. A number of laboratories have succeeded in isomerizing the 15(*R*) hydroxyl group to the active 15(*S*)-forms. For example, Spraggins (1972) reported that the treatment of the acetate, methyl diester of 15 *epi*-PGA_2 with a formic acid–potassium formate buffer at 25°, followed by hydrolysis, produced a biologically active PGA_2 methyl ester among other products. Bundy *et al.* (1972) demonstrated that (15*R*)-PGA_2 can be converted via several steps to PGE_2 methyl ester and the extremely active prostaglandin PGF_2 (**VII**).

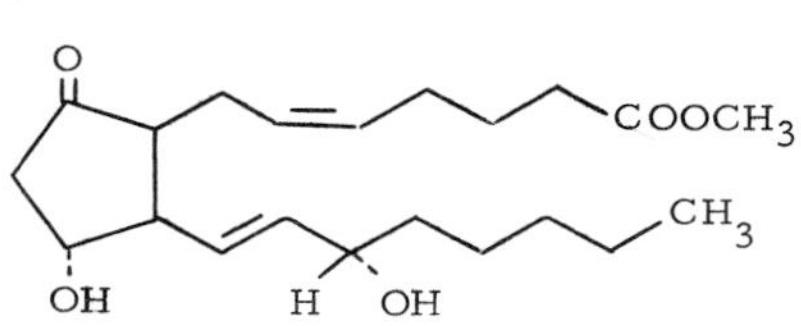

PGE = Methyl ester

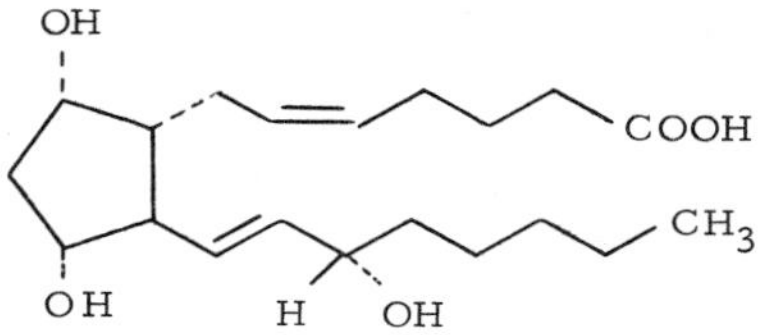

PGF_2
(**VII**)

Prostaglandin derivatives

In addition, Schneider *et al.* (1972) showed that some varieties of *P. homomalla* collected from various locations in the Caribbean contain the active prostaglandins, (15*S*)-PGE_2, (15*S*)-PGA_2 as their acetate methyl esters and the new 5,6-*trans* isomer of (15*S*)-PGA_2 acetate methyl ester (**VIII**). Light and Samuelsson (1972) also reported similar findings.

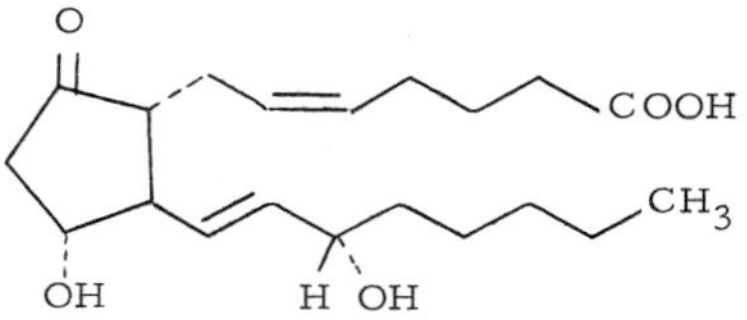

(15S) – PGE_2

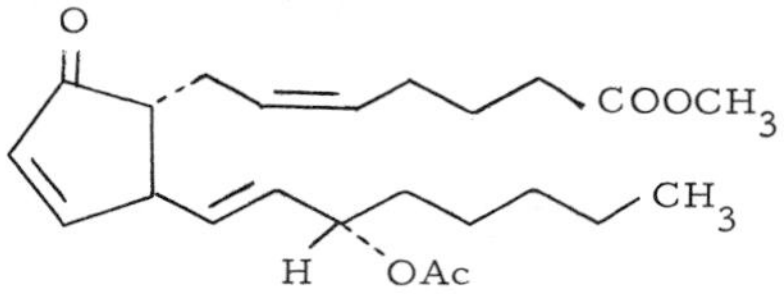

(15S)—PGA_2 acetate methyl ester

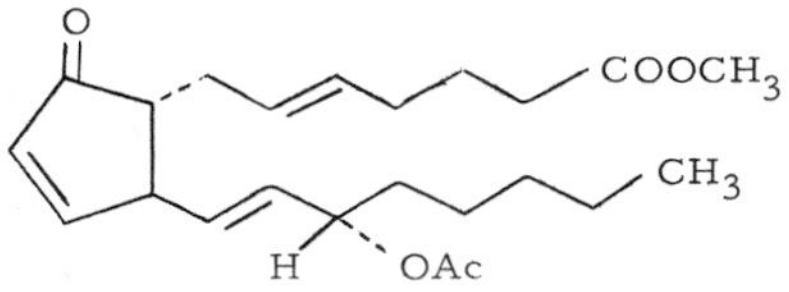

5,6-*trans*-(15S)-PGA_2 acetate methyl ester
(**VIII**)

Prostaglandins from Gorgonian, *Plexaura* corals

These findings indicate that coral can, in fact, be used as a raw material for the production of prostaglandins. Hinman (1972) has shown that with proper controls, extensive harvesting of coral for prostaglandins is economically feasible. Furthermore, he has demonstrated that the coral, *Plexaura homomalla*, will redevelop rapidly if at least 25% of the colony is left intact at time of harvest.

As noted above, the prostaglandins exhibit a wide variety of pharmacological effects. When injected, they affect many physiological processes, including blood pressure, lipolysis, gastric secretion and the function of circulating platelets. The specific effect is dependent upon the precise structure of the prostaglandin administered. The mechanism of the action is still under investigation, but it is said to be related to the production of cyclic 3′,5′-adenosine monophosphate (cAMP). The clinical areas in which prostaglandins have been implicated include induction of labour and abortion. An infusion of PGE_2 or PGF_2 is effective in terminating pregnancy or inducing labour in the pregnant woman. However, side effects, nausea, migraine, diarrhoea, inflammation etc., were also reported during prolonged infusion of the prostaglandin compounds used. Prostaglandins are also effective in alleviating bronchoconstriction in asthmatics and nasal decongestion and are also used as gastric antisecretory and antihypertensive agents.

Since the structures of the naturally occurring prostaglandins and their actions are so diverse, and because they are ubiquitously distributed in the body, these compounds are of considerable interest as potential therapeutic agents.

A group of bromine containing compounds, which are related to C_{16} straight-chain carboxylic acids, has been isolated from the red algae, *Laurencia* spp., by Japanese workers. Laureatin (**IX**) from *L. glandulifera* was shown to be a cyclic ether with a bromine atom and an acetylenic bond (Irie *et al.* 1968).

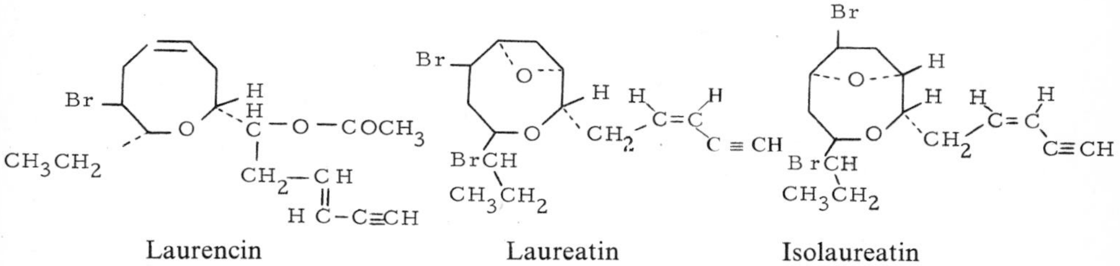

Laurencin Laureatin Isolaureatin

(IX)

A closely related algae, *L. nipponica*, yielded laureatin and isolaureatin of similar structures but with an additional bromine atom and epoxy group. (Irie, *et al.*, 1970). Little is known about their biological activities and future reports of results of their use as antibacterial, antiviral or other medicinal agents must be awaited.

A number of bromophenols isolated from marine algae have been shown to possess some antibacterial activity, although none has, as yet, reached the stage of clinical evaluation. In 1949, Mastaglin and Augier reported a bromophenol sulphate of composition $C_6Br_2(SO_3K)_2(OH)(COOH)$ in *Polysiphonia sp.*, but its structure was not established. Several years later Saito and Ando (1955) isolated another bromophenol from *Polysiphonia morrowii*, and showed its structure to be 5-bromo-3,4-dihydroxybenzaldehyde. Since that time, a number of other bromophenols have been isolated from marine algae. Craigie and his co-workers isolated the disulphate of 3,4-dibromo-5-hydroxymethylcatechol from *P. lanosa* (Craigie, 1965; Hodgkin, *et al.*, 1966). The same group also identified 3,4-dibromo-4-hydroxybenzyl alcohol in *Odonthalia dentata* and *Rhodomela conferoides* (Craigie and Gruenig, 1967). These phenolic compounds are considered to have a close connection with the antimicrobial activity found in algal extract (e.g. Burkholder *et al.*, 1960). The antifouling properties of brown algae are also attributed to polyphenols (Sieburth, 1964; Sieburth and Conover, 1965). Sargenin, a substance similar to these, and obtained from *Sargassum notans*, showed antibiotic activity toward many microorganisms (Martinez-Nadal, *et al.*, 1965).

23.4 Steroids and Terpenes

Generally speaking, marine animals contain cholesterol and related steroids similar to those found in land animals, but some of the marine animals of lower phyla contain specific compounds such as poriferasterol, spongosterol and gorgosterol which differ structurally from those found in terrestrial organisms. In some marine green algae, isofucosterol and sitosterol are the major sterol components. Fucosterol and its C-20 epimer sargasterol are the main steroids found in many marine brown algae, e.g. *Fucus*, *Sargassum* and *Laminaria* species. Surprisingly, the major steroid of red algae was shown to be cholesterol, a zoosterol seldom found in terrestrial plants (Tsuda *et al.*, 1958; Meunier *et al.*, 1969).

Reiner *et al.* (1962) reported that fucosterol and sargasterol and mixtures of these marine algal sterols produced a hypocholesterolemic action in chicks which were fed with diets containing cholesterol. The same group (1960) reported that an unsaponifiable fraction from clams also lowered blood cholesterol levels. This hypocholesterolemic property is probably due to the steroids such as 24-methyl(ene) and 24-ethyl(ene) cholesterols, which have extra carbons on the side-chain. The same action was observed to a lesser degree with a phytosterol, β-sitosterol, found in terrestrial plants. Recently, Idler and co-workers (1970) isolated 22-*trans*-24-nor-cholesta-5, 22-dien-3β-ol **(XI)** from the sea scallop, *Placoptecten magellanicus*. They

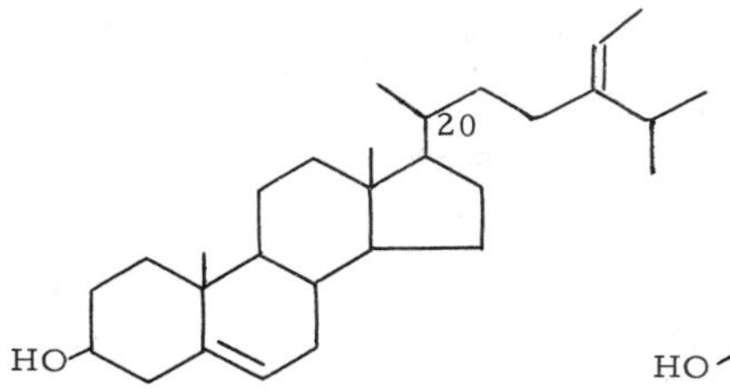

HO 20

Fucosterol (20β)
Sargasterol (20α)

(X)

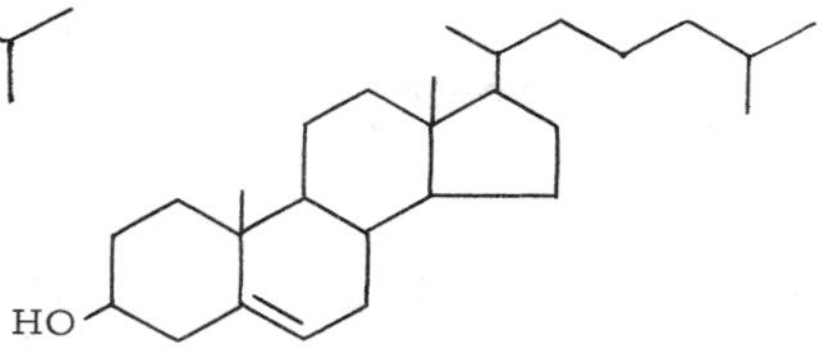

HO

Cholesterol

HO

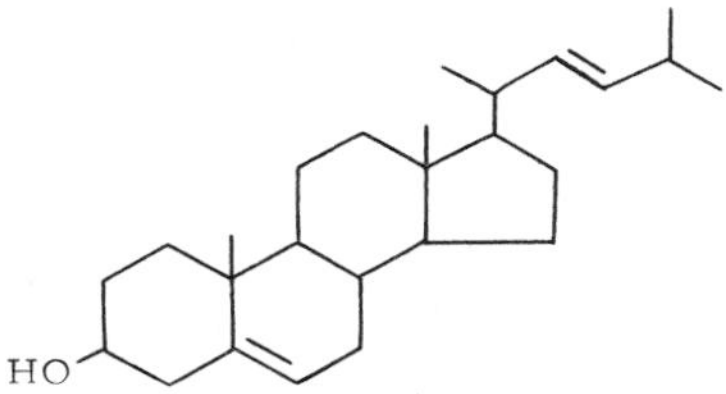

22-*trans*-24-norcholesta-5,22-dien-3β-ol

(XI)

also reported the presence of this unusual steroid in other molluscs, crustaceae and red algae. The structure of the compound was confirmed by synthesis (Fryberg *et al.*, 1971).

It is not yet known whether the 24-norsteroid has any hypocholesterolemic activity, or indeed whether it possesses any pharmacological activity at all.

Another class of steroids with an unusual cyclopropane side-chain has recently been reported. Gorgosterol, first isolated from a Gorgonia coral, *Plexaura flexuosa*, by Bergmann *et al.* (1943) was shown by Hale *et al.* (1970) to be a new type of C_{30} steroid. The unique feature of this structure is the extra carbon atom in the side chain forming a new type of cyclopropane ring (**XII**). Acansterol first recognized in the Crown of Thorns (*Acanthaster planci*) by Gupta and Scheuer (1968) was shown by Sheikh *et al.* (1971) to be the Δ^7 analogue of gorgosterol. Weinheimer and his coworkers examined the steroids from the Gorgonian, *Pseudopterogorgia americana*, and discovered a 9,11-secogorgosterol derivative which is a possible metabolite of gorgosterol (Weinheimer, 1971). 23-Desmethylgorgosterol was also isolated from Gorgonians by Schmitz and Pattabhiraman (1970).

Biological properties of these cyclopropane-containing steroids have not been reported. However, their presence in marine organisms suggests that

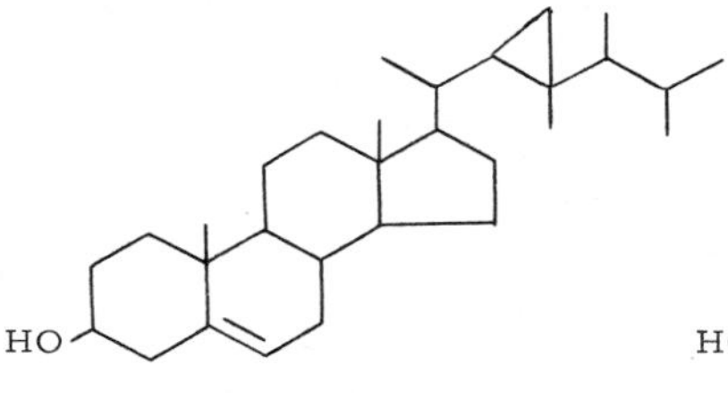

HO HO H

Gorgosterol 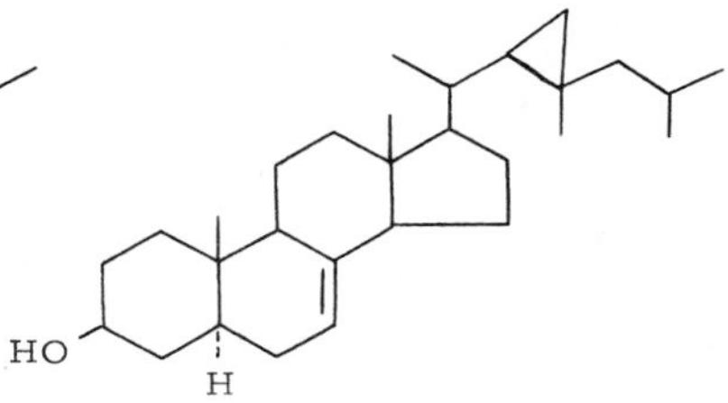 Acansterol

(**XII**)

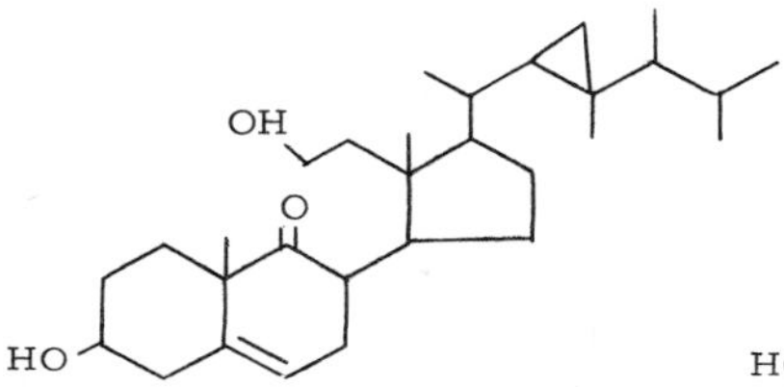

OH O HO HO

9,11-Secogorgost-5-en-3,11-diol-9-one 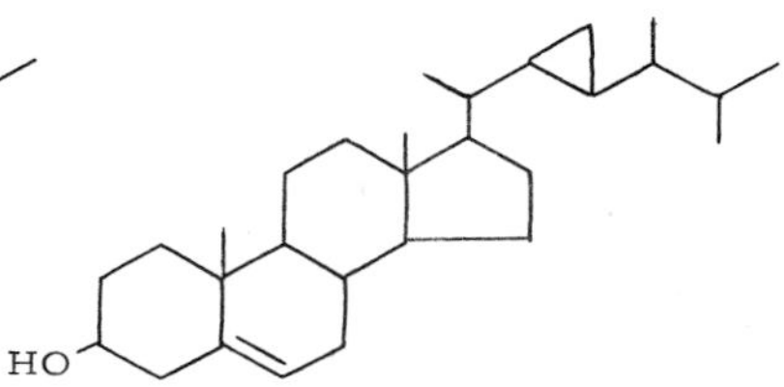 23-Desmethylgorgosterol

(**XIII**)

they are involved in some manner with the biosynthesis of steroids and in cellular metabolism (Hale *et al*., 1970). The possible use of marine steroids in producing hypocholesterolemic activity in humans awaits further clinical trial.

The oxygenated cholestanes, ecdysones, are extremely important and potentially useful compounds isolated from marine crustaceae. These hormone-like compounds are responsible for the moulting of insects, crayfishes and other crustaceae, and appear to play a role in insect metamorphosis. Hampshire and Horn (1966) isolated the active substance from the crayfish, *Jasus Islandei*, in 1966 and named it crustecdysone (=β-ecdysone, ecdysterone). Its structure was shown to be 20-hydroxyecdysone (Horn *et al.* 1968). Interestingly enough, crustecdysone has been found in some land plants in large amounts, although its presence in marine algae has so far not been detected (Takemoto *et al.* 1967).

Deoxyecdysone was isolated by an Australian group (Galbraith *et al*., 1968) from crayfish (200 μg $(1000\ kg)^{-1}$) and was found to be 2-deoxycrustecdysone, an isomer of ecdysone. Since ecdysones exhibit hormone-like activity in the moulting of crustaceans, e.g. crabs, shrimps and lobsters, their pharmacologic properties are the focus of further study, as is also their potential application in mariculture as growth regulators and in the control of the life cycles of insects.

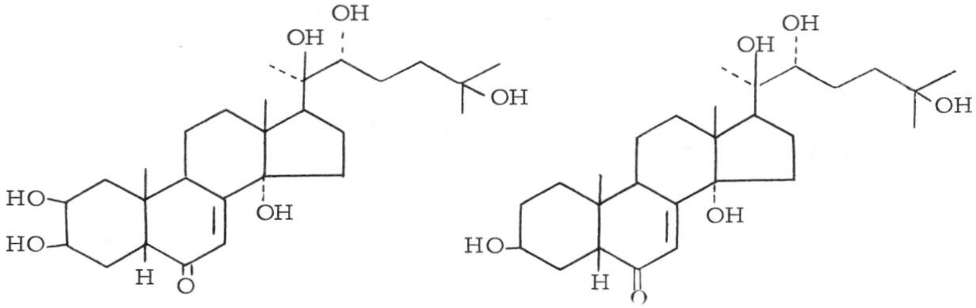

Crustecdysone Deoxycrustecdysone

(XIV)

The glycosidic steroids, holothurins, from sea cucumbers, and starfish saponins, asterosaponins, have a wide variety of pharmacological activities. The term "holothurin" was proposed by Yasumoto *et al.* (1967) for sea cucumber saponins in general and, therefore, does not identify a single constituent common to the sea cucumbers of which there are a number of genera and species. Species of *Holothuria*, *Actinopyga*, *Stichopus*, *Cucumaria* and *Pentacia* are among the common sea cucumbers used as sources for holothurin saponins.

Evidence for the presence of these steroid saponins in the body wall of the sea cucumber was originally presented by Yamanouchi (1955), Chanley *et al.* (1955) and Nigrelli *et al.* (1955). Later reports (Nigrelli and Jakowska, 1960; Sobotka *et al.*, 1964; Friess and Durant, 1965; Friess *et al.*, 1968; Shimada, 1969) have described the haemolytic, neuromuscular cytotoxic, antifungal, cardiotonic and antitumour effects of the holothurin components.

Chemically, the holothurin mixture is soluble in water and alcohol, is heat stable and nondialyzable; the aglycone (holothurinogenin) usually possesses a steroid structure to which are attached sulphate residues and numerous sugar moieties.

Holothurins from *Actinopyga agassizi* yield holothurin A, which contains a sulphate group and sugars moieties, which have been identified as D-glucose, 3-*O*-methylglucose, quinovose (6-desoxyglucose) and xylose. The aglycone holothurinogenins, have been assigned lanostane structure with 7,9(11)-diene and a lactone ring (Chanley *et al.* 1966).

The original holothurin A, however, lacked the diene system. The aglycones were considered to have a sulphate group at the 7-position and to be artifacts formed during acid hydrolysis. This structure, which can easily be converted

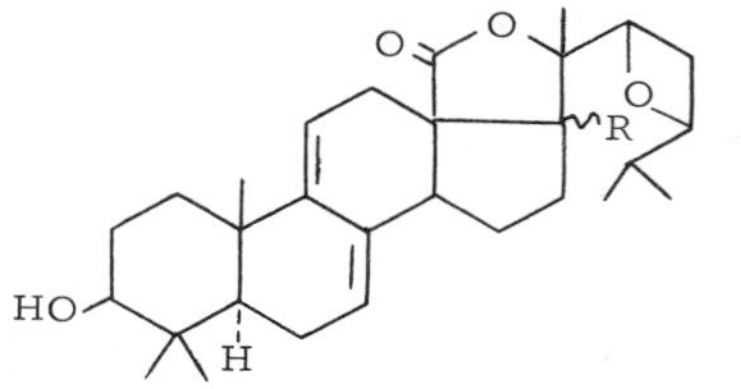

R = OH 22,25-Oxidoholothurinogenin
R = H 17-Deoxy-22,25-Oxidoholothurinogenin

(XV)

to the diene system, was tentatively assigned to the holothurins (see Baslow, 1969). Later, however, this was corrected by Chanley and Rossie (1969) who hydrolysed desulphated holothurin using a snail enzyme mixture to obtain 22,25-oxido-neoholothurinogenin. The neoholothurinogenin has an allylic alcohol structure, $\Delta^{9,11}$-12-ol, which upon acid treatment, readily gives the diene system.

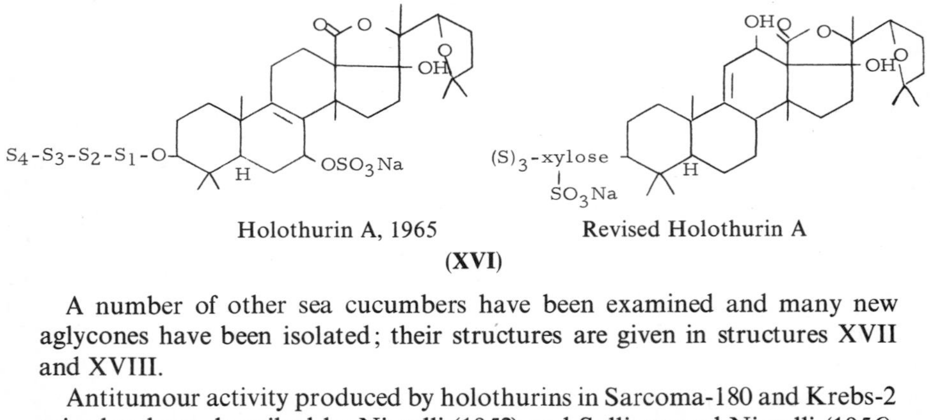

S_4-S_3-S_2-S_1-O— ... OSO_3Na (S)$_3$-xylose — SO_3Na

Holothurin A, 1965 Revised Holothurin A

(XVI)

A number of other sea cucumbers have been examined and many new aglycones have been isolated; their structures are given in structures XVII and XVIII.

Antitumour activity produced by holothurins in Sarcoma-180 and Krebs-2 mice has been described by Nigrelli (1952) and Sullivan and Nigrelli (1956). Shimada (1969) reported that "holotoxin" isolated from the body wall of a Japanese sea cucumber, *Stichopus japonicus*, possesses strong antifungal activity in the clinical treatment of fungal infections.

After the discovery of holothurins in sea cucumbers, Hashimoto and Yasumoto (1960) observed the presence of similar compounds (asterosaponins) in the starfish, *Asterias amurensis* and *Asterina pectinifera*. These compounds have pharmacological characteristics of saponins, i.e. haemolysis, toxicity to fish etc. (Rio *et al.*, 1965; Yasumoto *et al.*, 1964). Their cytotoxicity to tumour (KB) cells was observed by Nigrelli *et al.* (1967). Starfish and its

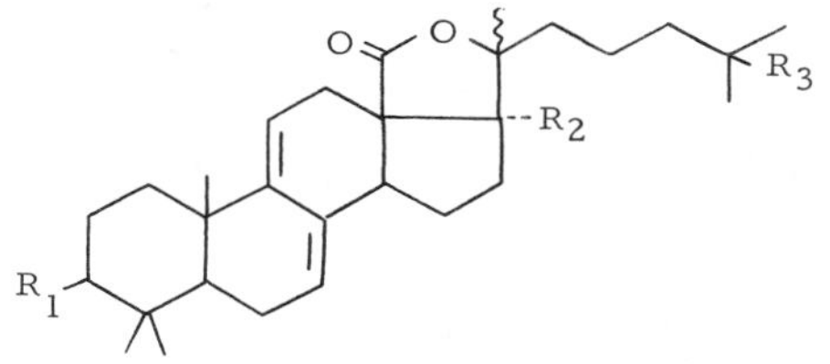

	Origin
Seychellogenin: R_1 = OH R_2 = H R_3 = H	*Bohadschia koellekari* (Roller *et al.*, 1969)
Koellikerigenin: R_1 = OH R_2 = H R_3 = OH	*B. koellekeri* (Roller *et al.*, 1969)
Tiernaygenin: R_1 = OH R_2 = H R_3 = OCH_3	*B. koellekeri* (Roller *et al.*, 1969)
Praslinogenin: R_1 = OH R_2 = OH R_3 = OCH_3	*B. koellekeri* (Tursch *et al.*, 1970)
Griseogenin: R_1 = OH, 22-OH	*Halodeima gaisea* (Tursch *et al.*, 1967)

(XVII)

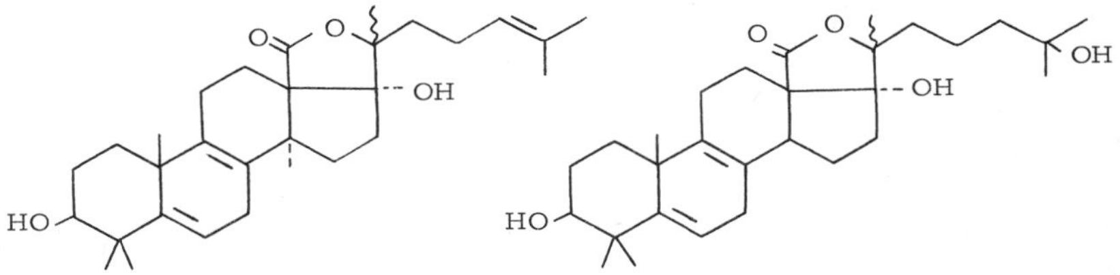

Stichopogenin A_2 *Stichopus japonicus* and Stichopogenin A_3 *S. japonicus* (Elyakov *et al.*, 1969)

(XVIII)

extracts have been used in parts of Japan to kill fly maggots, and Hashimoto and Yasumoto (1960) reported that these saponins act as surfactant agents in preventing the ecdysis of maggots.

The saponin from the starfish. *A. amurensis*, was fractionated to yield asterosaponins A and B, both of which contain a sulphate group, D-quinovose and fucose. The structure of Asterosaponin B has, in addition, D-xylose and D-galactose moieties (Yasumoto and Hashimoto, 1965). Similar findings were reported for other Pacific and Atlantic starfish species (Rio *et al.*, 1963, 1965; Ruggieri, 1965).

In an attempt to isolate the starfish compound which causes the avoidance reaction to molluscs (e.g. *Buccinum undatum*), Mackie and co-workers (1968) isolated "saponin-like" substances from *Asterias rubens* and *Marthasterias glacialis*. The same group partially characterized the compounds reporting the presence of glucose, quinovose, fucose, a sulphate group and a C_{27}

aglycone–marthasterone (Mackie and Turner, 1970). A tentative structure (**XIX**) containing a 23-keto group has recently been proposed for dihydro-marthasterone (Turner *et al.*, 1971).

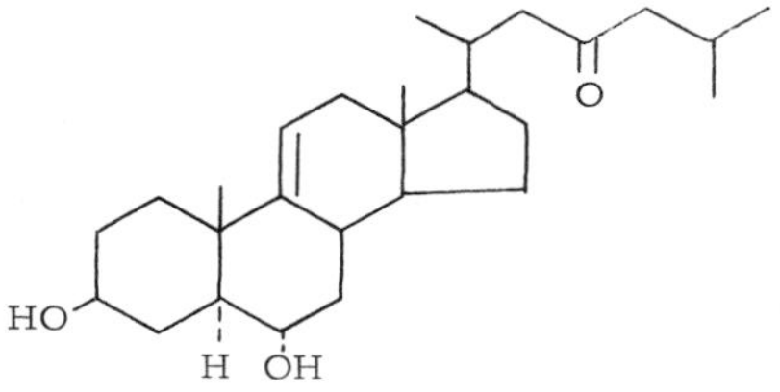

(Dihydro)marthasterone

(**XIX**)

Shimizu* (1970, 1971) found antiviral substances in the crude extracts of three starfish species, *Asterias forbesi* (Atlantic), *Acanthaster planci* (Pacific) and *Asterina pectinifera* (Japan Sea). These were active in the antiviral test using influenza virus B in the chicken embryo, and the compounds, when purified, were shown to be saponin-like substances. The chemical structure of one ASP-I from *A. pectinifera* possesses sugar moities, a sulphate group and a cholestane type aglycone, $C_{27}H_{46}O_4$ with four hydroxyl groups in the molecule.

The compound from *A. planci* (the Crown of Thorns found, for example, on the Great Barrier Reef of the South Pacific) was shown to have a sulphate group and a molecular weight of *ca.* 1500. Acid hydrolysis gave quinovose and fucose in about a 2:1 molar ratio, and a mixture of aglycones (Shimizu, 1972a). One of the aglycones has the molecular formula $C_{21}H_{32}O_3$. This compound was established to be 3β,6α-dihydroxy-5α-pregn-9(11)en-20-one almost simultaneously by four independent groups (Sheikh *et al.*, 1972a; Ikegami *et al.*, 1972a; Shimizu, 1972a; ApSimon and Buccini, 1972).

Sheikh and co-workers also isolated the saponin 3β,6α-dihydroxy-5α-cholesta-9(11), 20(22)-dien-23-one, a double bond isomer of marthasterone from the hydrosylate of extracts of *A. planci.* The same compound and its deconjugated form were also recognized in the hydrolysate of the saponin from *A. planci* and *Ast. forbesi* (Shimizu, 1972b). The unexpected occurrence of the pregnane derivative in the hydrolysate prompted various speculations about the origin of the compound. Shimizu (1972b) noted that the purified

* This research is part of an extensive *Drugs from the Sea* programme sponsored by the National Sea-Grant Program and conducted at the University of Rhode Island, Kingston, R.I., U.S.A.

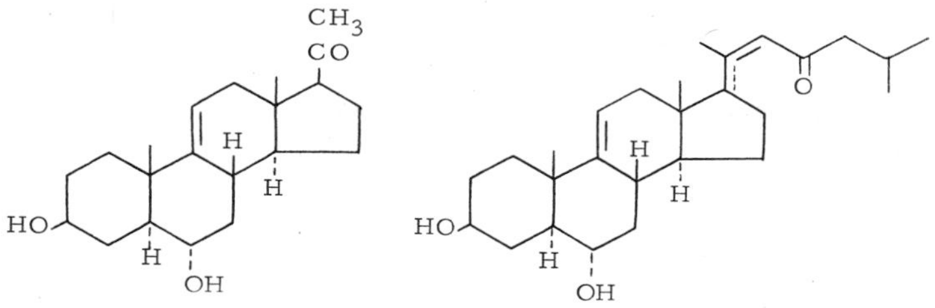

$3\beta,6\alpha$-dihydroxy-5α-pregn-9(11)-en-20-one

$3\beta,6\alpha$-dihydroxy-5α-cholesta-9(11), 20(22)-dien-23-one

(XX)

A. planci saponin lacks u.v. absorption maxima and that absorption appeared only after acid treatment. He suggested that this "preasterosapogenin" has the structure of 3β,6α,20-trihydroxy-5α-cholest-9(11)-en-23-one; upon treatment with acid it is dehydrated to the conjugated compound (or the deconjugated form) or undergoes a retro-aldol type cleavage to give the pregnane derivative.

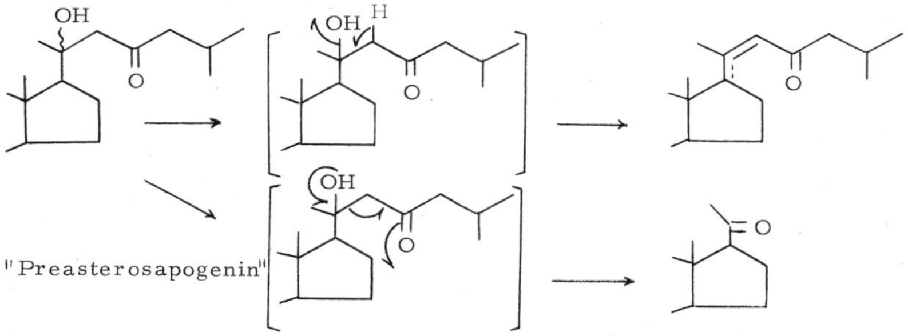

"Preasterosapogenin"

(XXI)

In support of the above speculation, two new aglycones, which were apparently formed by the dehydration of 20-hydroxyl groups, were subsequently isolated as the minor aglycones from *A. planci* (Sheikh *et al.*, 1972b).

Friess *et al.* (1968) reported that asterosaponins A and B produce an irreversible blockage of cholinergic neuromuscular transmission in mice. The saponins inhibit spawning by starfish by counteracting the spawning stimulating substances (GSS) (Ikegami *et al.*, 1972b). This finding is in agreement with the observation that, during summer months, the asterosaponins are found in high concentrations in starfish, whereas during winter, or at time of spawning, they are only present in extremely small amounts (Yasumoto *et al.*, 1966).

Di- and sesquiterpenes with antibacterial activities have been isolated from marine organisms. Among examples of these are eunicin, crassin, aplysin and laurencin. Eunicin is obtained from a gorgonian coral, *Eunicea mammosa* (Ciereszko *et al.*, 1960). A peculiar structure with a cembranolide skeleton, and 14-membered ring with an oxygen bridge (**XXII**), was reported for this compound by Weinheimer and co-workers (1968). This structure was confirmed by the X-ray crystallography of the iodoacetate (Hossain *et al.*, 1968).

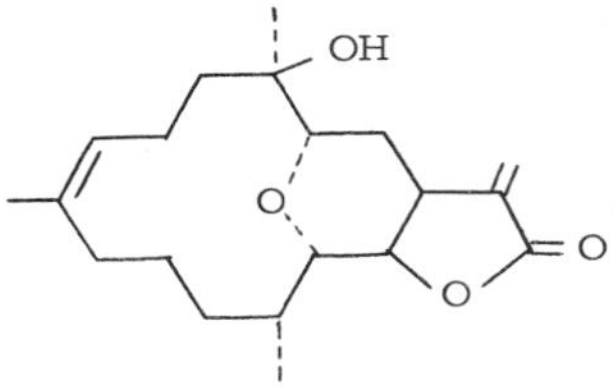

Eunicin

(**XXII**)

Crassin acetate, a diterpene unsaturated polycyclic lactone isolated from the horny corals of *Pseudoplexaura* species by Ciereszko *et al* (1960) also has antibacterial activity. Eunicin and crassin acetate are toxic to the organism, *Endamoeba histolytica,* which causes dysentery in humans. These compounds also inhibit the growth of *Staphylococcus aureus* and *Clostridium* species.

The sea hares, *Aplysia* sp, are known to excrete irritating substances and these have attracted attention because of their toxic neuromuscular effects in mice which result in paralysis. In 1963, Yamamura and co-workers isolated the bromine containing sesquiterpenes, Aplysin and Aplysinol, from a Japanese species, *Aplysia kurodai.* They also found a debrominated derivative, debromoaplysin, in the same animal.

Aplysin X = Br, Y = H
Aplysinol X = Br, Y = OH
Debromoaplysin X = Y = H

Laurinterol X = Br, R = H
Debromolaurinterol X = R = H

(**XXIII**)

Later, Irie and co-workers (1965) isolated a number of structurally related compounds from red algae of the *Laurencia* sp. These include the cyclopropane derivatives: laurinterol, isolaurentinol and debromolaurinterol from *L. intermedia* (Irie *et al.*, 1966). Aplysin, laurinterol, aplysinol and debromoaplysin have been isolated from *L. okamurai* (Irie *et al.*, 1969a) and laurenisol from *L. nipponica* (Irie *et al.*, 1969b).

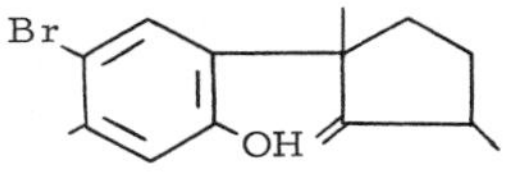

Isolaurentinol

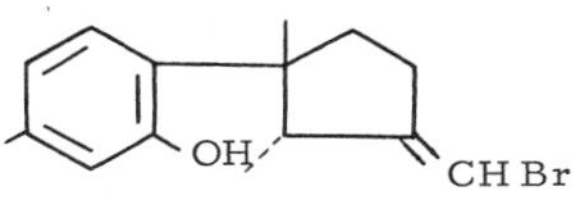

Laurenisol

(**XXIV**)

The chemical resemblance of the compounds to each other is obvious, and, moreover, some of the sea hares feed on the *Laurencia* species. The pharmacological activities of these terpenoids, other than their antibacterial effects, have not been fully established. *Aplysia* and *Laurencia* species are good sources of halogen containing compounds. For example, Matsuda *et al.* (1967) reported a bromine-containing diterpene, aplysin-20 (**XXV**) in the same species and in reasonable concentrations. However, the presence of this halogen in a drug molecule for antibacterial use could also cause undesirable effects in human tissues.

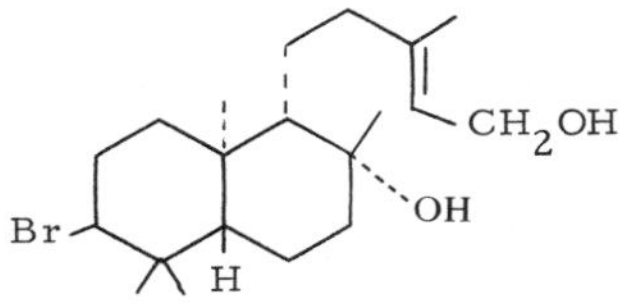

Aplysin-20

XXV

Two fungicidal components, zonarol and isozonarol, were isolated from the brown "iridescent" seaweed, *Dictyopteris zonarioides*, by Fenical *et al.* (1972). The structure of zonarol contains a hydroquinone moiety and a residue of terpenoidal origin. Isozonarol is the double bond isomer of zonarol.

The corresponding quinones, zonarone and isozonarone, were also isolated from the extract. All these compounds were found to be effective against the growth of fungi on plants, but to be devoid of significant antibacterial activity in humans.

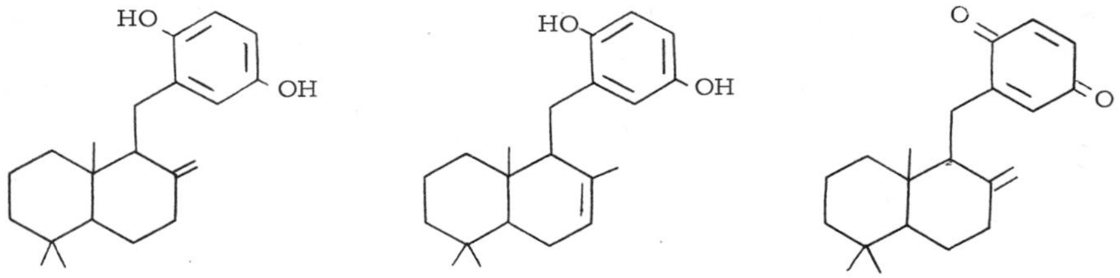

Zonarol Isozonarol Zonarone

(XXVI)

23.5 Nitrogenous Compounds

Like most natural products which contain nitrogen, those of marine origin exhibit many kinds of pharmacological activities and chemical structures.

A large number of simple amines, choline derivatives, betaines, creatinines, and guanidines have been found in marine animals from the lower invertebrates and anthropods to vertebrates. For example, the simplest quaternary amine, tetramethylammonium hydroxide (tetramine) was first discovered by Ackermann *et al.* (1923) in the sea anemone, *Actinia equina*, and was later found in many coelenterates. The strong paralytic activity produced by sea-anemones and some molluscs is associated with this compound.

Homarine is another significant quaternary amine found in many marine invertebrates. Details of its distribution and functions are discussed by Gasteiger *et al.* (1960). Homarine and related compounds are reported to decrease the heartbeat and, at high concentrations, cause cardiac arrest.

Another class of substances associated with the paralytic activity are the choline esters found in the hypobronchial glands of molluscs. Whittaker (1960) examined a number of species of molluscs and found urocanylcholine (murexine), senecioylcholine and arylcholine in extracts of these organisms. These are listed in Table 23.3.

Urocanylcholine has a strong paralytic activity (nicotinic and curare type) and is used as an experimental muscle relaxant.

Nereistoxin insecticide was first recognized by Nitta (1934) in a Japanese clam worm, *Lumbriconereis heteropoda.* The dead bodies of this worm have been known for centuries to repel and kill flies. Okaichi and Hashimoto (1962) reinvestigated the phenomenon and isolated nereistoxin, the active component, as a crystalline salt. This is a sulphur-containing simple amine derivative with the structure 4-*N*,*N*-dimethylamino-1,2-dithiolane.

Nereistoxin, as well as many of its analogues, can be synthesized by several

TABLE 23.3
Choline esters present in molluscs

Species	Choline Esters
Murex trunculus	Urocanylcholine (murexine)
M. brandaris	Urocanylcholine (murexine)
M. fulvescens	Urocanylcholine (murexine)
Toritonalia erinaceau	Urocanylcholine (murexine)
Urosulpinx cinereus	Urocanylcholine (murexine)
Thais lapillus	Urocanylcholine (murexine)
T. floridana	Senecioylcholine
Buccinum undatum	Acrylcholine

$(CH_3)_4N^+OH^-$

Tetramine

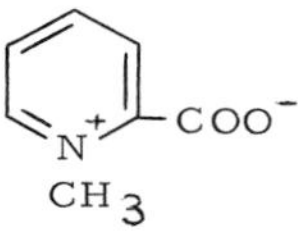

COO^-, N^+, CH_3

Homarine

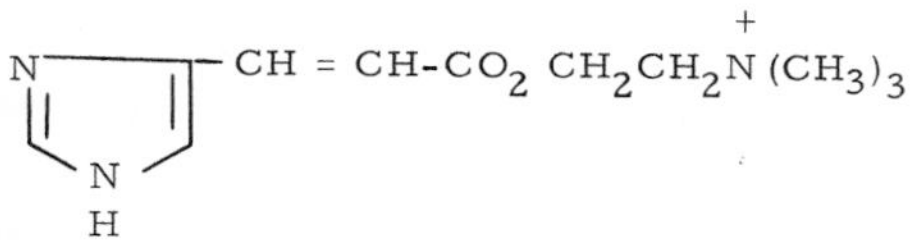

N, N, H, $CH=CH\text{-}CO_2CH_2CH_2\overset{+}{N}(CH_3)_3$

Urocanylcholine

H_3C, H_3C, $C=CH\,CO_2CH_2CH_2\overset{+}{N}(CH_3)_3$

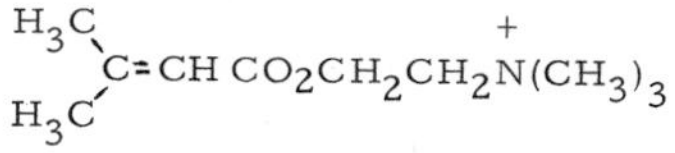

Senecioylcholine

$H_2C{=}CHCO_2CH_2CH_2\overset{+}{N}(CH_3)_3$

Acrylcholine

(XXVII)

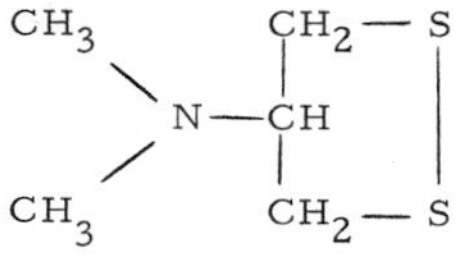

CH_3, CH_3, N—CH, CH_2—S, CH_2—S

Nereistoxin

CH_3, CH_3, N—CH, CH_2— S —CO NH_2, CH_2— S —CO NH_2

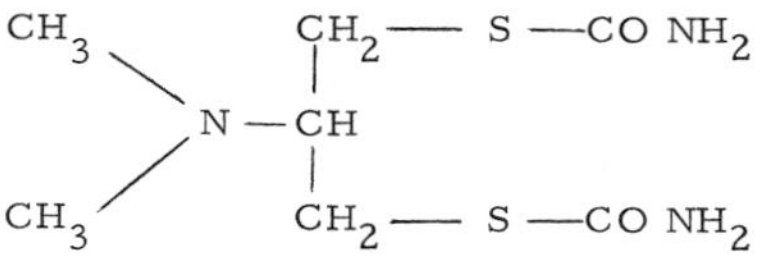

Cartap®, Padan®

(XXVIII)

routes (Hashimoto *et al.*, 1972a). One of the derivatives, dihydronereistoxin dicarbamate hydrochloride (1,3-bis(carbamoylthio)2-*N*,-dimethylamine propane) has been marketed in Japan as Cartap® and Padan® since 1967 as an insecticide to replace DDT and BHC.

Nereistoxin produces a marked ganglionic blocking action on the central nervous systems of insects (Sakai, 1967). It is used extensively in Asia and Japan against rice stem borers. However, the action of the compound as a cholinesterase inhibitor is weak, unlike that of chlorinated hydrocarbons or phosphate derivatives used as insecticides. The use of this compound as a neuroactive drug for humans is in the investigative stage.

Another example of an active compound, isolated from the marine annelid, is anabaseine, which is closely related to the tobacco alkaloids, nicotine and anabasine. This compound was first found in the annelid *Paranemertes*

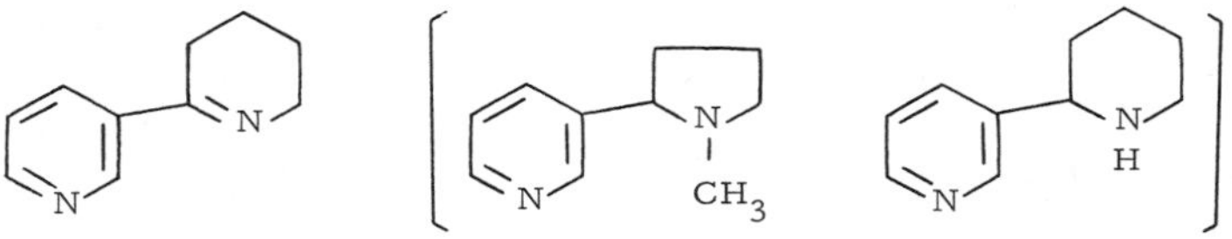

Anabaseine Nicotine Anabasine

(XXIX)

peregrina as a toxic component of the worm (Kem *et al.*, 1969). The presence of the same compound was also confirmed in *Hoplonemertinea* by Kem *et al.* (1971) and it is probably responsible for the insecticidal principle of these sea worms. The algae *Digenea* is also widely distributed in the subtropical and tropical waters of the Mediterranean, Red Sea and Indian Ocean. Kainic acid, its anthelmintic principle, was purified by Murakami *et al.* (1953) and the structure was shown to be that of the amino acid, L-α-kainic acid. Its name was derived from the Japanese name for the alga, "Kaininso."

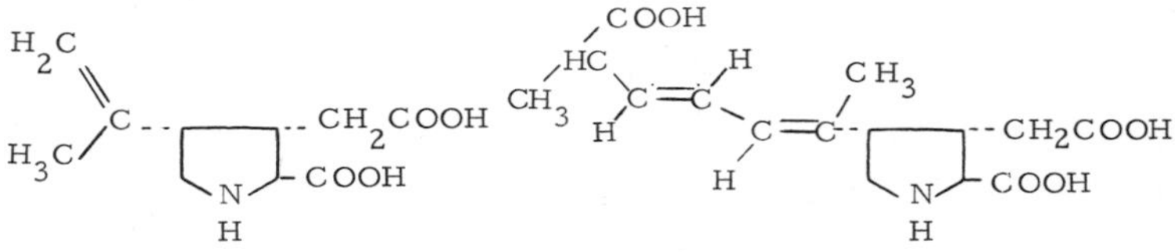

L-α Kainic acid Domoic acid

(XXX)

Chondria armata, an alga which belongs to the same family, has also been used for anthelmintic purposes in the southern part of Japan. Its active constituent is domoic acid, an analogue of kainic acid (Takemoto and Daigo, 1958; Daigo, 1959).

Some brown algae and their extracts have been used for centuries as hypotensive drugs in Oriental medicine. Takemoto and his co-workers (1964a) isolated the basic amino acid, laminine, as the active hypotensive principle of *Laminaria angustata.* This was accomplished by ion-exchange chromatography using amberlite IR-120, followed by crystallization of laminine as the oxalate. Laminine possesses a quaternary amino group (trimethyl ammonium) and a L-α-amino acid moiety. Takemoto *et al.* (1964a) assigned the structure of 5-amino-5-carboxypentyl-trimethyl ammonium trimethyl lysine oxalate to laminine oxalate. The basic compound was subsequently synthesized by methylation of the copper complex of lysine and N^2-acetyl-L-lysine (Takemoto *et al.* 1964b). The same group also

$$(CH_3)_3\overset{+}{N}CH_2CH_2CH_2CH_2\underset{^{+}NH_3}{\underset{|}{C}}HCOOH \qquad \left(2\begin{array}{c}COO^-\\|\\COOH\end{array}\right)$$

Laminine hydrogen oxalate

found that laminine is widely distributed in brown algae of the *Laminariaceae* family, in concentrations of up to 0·003% (dry weight). The hypotensive action of laminine is comparable with that of choline also found in extracts of *Laminaria* sp. Baslow (1969) concluded that although the hypotensive action of seaweeds such as *Laminaria* has not been fully elucidated, their choline and laminine constituents may have an important role in producing such action. These substances appear to be potentially useful pharmacological agents.

Pahutoxin from the Hawaiian boxfish "Pahu", *Ostracion lentiginosus*, is an example of a choline-like substance of marine origin. This is excreted by the boxfish as a toxic agent to other fish, probably as a defensive measure. Boylan and Scheuer (1967) isolated a crystalline substance, pahutoxin, from the same fish and established its structure as 3-acetoxyhexadecanoyl choline. This compound is a surfactant and has haemolytic properties.

$$CH_3(CH_2)_{12}\overset{OAc}{\overset{/}{C}}HCH_2COOCH_2CH_2\overset{+}{N}(CH_3)_3$$

Pahutoxin

The soapfish, *Rypticus saponaccus* (*Grammistedae*), and the Pacific bass, *Grammistes sexlineatus*, secrete toxins from mucus producing glands. These toxins differ from those of the boxfish because of their peptide structures. Hashimoto and Oshima (1972) reported the fractionation of the toxins from soapfish, *Pogonoperca punctata* into three peptides, Grammistins A, B and C. The amino acids leucine and phenylalanine were identified as components of these peptides. The distribution in fish and the properties of grammistins are described by Randall *et al.* (1971). The toxins are excreted when the fish is frightened whereupon they produce toxic haemolytic and stuporific effects in predators including other fish and mammals.

It has already been mentioned that low molecular weight amines, such as tetramine, are recognized as mulluxan toxins. Ghiretti (1953, 1960) reported the presence of a larger molecular toxin, cephalotoxin, in *Octopus vulgaris*. Cephalotoxin is acetone soluble and is probably a kind of glucoprotein (Ghiretti, 1960). In addition, a biologically active peptide, eledoisin, was isolated from the salivary glands of octopus, *Eledone moschata* and *E. aldrovandi* (Anastasi and Erspamer, 1963). Eledoisin is composed of eleven amino acids, and the sequence and the composition were shown to be H-Pyr-Pro-Ser-Lys-Asp(OH)-Ala-Phe-Ile-Gly-Leu-Met-NH_2. Eledoisin has, together with other activities, a hypotensive action and a large number of the homologues have been synthesized and tested (Bernardi *et al.*, 1964).

Bergmann and his coworkers have discovered several unusual nucleosides in sponges, e.g. *Cryptotethya crypta* (Bergmann and Feeney, 1950, 1951; Bergmann and Burke, 1955, 1956; Bergmann and Stempien, 1957).

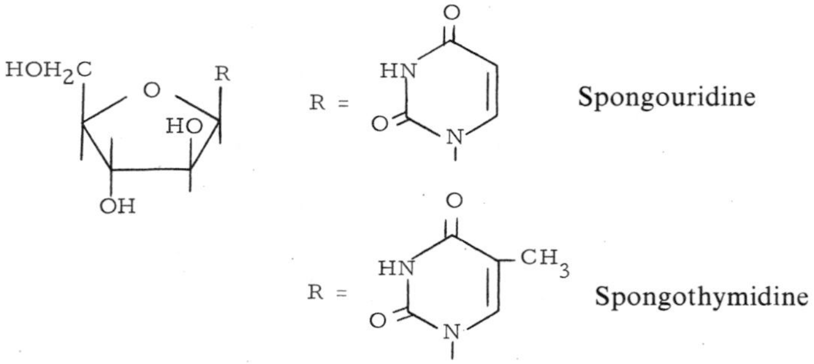

HOH_2C R O HO OH

R = Spongouridine

R = CH_3 Spongothymidine

(XXXI)

These include, spongouridine and spongothymidine which are arabinose-containing pyrimidine nucleosides. These are considered to be the 2′-epimers of the cellular nucleosides. These nucleosides provided models for antitumour nucleic acid antagonists, e.g. arabinosylcytosine.

When sulphated and standardized the protamine of various fish constitutes a useful chemical agent which can be employed as an antidote for the overdosage of the anticoagulant drug *heparin*, and in the preparation of Protamine Zinc Insulin, a long acting antidiabetic drug. This strongly basic simple protein is obtained from the sex cells of fish such as the salmon, trout, herring, carp etc. Fish sperm and mature testes yield protamine in amounts practical for isolation and purification. The amino acid composition of this protein and its molecular weight (ranging from 1700 to 8000) is characteristic of the type of fish used. Usually the name given to the protamine depends upon its origin, e.g. cyprinine from the carp, scombrine from mackerel, staurine from sturgeon, iridine from trout and salmine from salmon.

Because this drug is used intravenously and because of its protamine content, considerable care must be observed in standardizing its preparation. For example, the limits of its nitrogen equivalent content are usually set at 22·5 to 25·5%. The preparation used for injection is assayed on the basis of the amount of the drug heparin sodium, the anticoagulant effect of which is neutralized by a specific volume of standardized reference protamine sulphate injection. A specified amount of phenol, i.e. 0·25%, is permitted in the injection product for bacteriostatic purposes.

Protamine, used in preparing Protamine Zinc Insulin, an official drug described in the U.S., British, and International Pharmacopoeias, is usually from the same source, i.e. trout of the genus *Oncorhynchus* or *Trutta* and/or salmon, *Salmo*. Usually the insulin drug in this preparation contains from 1·0 to 1·5 mg of protamine for each 100 units of Insulin.

As an antidote, protamine, in controlled dosage, combines with heparin so as to render it biologically inactive in so far as its anticoagulant property is concerned. This property of protamine sulphate is also made use of in other haemorrhage conditions where the anticoagulant properties of protamine itself can be used, e.g. as an antithromboplastic or antiprothrombic agent, or as an agent capable of precipitating fibrinogen.

23.6 Antibiotic Compounds From Sponges

Although for many years antibiotic substances derived from sponges have been of interest thus far none has reached the stage of extensive clinical use. Baslow (1969) who reviewed the status of this research up to 1969 concluded that the chemical agents reported to be present in sponges would seem to have no potential use as antibiotics.

There are probably two reasons for this; chemically, the compounds thus far isolated for antibiotic tests are brominated cyclohexadienes and polyhydroxybrominated phenols; some are acetamide derivatives and dihydroxyindoles which possess potential toxicity in humans, and tests on agar plate discs show that although the compounds are often active against a wide range of test organisms *in vitro*, they do not appear to show appreciable inhibition compared with standard antibiotic agents.

Nevertheless, this chapter would hardly present a true picture of the efforts of many investigators into the area of marine drugs if a brief account was not given of certain antibiotic substances isolated from the sponges. The resistance of sponges to bacterial decomposition has long stimulated investigations of possible antibacterial compounds in them. These organisms possess an efficient filtering canal system which enables them to extract chemical substances from the vast amounts of sea water which pass into them to be absorbed onto cell surfaces. These can be subsequently extracted for chemical and biological testing.

Specific compounds with some degree of antibacterial activity *in vitro* and *in vivo* which have been extracted from sponges include ectyonin, aerophysinin, aerothionin, homoaerothionin, dibromophakellin and oroidin. The extracts of more than a hundred sponge species which have been subjected to screening have given positive results (Jakowska and Nigrelli, 1960). Stempien *et al.* (1969) reported the presence of antibacterial substances in extracts of 23 or 125 sponge species collected from Caribbean waters and the Bahama Islands. Several of these showed antibiotic activity to a rather wide spectrum of gram-negative and gram-positive organisms. Antibiotic compounds, such as quinol containing bromine and 2-(3′,5′-dibromo-1′-hydroxy-4′-ketocyclohexadien-1′-yl) acetamide, and its dimethyl ketal were isolated from *Verongia cauliformis* and related species of Jamaican sponges (Sharma and Burkholder, 1967).

A similar compound, which is a dihydrobenzyl cyanide derivative, Aeroplysinin-I was isolated in good yields from a Mediterranean species of sponge, *Verongia* (*Aplysina*) *aerophoba* (Fattorusso *et al.*, 1970, 1972). The compound surprisingly constitutes 3% of the total dry weight of the sponge. It is optically active ($[\alpha]_D + 186°$) and, interestingly, its antipode, (($-$)-form) was isolated from a Caribbean species, *Ianthella ardis* (Fulmer *et al.*, 1970). The structure was established by X-ray crystallography by Cosulich and Lovell (1971).

Aeroplysinin-I showed antibiotic activity to *Staphylococcus alba*, *Bacillus cereus* and *B. subtilis* (Fattorusso *et al.*, 1972).

Verongia aerophoba also yielded an interesting compound, Aerothionin, which constitutes 10% of the dried sponge. The structure of aerothionin was shown to be two spirocyclohexadienylisoxazoles connected through 1,4-diaminobutane.

This finding was made by a combined British and Italian group (Moody

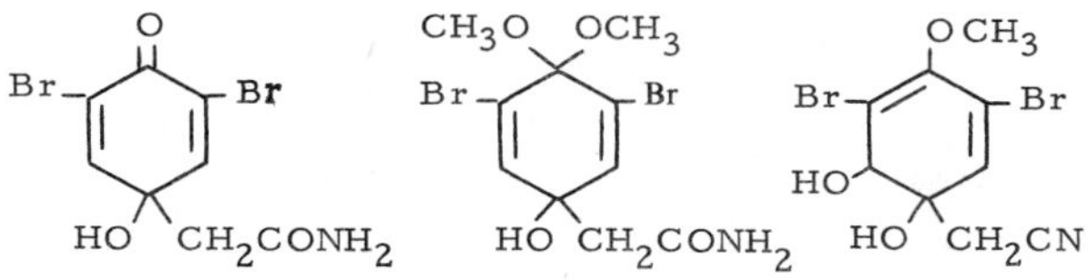

4′-keto*cyclo*hexadien-1′-ylacetamides Aeroplysinin-I from sponges

(XXXII)

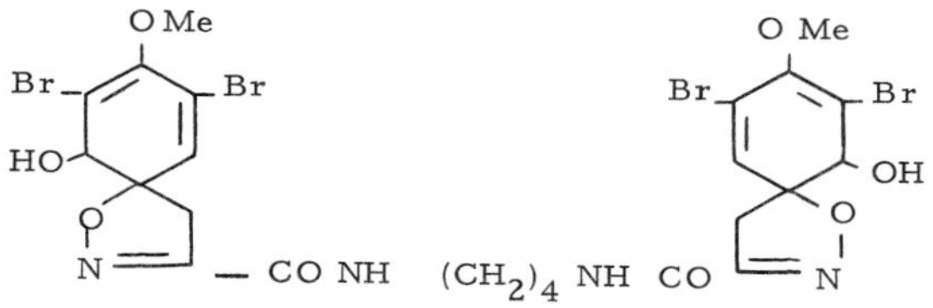

Aerothionin

(XXXIII)

et al., 1972). Aerothionin was also found in a Californian species, *V. thiona* (Moody *et al.*, 1972). Homoaerothionin, which contains 1,5-diaminopentane instead of diaminobutane, was also isolated from *V. thiona*.

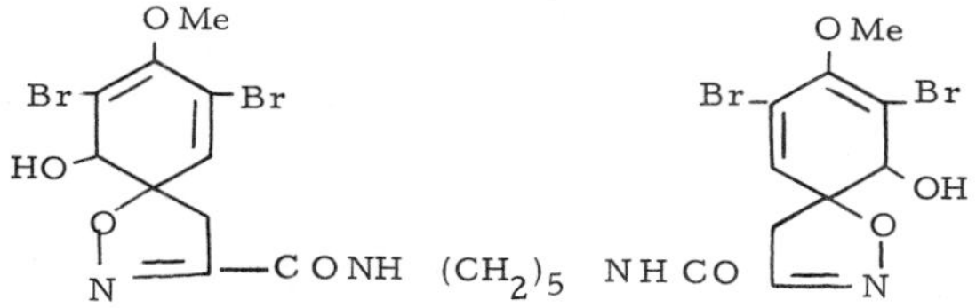

Homoaerothionin

(XXXIV)

Some sponges were also found to yield several bromo-compounds with a pyrrole ring system. One compound isolated from *Phakellia flabellata*, dibromophakellin, was shown to have a complex ring system which included an imidazole ring (Sharma and Burkholder, 1971).

The compound, oroidin, was isolated from a Mediterranean sponge, *Agelas oroides* and its structure is considered to be related to that of phakellin. The molecule is made up of two ring structures, pyrrole and imidazole, connected by an amide linkage (Forrenza *et al.*, 1971). Other compounds which have only the pyrrole ring were also isolated from the same organism.

An interesting non-nitrogenous bromophenol with antibiotic activity has

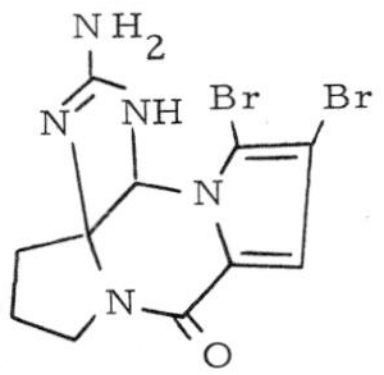

Dibromophakellin 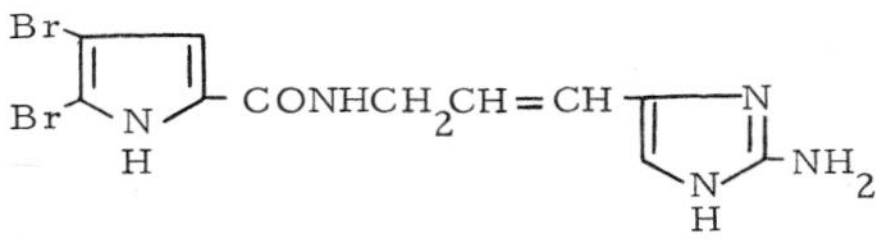 Oroidin

(XXXV)

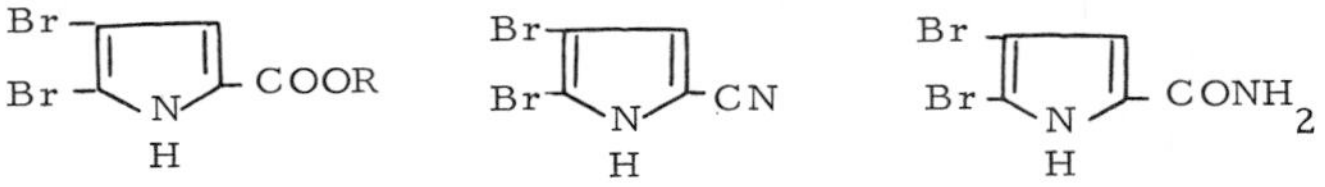

Pyrroles from *A. oroides*

(XXXVI)

been isolated from the Caribbean sponge, *Dysidea herbacea*, by Sharma *et al.* (1969) and Burkholder and Sharma (1969). These workers recognized the antibiotic activity of bromophenolic substances from sponges in general to gram-negative and gram-positive organisms. Sharma and Vig (1972) subsequently purified two of these sponge substances (A and B) and established them to be the following polybrominated diphenyl ethers (**XXXVII**).

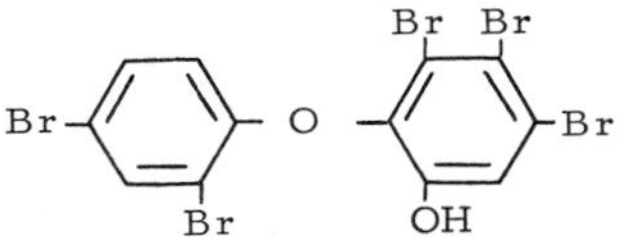

Substance A 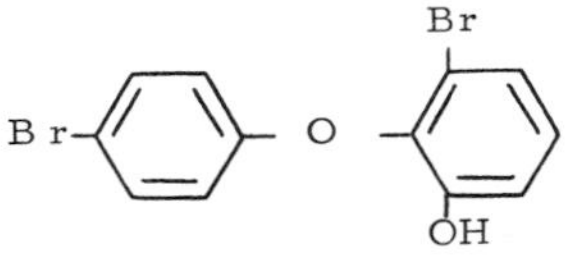 Substance B

Bromodiphenyl ethers from *D. herbacea*.

(XXXVII)

23.7 Cephalosporins

The cephalosporium fungus of Mediterranean origin is thus far the most significant source of an antibiotic from the oceans. In the middle 1940's Brotzu examined the microbial flora of sea water near a sewage outlet at Cagliari, Sardinia, in the Mediterranean, reasoning that the observed self-purification properties of the water might result from the antibacterial substances produced by other micro-organisms in it. He subsequently isolated a fungus, *Cephalosporium acremonium* and demonstrated that it inhibited the growth of gram-positive and gram-negative bacteria (Abraham and Loder, 1972). Later, the culture filtrate of the organism was found to be an active antibiotic in clinical tests conducted in Sardinia.

The chemical isolation of *Cephalosporium* active principles was first made by Crawford *et al.* (1952) in Oxford, England. These workers described a penicillinase sensitive substance and named it Antibiotic N because of its activity against gram-negative bacteria. Other cephalosporin substances active against gram-positive bacteria were then named cephalosporins P (Crawford *et al.*, 1952; Burton *et al.*, 1956). In 1956, Newton and Abraham identified and purified the most active compound, cephalosporin C, an antibiotic which possesses activity over a broad spectrum of both gram-negative and gram-positive bacteria, similar to that shown by pencillin but differing from penicillin in its resistance to penicillinase, an enzyme which destroys the effectiveness of penicillin.

Abraham and Newton (1961) suggested that cephalosporin C contains a β-lactam-dihydrothiazine ring system instead of the β-lactam-thiazolidine ring of the penicillins. The structure was later confirmed by X-ray crystallography (Hodgkin and Maslen, 1961).

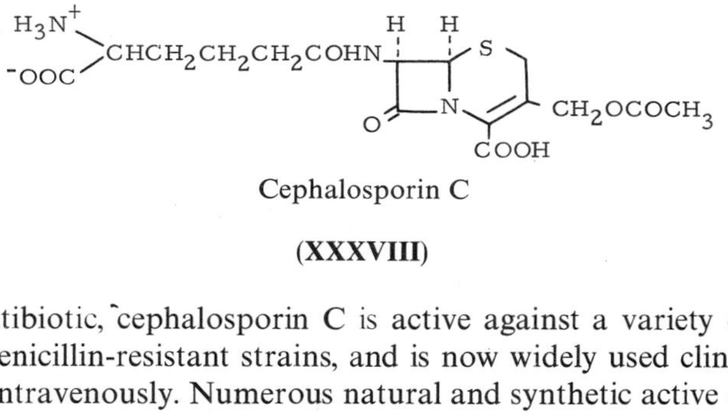

Cephalosporin C

(XXXVIII)

As an antibiotic, cephalosporin C is active against a variety of bacteria including penicillin-resistant strains, and is now widely used clinically both orally and intravenously. Numerous natural and synthetic active derivatives of this dihydrothiazine compound are also available for clinical use. (For an extensive review of the cephalosporins, see Abraham and Loder, 1972.)

23.8 Fish and Shellfish Toxins

Severe toxic effects, including muscle paralysis, have been reported following human ingestion of certain shellfish, molluscs and vertebrate fishes. Nitrogenous compounds such as tetrodotoxin, saxitoxin, gonyaulax toxin and ciguatoxin are known to cause these poisonings, and they have been isolated, chemically characterized and subjected to pharmacological evaluation. Algae, especially dinoflagellates of the fish food chain, are most likely to be involved in the biological synthesis of these compounds, and when herbivorous species feed on these algae, their toxins are stored in fish tissues or serve as chemical intermediates in the production of the fish toxins. When carnivorous species feed on these herbivores the toxin accumulates or, in some manner is concentrated, and may be passed on to other fish and to those who ingest them. Thus, a biochemical cycle, which includes plant and animal life, makes the toxins an ever present threat to sea predators, including humans, who consume them.

These toxic substances possess marked effects on the central nervous system, and muscle and/or respiratory paralysis often results. The severity of the poisoning is not only dependent upon the amount of pure toxin consumed, but also on the presence of nitrogenous substances, e.g. toxic amines and peptides, in the algal dinoflagellates of the food chain of the fish.

Tetrodotoxin, puffer or fugu poison, a complex polyhydroxylated perhydromethylquinazoline, is found in Porcupine fish, *Diodontidae*, Sunfish *Molidae* or Puffers, *Tetrodontidae* and Fugu fish. Its LD_{50} in mice is 8–20 μg per kg body weight. Cysteine is reported by Fujii *et al.* (1967) to be a useful antidote for tetrodotoxin poisoning if used 10 to 30 minutes after intoxication.

Russell (1967), Halstead (1967) and Kao (1966) have reviewed the pharmacological actions of the toxin, and report that basically it blocks nerve conduction by acting on the nerve motor axons and muscle cell membranes and thus prevents membranes being permeable to sodium. In addition, it blocks the excitability of skeletal muscle fibres, thus producing skeletal muscle paralysis. As a result of the chemical and pharmacological interest in this compound, it was finally synthesized by Kishi and his group in 1971.

Tetrodotoxin* is commercially available from Sankyo Co., Japan, and Calbiochem, U.S., and in carefully controlled doses is being used clinically as a muscle relaxant and pain killer in neurogenic leprosy and terminal cancer. It has also been used with some success as a local anaesthetic in Japan (Ogura and Mori, 1968).

Unlike tetrodotoxin, the toxic principles saxitoxin and gonyaulax toxins which can be isolated from Alaska butter clams, *Saxidomas giganteus*, mussels,

* Brown and Mosher (1963) isolated a crystalline neurotoxin from the eggs of the common Californian newt (*Taricha torosa*) and named it tarichatoxin. This toxin was found to have an LD_{50} in mice similar to tetrodotoxin, and was subsequently identified as the same by direct comparison (Buchwald *et al.*, 1964). Hashimoto and Noguchi (1971) discovered tetrodotoxin in the Goby fish of Japan, *Gobius criniger*. The content of the toxin in the goby fish, which is also found in the Philippines and Taiwan, is about 60 000 mouse units per kg.

e.g. *Mytilus sp.*, and the dinoflagellates of *Gonyaulax sp.* have yet to find clinical use. Nevertheless, these exhibit similar chemical and pharmacological properties and produce marked neuromuscular paralysis when ingested. The action is apparently caused by a blockage of membrane permeability similar to that caused by tetrodotoxin (Evans, 1964; Kao, 1966).

Sommer and Meyer (1937) first linked the cause of shellfish poisoning to the microalga, *Gonyaulax catenella*, during blooms of the latter dinoflagellate off the California coast. Later Schantz *et al.*, (1957) isolated, and in 1966 identified the same toxin from Alaska butter clams as saxitoxin, sometimes called gonyaulax toxin.

Purified saxitoxin is a non-crystalline powder and has a potency of about 5 500 mouse units* per milligram. Wong *et al.* (1971b), confirmed its empirical formula to be $C_{10}H_{15}N_7O_3 2HCl$ and showed that it is a strongly basic compound with two guanidium groups each containing a chlorine atom. It belongs to the perhydropurines. Rapoport's group synthesized degradation products of the molecule (Wong, *et al.*, 1971a) and proposed the following structure (Wong, *et al.*, 1971b).

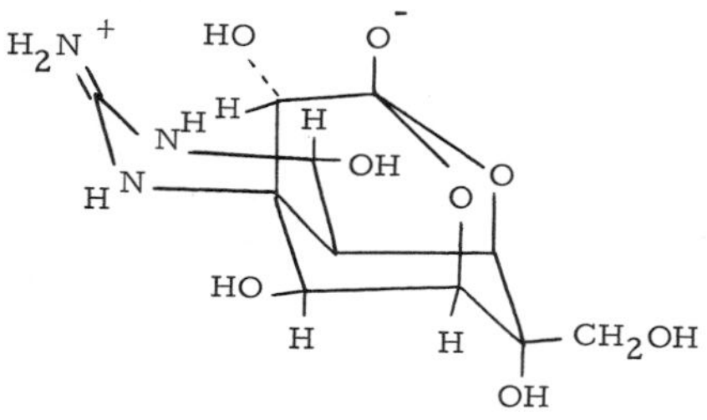

Tetrodotoxin

(XXXIX)

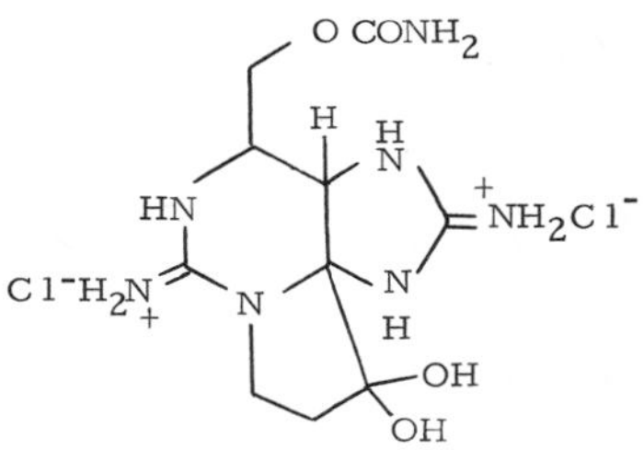

Saxitoxin

(XL)

Saxitoxin was also discovered in toxic crabs from islands off south east Japan. Hashimoto *et al.* (1967) recognized the presence of a toxin in two kinds of crabs, *Zosimus aereus* and *Platypodia granulosa*. It was subsequently purified and found to be saxitoxin (Noguchi *et al.*, 1969). As with tetrodotoxin poisoning, occurrences of saxitoxin poisoning from shellfish are worldwide.

Although the *Gonyaulax* dinoflagellate, *G. catenella*, which produces saxitoxin, is distributed along the Pacific coast of the United States, Canada, Alaska and the Aleutians to Japan, a related species, *G. tamarensis*, is the principal cause of shellfish poisoning along the Atlantic coasts of New

* One mouse unit is defined as an amount of toxin required to kill a 20 gram mouse in 15 minutes.

TABLE 23.4

Types of pharmacological activity and potential drugs from marine organisms

Taxonomic groups (Phyla)	Type of pharmacological activity observed in various organisms[h] CNS[a]	RS[b]	NMS[c]	ANS[d]	CVS[e]	GI[f]	Local[g]	Antibiotic activity	Other activity	Nature of toxin and toxicity	Potential pharmacological-drug use
Monerans											
Schizophyta											
Marine bacteria											
Bacillus spp.	—	—	—	—	—	—	—	+		Antibacterial principles	Potential source of antibiotics
Micrococcus spp.	—	—	—	—	—	—	—	+			
Chromobacterium spp.	—	—	—	—	—	—	—	+	Antifungal and antiyeast activity	Antibacterial, antifungal, and antiyeast principles	Antibiotics with control against yeast and fungal pathogens
Aeromonas spp.	—	—	—	—	—	—	—	+			
Pseudomonas spp.	—	—	—	—	—	—	—	+			
Vibrio spp.	—	—	—	—	—	—	—	+			
Flavobacterium spp.	—	—	—	—	—	—	—	+			
Alcaligenes spp.	—	—	—	—	—	—	—	+			
Flavobacterium piscicida	+	—	—	—	—	—	—	+		Toxin	CNS drug
Marine Actinomycetes											
Nocardia spp.	—	—	—	—	—	—	—	+		Antibacterial	Antibiotic
Cyanophyta (blue-green algae)											
Lyngbya majuscula	—	—	—	—	—	—	+	+	Toxic	Antibacterial	Antibiotic
Microcystis aeruginosa (freshwater)	+	—	—	—	—	—	—	—	Toxic	Polypeptide endotoxin	CNS drug
Anabaena flos-aquae (freshwater)	+	—	—	—	—	—	—	—	Toxic	Polypeptide	CNS drug
Phormidium spp.	—	—	—	—	—	—	+	—	Stimulates growth of bacteria, plant, animal, algae cultures	Growth stimulant	Wound healing
Nostic rivulare	—	—	—	—	—	—	+	—	Carcinogenic	Unknown	Study of cancers

Protistans											
Rhodophyta (red algae)											
Digenea simplex	—	—	—	—	—	—	+	—	Antihelminthic	Kainic acid	Against parasitic intestinal worms (ascarid)
Chondrus crispus	—	—	—	—	—	—	—	—	Antiviral	Carrageenan	Antiviral drug
Gelidium cartilagenium	—	—	—	—	—	—	—	—	Antiviral	Polysaccharide	Antiviral drug
Phaeophyta (brown algae)											
Rhodomela larix	—	—	—	—	—	—	—	+		Brominated phenolic compound	Antibiotic
Laminaria spp.	—	—	—	—	—	—	—	—	Blood anticoagulant	Laminarin	Anticoagulant
Chlorophyta (green algae)											
Chlorella spp.	—	+	+	—	—	+	—	+	Toxic	Unknown, oxidation products of fatty acids	NMS studies, antibiotics
Chlamydomonas reinhardtii	—	—	—	—	—	—	—	+		Fatty acids	Antibiotic
Chrysophyta (diatoms)											
Ochromonas spp.	+	—	+	—	—	—	—	—	Ichthyotoxic	Unknown	CNS and neuromuscular drugs
Prymnesium parvum	+	—	+	—	—	—	—	—	Ichthyotoxic, haemolytic, cytolytic, antispasmodic activities	Prymnesin	CNS and neuromuscular drugs
Phaeocystis pouchetii	—	—	—	—	—	+	—	+		Acrylic acid	Broad spectrum antibiotic for GI tract
Pyrrophyta (dinoflagellates)											
Gymnodinium spp.	+	+	+	+	+	+	+	—		Alkylguanidine compounds	CNS drugs
Gonyaulax spp.	+	+	+	+	+	+	+	—		Alkylguanidine compounds	CNS drugs
Invertebrates											
Porifera (sponges)											
Tedania toxicalis	—	—	—	—	—	—	+	+		Unknown	Antibiotic
Suberites domunculus	—	+	—	—	+	+	+	—		Unknown	?
Microciona prolifera	—	—	—	—	—	—	—	+		Ectyonin	Antibiotic

L

Table 23.4—*continued*

Taxonomic groups (Phyla)	Type of pharmacological activity observed in various organisms[h]								Other activity	Nature of toxin and toxicity	Potetnial pharmacological-drug use
	CNS[a]	RS[b]	NMS[c]	ANS[d]	CVS[e]	GI[f]	Local[g]	Antibiotic activity			
Haliclona variabilis	—	—	—	—	—	—	—	—	Aggregation factor	Protein	Study of healing process
Cryptotethya crypta	—	—	—	—	—	—	—	—	Growth regulators	Nucleosides, spongothymidine, and spongouridine	Antagonists in nucleic acid metabolism
Coelenterata (*cnidaria*)											
Hydroids											
Physalia physalis	+	+	+	+	+	+	+	—		5-HT, low-molecular weight proteins and peptides	Cardioactive and neuromuscular drugs
Jellyfish											
Chironex fleckeri	+	+	+	+	+	+	+	+		5-HT, low-mol. wt. proteins and peptides	Cardioactive and neuromuscular drugs
Aurelia aurita	—	—	—	—	—	—	—	—	Neurohumoral compounds	Extract affects neurofunction of related species	CNS drug
Sea anemones											
Actinea equina	+	—	+	—	—	+	+	—		Tetramine, homarine	Cardioactive and neuromuscular drug
Rhodactis howesii	+	—	—	—	—	—	—	—	Anticoagulant	—	CNS drug, anticoagulant
Corals											
Acropora palmata	+	—	—	—	—	—	+	—		Unknown	?
Plexaura crassa	—	—	—	—	—	—	—	+	Toxic	Crassin	Antibiotic

Echinodermata											
Starfish											
Asterias spp.	—	—	—	—	+	—	—	—	Haemolysis, toxic sperm immobilization, induces egg and sperm shedding	Automizing toxin, saponins, asterotoxin	Tissue regeneration studies, sperm inactivation studies egg maturation studies
Sea urchins											
Paracentrotus lividus	+	+	+	+	—	—	+	—	Toxin in ovaries and spines	Unknown	CNS drug, NMS drug
Tripneustes gratilla	+	+	+	+	+	—	+	—	Toxin in ovaries and spines	Toxic protein (acetylcholine-like)	Neuromuscular blocking agents
Sea cucumbers											
Actinopyga agassizi	—	—	+	—	—	—	+	—	Haemolytic, toxic, cytotoxic, antitumor activity	Holothurin (complex of steroidal glycosides, salts and polypeptides)	Neuroactive drug
Mollusca											
Gastropods											
Haliotis spp.	—	—	—	—	—	—	—	+	Antiviral	Paolin I is antimicrobial and paolin II is antiviral: both are protein	Antiviral drugs
Conus spp.	+	+	+	+	+	—	+	—	Toxic	Venom contains mixture of peptides and ammonium Compds. (homarine, *N*-methyl-*N*-methylpyridinium, γ-butyrobetaine) with protein	Neuromuscular and CNS drugs
Neptunea arthritica	—	+	+	—	+	—	—	—		Saliva toxin (tetramine)	Neuromuscular and CNS drugs
Murex spp.	+	+	+	+	+	—	—	—		Murexine	Neuromuscular and CNS drugs

Table 23.4—*continued*

Taxonomic groups (Phyla)	Type of pharmacological activity observed in various organisms[h]								Other activity	Nature of toxin and toxicity	Potential pharmacological-drug use
	CNS[a]	RS[b]	NMS[c]	ANS[d]	CVS[e]	GI[f]	Local[g]	Antibiotic activity			
Bivalves											
Mytilus spp.	—	+	+	+	+	+	+	—		Poisoning caused by ingestion of toxic dinoflagellates	Neuroactive drugs
Mercenaria mercenaria	—	—	—	—	—	—	—	—	Growth inhibitor	Mercenene	Antitumor drug
Octopuses											
Octopus spp.	+	+	+	+	+	+	+	—	Haemolysis	Toxic venom in saliva; tyramine, octopamine, 5-HT, histamine, protein (cephalotoxin)	CNS drugs
Annelida (segmented worms)											
Lumbriconereis heteropoda	—	—	+	+	+	+	+	—	Anaesthetic for insects	Nereistoxin	Neuroactive drug
Anthropoda (joint-footed marine animals)											
Carcinoscorpius notundicauda	+	—	+	—	+	+	+	—		Unknown	Neuromuscular drugs
Carcinus maenas	—	—	—	—	+	—	—	—		6-HT, a mucopeptide	Cardiac drug
Vertebrates											
Fish											
Agnatha (jawless fish)											
Eptatretus stoutii	—	—	—	—	+	—	—	—		Eptatretin (low-mol. wt. amine) obtained from heart	Cardioactive agent (hypersensive)

Chondrichthyes (cartilaginous fish)											
Sharks											
Somniosus micro cephalus	+	—	+	—	—	+	—	—	Visual disturbances	Form of ciguatera toxin (?) due to ingestion of livers and musculature	CNS drug
Hexanchus grisseus	+	—	+	—	—	+	—	—	Visual disturbances		
Stingrays											
Dasyatis pastinacus	+	+	—	+	+	+	+	—	Toxic	Sting venom is protein in nature	CNS or cardioactive drugs
Urobatis halleri	+	+	—	+	+	+	+	—	Toxic		
Osteichthyes (true or bony fish)											
A. Ichthyosarcotoxic (poison in musculature, viscera, skin)											
1. Ciguatera group (fish poisoning characterized by GI and neurological effects) Includes over 300 species in 12 families, *e.g.*, sturgeon fish, sea basses, snappers, barracudas, etc.	+	+	+	+	+	+	+	—		Herbivorous spp. may feed on toxic blue-green algae (cyanophyta). Carnivorous spp. feed on toxic herbivorous spp. and accumulate and concentrate the toxin(s). Toxin probably a mixture. Anticholinesterase implicated.	Neuroactive or gastrointestinal drugs
2. Tetraodon group (puffer or fugu poisoning of neurotoxic type)											

Table 23.4—*continued*

Taxonomic groups (Phyla)	Type of pharmacological activity observed in various organisms[h]								Other activity	Nature of toxin and toxicity	Potential pharmacological-drug use
	CNS[a]	RS[b]	NMS[c]	ANS[d]	CVS[e]	GI[f]	Local[g]	Antibiotic activity			
Diodontidae (porcupine fish, 10 spp.)	+	+	+	+	+	+	+	—	Toxic	Toxin most toxic of ichthyosarcotoxic types.	Neuroactive drug
Molidae (sunfish, 1 spp.)	+	+	+	+	+	+	+	—	Toxic	Toxin concentrated in ovaries or testes, liver and intestines, Musculature free of poison.	Neuroactive drug
Tetradontidae (puffers, 40 spp.)	+	+	+	+	+	+	+	—	Toxic	Toxin known as tetrodotoxin. ($C_{11}H_{17}N_3O_8$)	Neuroactive drug
3. Scombroid group (mackerel-like fish, tunas, skipjacks, and bonitos)	+	—	—	—	—	+	+	—		Toxin called "saurine." Forms if fish are inadequately preserved. Has histamine-like properties.	?
4. Clupeoid group (herring-like fish of tropical Pacific)	—	—	—	—	—	—	—	—		Toxin produces symptoms somewhat like ciguatera poisoning	?
5. Hallucinogenic group [mullet and surmullet (goatfish)]											

Mugil cephalus	+	—	+	—	—	+	+	—	Light-headedness	Unknown	CNS drugs
Neomyxuc chaptalli	+	—	+	—	—	—	+	—	Hallucinations		
Parapenus chryserydros	+	—	+	—	—	—	+	—	Depression		
Upeneus arge	+	—	+	—	—	—	+	—	Violent nightmares.		
B. Ichthyootoxic fish (poison in gonads)											
Scorpaenichthys marmoratus	+	—	—	—	—	+	—	—		Unknown	?
C. Ichthyohaemotoxic fish (poison in blood)											
Muraena helena	+	—	+	—	—	+	—	—		Toxic protein (?) in ingested blood	?
D. Venomous fish											
1. Weeverfish											
Trachinus draco	+	—	—	+	+	—	+	—		Envenomation on mishandling venomous spines	Cardiovascular drugs
Trachinus vipera	+	—	—	+	+	—	+	—		Venom contains protein, 5-HT, histamine releaser	
2. Scorpion fish (rock fish)											
a. Zebrafish											
Pterois spp.	+	+	+	—	+	—	+	—		Venomous spines	Cardioactive drugs
Dendrochirus spp.	+	+	+	—	+	—	+	—		Venom protein in nature	
b. Scorpaena											
Scorpaena guttata	+	—	—	—	+	+	+	—		Venomous spines	Cardioactive drugs
Urolophus halleri	+	—	—	—	—	—	+	—		Venom protein in nature	
c. Stone fish											
Synanceja horrida	+	+	+	—	+	—	+	—		Venom protein in nature	Cardioactive, muscle-relaxing drugs

Table 23.4—*continued*

Taxonomic groups (Phyla)	Type of pharmacological activity observed in various organisms[h]								Other activity	Nature of toxin and toxicity	Potential pharmacological-drug use
	CNS*	RS[b]	NMS[c]	ANS[e]	CVS[e]	GI[f]	Local[g]	Antibiotic activity			
Amphibians											
Salamandridae (true newts)											
Taricha torosa	+	+	+	+	+	+	+	—	Toxic	Toxin concentrated in skin, muscles, and blood. Toxin identified with tetrodotoxin	Neuroactive drug
Reptiles											
Marine turtles											
Chelonia mydas	+	—	—	—	—	+	+	—		Toxic on ingestion	?
Eretmochelys imbricata	+	—	—	—	—	+	+	—		Toxin unknown	
Dermochelys corialea	+	—	—	—	—	+	+	—			
Sea snakes											
Pelamis platunas	+	+	+	+	—	—	+	—		Venomous fangs	Neuroactive ,CNS
Enhydrina schistosa	+	+	+	+	—	—	+	—			
Hydrophis caerulescens	+	+	+	+	—	—	+	—			

[a] CNS = Central Nervous System (nausea, headache, confusion, visual disturbances, nervousness, drowsiness etc.). [b] RS = Respiratory System (depression, distress, syncope, dyspnea etc.). [c] NMS = Neuromuscular System (muscle weakness, incoordination, spasms, curare-like action, paralysis etc.). [d] ANS = Autonomic Nervous System (pupil dilation, anticholinesterase activity, parasympathetic action etc.). [e] CVS = Cardiovascular System (cardiac stimulation, bradycardia, congestion, myocardial ischemia, etc.). [f] GI = Gastrointestinal (vomiting, diarrhoea, abdominal pain etc.). [g] Local = pruritis, parasthesias, pain, necrosis, oedema etc. [h] + = activity; – = no activity.

Reprinted by permission from Der Marderosian, A., *J. Pharm. Sci.* (1969) Vol. 58, No. 1, 6–9, 2215 Constitution Avenue, N.W., Washington, D.C., U.S.A.

England and Canada and in the North Sea areas of Britain and Europe. The paralytic toxic principle of this organism apparently, however, is not saxitoxin (Schantz, 1960; Evans, 1970; Shimizu, 1972c).

The Japanese gastropod, "bai" (*Babylonia japonica*) is poisonous in certain areas. The poisoning symptoms resemble those of atropine and include amblyopia, mydriasis, thirst and aphasia (Hashimoto *et al.*, 1967). Shibota and Hashimoto (1970) purified the toxin by gel filtration through Sephadex G-25. They also reported mydriasis in mice (see also Shibota and Hashimoto, 1970, 1971) and this can be used to test the toxicity. Kosuge and co-workers (1971) isolated the crystalline toxin, which they named surugatoxin, using essentially the method developed by Shibota and Hashimoto. Its structure was established by X-ray crystallography, taking advantage of the bromine atom in the molecule (Kosuge *et al.*, 1971). It contains a bromoindole, *myo*-inositol moiety connected to a pteridine nucleus through a spiro linkage. The chemical and preliminary pharmacological properties of this compound warrant further investigation in view of its effects on the nervous system.

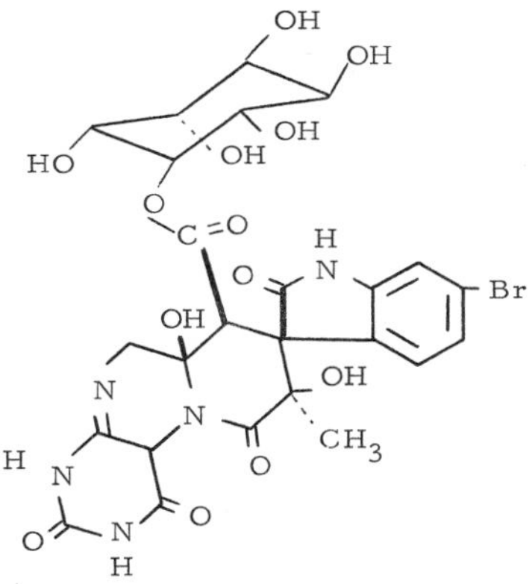

Surugatoxin

(XLI)

Ciguatera (ciguatoxin) toxin, responsible for "cigua" poisoning, was first linked to the ingestion of a marine snail native to the Caribbean areas. This toxin is now considered to be the cause of neurointoxication in humans following the ingestion of sea food from several different species of marine fish and shellfish of tropical and temperate regions. As many as 400 species of fish have been implicated in this poisoning, and as is true for other similar toxic fishes, marine microalgae of the food chain, particularly those of

Gymnodinium species upon which fish feed, are probably original biosynthetic sources of ciguatera toxin (Helfrich and Banner, 1963; Banner, 1967).

Scheur *et al.* (1967) describe ciguatoxin as an unstable lipid compound with a quaternary nitrogen atom and possessing the empirical formula $C_{35}H_{65}NO_8$.

Other toxic fish and shellfish components of interest, but for which little or no complete chemical and pharmacological data are available, are the proteinaceous toxin of the eel, *Anguilla vulgaris* (Rocca and Ghiretti, 1964), lipoproteins in the roe of the blenny, *Stichaeus* species (Hatano *et al.*, 1964) and samples of unidentified toxins of sharks, tuna fish and mullets. These have not been studied sufficiently to permit any conclusions to be drawn at present about their potential medical applications. However, the potent effects of fish toxins in general on the neurophysiology of animals, including man, prompts the medicinal chemist to use them as models for the synthesis of newer analogues which are useful as neuromuscular relaxants, local anaesthetics or similar agents but which have less toxicity than the natural materials. These efforts are the subject of continuing research.

23.9 Conclusions

Table 23.4 summarizes the pertinent findings concerning many types of chemical compounds present in marine organisms, together with their pharmacological activities. The potential use of many of these compounds as drugs should act as an incentive for the intensification of research into these compounds and their synthetic analogues.

References

Abraham, E. P. and Newton, G. G. F. (1961). *Biochem. J.* **79**, 377.
Abraham, E. P. and Loder, P. B. (1972). "Cephalosporins and Penicillins" (E. H. Flynn, ed.), p. 2. Academic Press, London.
Ackermann, D., Holtz, F. and Reinwein, H. (1923). *Z. Biol.* **79**, 113.
Anastasi, A. and Erspamer, V. (1963). *Arch. Biochem. Biophys.* **101**, 56.
Anderson, W. and Duncan, J. G. C. (1965). *J. Pharm. Pharmacol.* **17**, 647.
ApSimon, J. W. and Buccini, J. (1972). Third Food-Drugs from the Sea Conference, Abstract of Papers, p. 4. University of Rhode Island, Kingston, Rhode Island, U.S.A.
Banner, A. H. (1967). "Animal Toxins" (F. E. Russell and P. R. Saunders, eds.), p. 157. Pergamon, New York.
Baslow, M. H. (1969). "Marine Pharmacology". Williams & Wilkins, Baltimore.
Bergmann, W., McLean, M. J. and Lester, D. J. (1943). *J. Org. Chem.* **8**, 271.
Bergmann, W. and Feeney, R. J. (1950). *J. Amer. Chem. Soc.* **72**, 2809.

Bergmann, W. and Feeney, R. J. (1951). *J. Org. Chem.* **16**, 981.
Bergmann, W. and Stempien, M. F., Jr. (1957). *J. Org. Chem.* **22**, 1575.
Bergmann, W. and Burke, D. C. (1955). *J. Org. Chem.* **20**, 1501.
Bergmann, W. and Burke, D. C. (1956). *J. Org. Chem.* **20**, 1501.
Bergstrom, S. and Sjovall, J. (1960). *Acta Chim. Scand.* **14**, 1701, 1693.
Bernardi, L., Bosisio, G., Chiliem, F., DeCaro, G., DeCastiglinone, R., Erspamer, V., Gleassner, A. and Goffredo, O. (1964). *Experientia*, **20**, 306.
Besterman, E. and Evans, J. (1957). *Brit. Med. J.* 310.
Boylan, D. B. and Scheuer, P. J. (1967). *Science, N.Y.* **155**, 52.
Brown, M. S. and Mosher, H. S. (1963). *Science, N.Y.* **140**, 295.
Buchwald, H. D., Durham, L., Fischer, H. G., Harada, R., Mosher, H. S., Kao, C. Y. and Fuhrman, F. A. (1964). *Science, N.Y.* **143**, 464.
Bundy, G. L., Schneider, W. P., Lincoln, F. H. and Pike, J. E. (1972). *J. Amer. Chem. Soc.* **94**, 2123.
Burkholder, P. R., Burkholder, L. M. and Almodovar, L. R. (1960). *Bot. Mar.* **2**, 149.
Burkholder, P. R. and Sharma, G. M. (1969). *Lloydia*, **32**, 466.
Burton, H. S., Abraham, E. P. and Cardwell, H. M. E. (1956). *Biochem. J.* **62**, 171.
Cantoni, G. L. and Anderson, D. G. (1956). *J. Biol. Chem.* **222**, 171.
Challenger, R. and Simpson, M. I. (1948). *J. Chem. Soc.* 1591.
Chanley, J. D., Kohn, S. K., Nigrelli, R. F., and Sobotka, H. (1955). *Zoologica*, **40**, 99.
Chanley, J. D., Mezzetti, P. and Sobotka, H. (1966). *Tetrahedron*, **22**, 1857.
Chanley, J. D. and Rossi, C. (1969). *Tetrahedron*, **25**, 1897, 1911.
Ciereszko, L. S., Sifford, D. H. and Weinheimer, A. J. (1960). *Ann. N.Y. Acad. Sci.* **90**, 917.
Corey, E. J., Anderson, N. H., Carlson, R. M., Paust, J., Vedejs, E., Vlattas, J. and Winter, R. E. K. (1968). *J. Amer. Chem. Soc.* **90**, 3245.
Cosulich, D. B. and Livello, F. M. (1971). *Chem. Comm.* 397.
Craigie, J. S. (1965). *Proc. Soc. Plant Physiol.* **6**, 16.
Craigie, J. S. and Gruenig, D. E. (1967). *Science, N.Y.* 1058.
Crawford, K., Heatley, N. G., Boyd, P. F., Hale, C. W., Kelly, B. K., Miller, G. A. and Smith, N. (1952). *J. Gen. Microbiol.* **2**, 361.
Daigo, K. (1959). *Yakugaku Zasshi*, **79**, 353, 356.
Der Marderosian, A. (1969). *J. Pharm. Sci.* **58**, 1.
Dewar, E. J. (1956). "Second Seaweed Symposium" (T. Braarud and N. A. Sorensen, eds.), p. 55. Pergamon Press, New York.
Elyakov, G. B., Kuznetsova, Dziezenko, A. K. and Elkin, Y. N. (1969). *Tetrahedron Lett.* 1151.
Evans, M. H. (1964). *Brit. J. Pharmacol.* **22**, 478.
Evans, M. H. (1970). *Mar. Pollut. Bull.* **1**, 184.
Fattorusso, E., Minale, L. and Sedano, G. (1970). *Chem. Comm.* 751.
Fattorusso, E., Minale, L. and Sedano, G. (1972). *J. Chem. Soc. Perkin I.* 16.
Faulkner, D. J. and Andersen, R. J. (1974). *In* "The Sea" (Goldberg, E. D. ed.) Vol 5, p 679, Wiley, New York.
Fenical, W., Sims, J. J., Radlick, P. and Wing, R. M. (1972). Third Food-Drugs from the Sea Conference, Abstract of Papers, p. 17. University of Rhode Island, Kingston, Rhode Island, U.S.A.
Forenza, S., Minale, L. and Riccio, R. (1971). *Chem. Comm.*, 1129.
Foster, A. B. and Webber, J. M. (1960). *Advan. Carbohyd. Chem.* **15**, 371.
Friess, S. L. and Durant R. C. (1965). *Toxicol. Appl. Pharmacol.* **7**, 373.

Friess, S. L., Durant, R. C. and Chanley, J. D. (1968). *Toxicon*, **6**, 81.
Fryberg, M., Ochschlazer, A. C. and Unrau, A. M. (1971). *Chem. Comm.* 1194.
Fujii, M., Harada, K. and Matsuda, M. (1967). *Chem. Abstr. Jap.* **69**, 2396.
Fulmer, W., Van Lear, G. E., Morton, G. O. and Mills, R. D. (1970). *Tetrahedron Lett.* 4551.
Galbraith, M. N., Horn, D. H. S., Middleton, E. J. and Hackney, R. J. (1968). *Chem. Comm.* 83.
Gasteiger, E. L., Haake, P. C. and Gergen, J. A. (1960). *Ann. N.Y. Acad. Sci.* **90**, 622.
Ghiretti, F. (1953). *Arch. Sci. Biol.* **37**, 435.
Ghiretti, F. (1960). *Ann. N.Y. Acad. Sci.* **90**, 726.
Gupta, K. C. and Sheuer, P. J. (1968). *Tetrahedron*, **24**, 5831.
Hale, R. L., Leclerg, J., Tursch, B., Djerassi, C., Cross, R. A., Weinheimer, A. J., Gupta, K. C. and Sheuer, P. J. (1970). *J. Amer. Chem. Soc.* **92**, 2119.
Halstead, B. W. (1967). "Poisonous and Venomous Marine Animals of the World" Vol. 2. U.S. Govt. Print Office, Washington, D.C.
Hampshire, F. and Horn, D. H. S. (1966). *Chem. Comm.*, 37.
Hashimoto, Y. and Yasumoto, T. (1960). *Bull. Jap. Soc. Sci. Fish.* **26**, 1132.
Hashimoto, Y., Konosu, S., Yasumoto, T., Inoue, A. and Noguchi, T. (1967). *Toxicon*, **5**, 85.
Hashimoto, Y., Miyazawa, K., Kamiya, H. and Shibota, M. (1967). *Bull. Jap. Soc. Sci. Fish.* **33**, 661.
Hashimoto, Y. and Noguchi, T. (1971). *Toxicon*, **9**, 79.
Hashimoto, Y. and Oshima, Y. (1972). *Toxicon*, **10**, 279.
Hashimoto, Y., Sakai, M. and Konishi, K. (1972). Third Food-Drugs from the Sea Conference, Abstract of Papers, p. 4. University of Rhode Island, Kingston, Rhode Island, U.S.A.
Hatano, M., Zama, F., Takama, K., Sakai, M. and Igarashi, H. (1964). *Bull. Fac. Fish. Hokkaido Univ.* **15**, 138.
Helfrich, P. and Banner, A. H. (1963). *Nature, Lond.* **197**, 1025.
Hesp, R. and Ramsbottom, B. (1965). *Nature, Lond.* **208**, 1341.
Hinman, J. W. (1972). Third Food-Drugs from the Sea Conference, Abstract of Papers, p. 15. University of Rhode Island, Kingston, Rhode Island, U.S.A.
Hirst, E. L. and Rees, D. A. (1965). *J. Chem. Soc.* 1182.
Hodgkin, D. C. and Maslen, E. N. (1961). *Biochem. J.* **79**, 393.
Hodgkin, J. H., Craigie, J. S. and McInnes, A. G. (1966). *Can. J. Chem.* **44**, 74.
Horn, D. H. S., Fabbri, S., Hampshire, F. and Lowe, M. E. (1968). *Biochem. J.* **109**, 399.
Hossain, M. B., Nicholas, A. F. and van der Helm, D. (1968). *Chem. Comm.* 385.
Idler, D. R., Saito, A. and Wiseman, P. (1970). *Steroids*, **11**, 465.
Ikegami, S., Kamiya, Y. and Tamura, S. (1972b). *Agr. Biol. Chem.* (*Jap.*.), **36**, 1087.
Ikegami, S., Kamiya, Y. and Tamura, S. (1972a). *Tetrahedron Lett.* 1601.
Irie, T., Yasunari, Y., Suzuki, T., Imai, M., Kurosawa, E. and Masamune, T. (1965). *Tetrahedron Lett.* 3619.
Irie, T., Suzuki, M. and Masamune, T. (1966). *Tetrahedron Lett.* 1837.
Irie, T., Suzuki, M. and Masamune, T. (1968). *Tetrahedron*, **24**, 4193.
Irie, T., Suzuki, M. and Hayakawa, Y. (1969a). *Bull. Chem. Soc.* (*Jap.*), **42**, 843.
Irie, T., Fukazawa, A., Izawa, M. and Kurosawa, E. (1969b). *Tetrahedron Lett.* 1343.
Irie, T., Izawa, M. and Kurosawa, E. (1970). *Tetrahedron*, **26**, 851.
Jakowska, S. and Nigrelli, R. F. (1960). *Ann. N.Y. Acad. Sci.* **90**, 913.

Kao, C. Y. (1966). *Pharmacol Rev.* **18**, 997.
Katayama, T. (1962). "Physiology and Biochemistry of Algae" (R. A. Lewin, ed.). Academic Press, New York.
Kathan, R. H. (1965). *Ann. N.Y. Acad. Sci.* **130**, 390.
Katz, F., Fishman, L. and Levy, M. (1972). *Chem. Eng. News*, Sept. 4, 16.
Kem, W. R., Coates, R. M. and Abbott, B. C. (1969). *Fed. Proc.* **28**, 610.
Kem, R. K., Abbott, B. C. and Coates, R. M. (1971). *Toxicon*, **9**, 15.
Kishi, Y., Araya, M., Tanino, H., Fukuyama, T., Nakatsubo, F., Goto, T., Inoue, S., Sugihara, S. and Kakoi, H. (1971). Symposium Papers, 15th Symposium on the Chemistry of Natural Products, Nagoya, Japan, p. 131.
Kosuge, T., Zenda, H., Ochiai, A., Masaki, N., Noguchi, M., Kimura, S. and Narita, H. (1971). Symposium Papers, 15th Symposium on the Chemistry of Natural Products, Nagoya, Japan, p. 145.
Kurzrok, R. and Lieb, C. C. (1930). *Proc. Soc. Exp. Biol. Med.* **28**, 268.
Light, R. J. and Samuelsson, B. (1972). *Eur. J. Biochem.* **28**, 232.
Mackie, A. M., Lasker, R. and Grant, P. T. (1968). *Comp. Biochem. Physiol.* **26**, 415.
Mackie, A. M. and Turner, A. B. (1970). *Biochem. J.* **117**, 543.
Martinez-Nadal, N. G., Rodriguez, L. V. and Casillas, C. (1965). *Antimicrob. Ag. Chemother.* 131.
Mastaglin, P. and Augier, J. (1949). *Compt. Rend.* **229**, 775.
Matsuda, H., Tomiie, Y., Yamamura, S. and Hirata, Y. (1967). *Chem. Comm.* 989.
Merck Index, Eighth Edition, (Paul G. Stecher, ed.). Merck & Co., New Jersey, 1968.
Meunier, H., Zelenski, S. and Worthen, L. R. (1969). Food-Drugs from the Sea Proceedings (H. W. Youngken, Jr., ed.). Marine Technology Society, Washington, D.C., p. 319.
Moody, K., Thompson, R. H., Fattorusso, E., Minale, L. and Sodano, G. (1972). *J. Chem. Soc. Perkin I*, 18.
Murakami, S., Takemoto, T. and Shimizu, Z. (1953). *Yakugaku Zasshi*, **73**, 1026.
Newton, G. C. F. and Abraham, E. P. (1956). *Biochem. J.* **62**, 651.
Nigrelli, R. F. (1952). *Zoologica*, **37**, 89.
Nigrelli, R. F., Chanley, J. D., Kohn, S. K. and Sobotka, H. (1955). *Zoologica*, **40**, 47.
Nigrelli, R. F. and Jakowska, S. (1960). *Ann. N.Y. Acad. Sci.* **90**, 884.
Nigrelli, R. F., Stempien, M. F., Jr., Ruggieri, G. D., Ligouri, V. R. and Cecil, J. F. (1967). *Fed. Proc.* **26**, 1197.
Nitta, S. (1934). *Yakugaku Zasshi*, **54**, 648.
Noguchi, T., Konosu, S. and Hashimoto, Y. (1969). *Toxicon*, **7**, 325.
Ogura, Y. and Mori, Y. (1968). *Europ. J. Pharmacol.* **3**, 58.
Okaichi, T. and Hashimoto, Y. (1962). *Agr. Biol. Chem.* (*Jap.*), **81**, 930.
O'Neill, A. N. (1955). *J. Amer. Chem. Soc.* **77**, 2037, 6324.
Organic Syntheses, Collective Volume 3. (1955). (E. C. Horning, ed.), p. 430. John Wiley & Sons, New York.
Peifer, J. J., Janssen, P. A., Cox, W. and Lundberg, W. O. (1960). *Arch. Biochem. Biophys.* **86**, 302.
Pratt, R., Mautner, R. H., Gardner, G. M., Sha, Y. and Dutrenoy, J. (1951). *J. Amer. Pharm. Assoc.* **40**, 575.
Preston, R. D. (1962). *Nature, Lond.* **196**, 1962.
Proctor, V. W. (1957). *Limmol. Oceanogr.*, **2**, 125.
Randall, J. E., Aida, K., Hibiya, T., Mitsuura, N., Kamiya, H. and Hashimoto, Y. (1971). *Publ. Seto Mar. Biol. Lab.*, **19**, 157.

Reiner, E., Idler, D. R. and Wood, J. D. (1960). *Can. J. Biochem. Physiol.* **38**, 1499.
Reiner, E., Topliff, J. and Wood, J. D. (1962). *Can. J. Biochem. Physiol.* **40**, 1401.
Rio, G. J., Ruggieri, G. D., Stempien, M. F., Jr. and Nigrelli, R. F. (1963). *Amer. Zool.* **3**, 554.
Rio, G. J., Stempien, M. F., Jr., Nigrelli, R. F. and Ruggieri, G. D. (1965). *Toxicon*, **3**, 147.
Rocca, E. and Ghiretti, F. (1964). *Toxicon*, **2**, 79.
Roller, P., Djerassi, C., Cloetens, R. and Tursch, B. (1969). *J. Amer. Chem. Soc.* **91**, 4918.
Ruggieri, G. D. (1965). *Toxicon*, **3**, 157.
Russell, F. E. (1967). *Fed. Proc.* **26**, 1206.
Saito, T. and Ando, Y. (1955). *J. Chem. Soc.* (*Jap.*), **76**, 478.
Sakai, M. (1967) *Botyu-Kagaku*, **32**, 21.
Schantz, E. J., Mold, J. D., Stanger, D. W., Sharel, J., Riel, F. J., Bowden, J. P., Lynch, J. M., Wyler, R. S., Riegel, B. and Sommer, H. (1957). *J. Amer. Chem. Soc.*, **79**, 5230.
Schantz, E. J. (1960). *Ann. N.Y. Acad. Sci.* **90**, 843.
Schantz, E. J., Lynch, J. M., Vayvada, G., Matsumoto, K. and Rapoport, H. (1966). *Biochemistry*, **5**, 1191, 2117.
Scheuer, P. J. (1973). "Marine Natural Products". Academic Press, New York.
Scheuer, P. J., Takahashi, W., Tsutsumi, J. and Yoshia, T. (1967). *Science, N.Y.* **155**, 1267.
Schmitz, F. J. and Pattabhiraman, T. (1970). *J. Amer. Chem. Soc.* **92**, 6073.
Schneider, W. P., Axen, U., Lincoln, F. H., Pike, J. E. and Thompson, J. L. (1968). *J. Amer. Chem. Soc.* **90**, 5896.
Schneider, W. P., Hamilton, R. D. and Ruhland, L. E. (1972). *J. Amer. Chem. Soc.* **94**, 2122.
Seno, N. (1961). *Seikagaku* (*Jap.*), **33**, 461, 465, 471.
Sharma, G. M., and Burkholder, P. R. (1967). *Tetrahedron Lett.* 4147.
Sharma, J. M., Vig, B. and Burkholder, P. R. (1969). "Food-Drugs from the Sea Proceedings" (H. W. Youngken, Jr., ed.). Marine Technology Society, p. 207.
Sharma, G. M. and Burkholder, P. R. (1971). *Chem. Comm.* 151.
Sharma, G. M. and Vig, B. (1972). *Tetrahedron Lett.* 1715.
Sheikh, Y. M., Djerassi, C. and Tursch, B. (1971). *Chem. Comm.* 217.
Sheikh, Y. M., Tursch, B. and Djerassi, C. (1972a). *J. Amer. Chem. Soc.* **94**, 3278.
Sheikh, Y. M., Tursch, B. and Djerassi, C. (1972b). *Tetrahedron Lett.* 3721.
Shibota, M. and Hashimoto, Y. (1970). *Bull. Jap. Soc. Sci. Fish.* **36**, 115.
Shibota, M. and Hashimoto, Y. (1971). *Bull. Jap. Soc. Sci. Fish.* **37**, 936.
Shimada, S. (1969). *Science, N.Y.* **163**, 1462.
Shimizu, Y. (1970). AAAS Meeting, Symposium Lectures, Chicago.
Shimizu, Y. (1971). *Experientia*, **27**, 1188.
Shimizu, Y. (1972a). *J. Amer. Chem. Soc.* **94**, 4051.
Shimizu, Y. (1972b). Third Food-Drugs from the Sea Conference, Abstract of Papers, p. 29. University of Rhode Island, Kingston, Rhode Island, U.S.A.
Shimizu, Y. (1972c). Unpublished data.
Sieburth, J. McN. (1960). *Science, N.Y.* **132**, 676.
Sieburth, J. McN. (1961). *J. Bacteriol.* **82**, 72.
Sieburth, J. McN. (1964). *Develop. Ind. Microbiol.* **5**, 124.
Sieburth, J. McN. and Conover, J. T. (1965). *Nature, Lond.* **208**, 52.
Sobotka, H., Friess, S. L. and Chanley, J. D. (1964). "Comparative Neurochemistry" (D. Richter, ed.). Oxford University Press, p. 471.

Sommer, H. and Meyer, K. F. (1937). *Amer. Med. Assoc. Arch. Pathol.* **24**, 560.
Spraggins, R. L. (1972). Third Food-Drugs from the Sea Conference, Abstract of Papers, p. 12. University of Rhode Island, Kingston, Rhode Island, U.S.A.
Stempien, M. F., Jr., Ruggieri, G. D., Nigrelli, R. F. and Cecil, J. T. (1969). "Food-Drugs from the Sea Proceedings" (H. W. Youngken, Jr., ed.). Marine Technology Society, p. 295.
Sullivan, J. D. and Nigrelli, R. F. (1956). *Proc. Amer. Assoc. Cancer Res.* **2**, 151.
Sutton, A. (1967). *Nature, Lond.* **216**, 1005.
Tagawa, S. (1968). *J. Shimonoseki Univ. Fish.* (*Jap.*), **17**, 35.
Takemoto, K. K. and Spicer, S. S. (1965). *Ann. N.Y. Acad. Sci.* **130**, 365.
Takemoto, T. and Daigo, K. (1958). *Chem. Pharm. Bull.* **6**, 578.
Takemoto, T., Daigo, K. and Takagi, N. (1964a). *Yakugaku Zasshi*, **84**, 1176.
Takemoto, T., Daigo, K. and Takagi, N. (1964b). *Yakugaku Zasshi*, **84**, 1180.
Takemoto, T., Ogawa, S., Nishimoto, N., Arihara, S. and Bue, K. (1967). *Yakugaku Zasshi*, **87**, 1414.
Tanaka, Y. (1969). "Food-Drugs from the Sea Proceedings" (H. W. Youngken, Jr., ed.), p. 311. Marine Technology Society, Washington, D.C.
Tsuchiya, Y. and Suzuki, Y. (1962). referred to in Y. Tsuchiya "Marine Chemistry". Koseisha, Tokyo.
Tsuda, K., Akagi, S., Kishida, Y., Hayatsu, R. and Sakai, K. (1958). *Chem. Pharm. Bull.* **6**, 724.
Turner, A. B., Smith, D. S. H. and Mackie, A. M. (1971). *Nature, Lond.* **233**, 209.
Tursch, B., DeSouza Guimaraes, I. S., Gilbert, B., Aplin, R. T., Duffield, A. M. and Djerassi, C. (1967). *Tetrahedron*, **23**, 761.
Tursch, B., Cloetens, R., and Djerassi, C. (1970). *Tetrahedron Lett.* 467.
Waldron-Edwards, D. (1968). "Drugs from the Sea" (H. D. Freudenthal, ed.), p. 267. Marine Technology Society, Washington, D.C.
Wagner, H. and Friedlich, H. (1964). *Naturwissenschaften*, **51**, 163.
Weinheimer, A. J., Middlebrook, R. E., Bledsoe, J. O., Marsice, W. E. and Karns, T. K. B. (1968). *Chem. Comm.* 384.
Weinheimer, A. J. and Spraggins, R. L. (1969). *Tetrahedron Lett.* 5185, and "Food-Drugs from the Sea Proceedings" (H. W. Youngken, Jr., ed.), p. 311. Marine Technology Society, Washington, D.C.
Weinheimer, A. J. (1971). In Symposium Lecture of Physiologically Active Marine Substances of Marine Origins, St. Petersburg, Florida.
Whittaker, V. P. (1960). *Ann. N.Y. Acad. Sci.* **90**, 695.
Wong, J. L., Brown, M. S., Matsumoto, K., Oesterlin, R. and Rapoport, H. (1971a). *J. Amer. Chem. Soc.* **93**, 4633.
Wong, J. L., Oesterlin, R. and Rapoport, H. (1971b). *J. Amer. Chem. Soc.* **93**, 7344.
Yamamura, S. and Hirata, Y. (1963). *Tetrahedron*, **19**, 1485.
Yamanouchi, T. (1955). *Publ. Seto Mar. Biol. Lab.* **4**, 183.
Yaphe, W. (1959). *Can. J. Bot.* **37**, 751.
Yasumoto, T., Watanabe, T. and Hashimoto, Y. (1964). *Bull. Jap. Soc. Sci. Fish.* **30**, 357.
Yasumoto, T. and Hashimoto, Y. (1965). *Bull. Agr. Biol. Chem.* (*Jap.*), **29**, 804.
Yasumoto, T., Watanabe, M. and Hashimoto, Y. (1966). *Bull. Jap. Soc. Sci. Fish.* **32**, 673.
Yasumoto, T., Nakamura, K. and Hashimoto, Y. (1967). *Agr. Biol. Chem.* **31**, 7.
Youngken, H. W., Jr. (1969). *Lloydia*, **32**, 407.

Appendix

Tables of physical and chemical constants relevant to marine chemistry

TABLE 1

Some physical properties of pure water (after Dorsey, 1940)

Molecular weight	18·0153
Heat of formation	285·89 $kJmol^{-1}$ (at 25°C and 1 atm)
Ionic dissociation constant	10^{-4} M^{-1} (at 25°C and 1 atm)
Heat of ionization	55·71 $kJmol^{-1}$ (at 25°C and 1 atm)
Viscosity	8·949 mP (at 25°C and 1 atm)
Velocity of sound	1496·3 ms^{-1} (at 25°C and 1 atm)
Density	0·9979751 g cm^{-3} (at 25°C and 1 atm)
Freezing point	0°C (at 1 atm)
Boiling point	100°C (at 1 atm)
Isothermal compressibility	45·6 × 10^{-6} atm^{-1} (at 25°C over the range 1–10 atm)
Specific heat at constant volume	4·1786 int.J $(g°C)^{-1}$ (at 25°C and 1 atm)
Thermal conductivity	0·00598 W cm^{-1} $°C^{-1}$ (at 20°C and 1 atm)
Temperature of maximum density	3·98°C (at 1 atm)
Dielectric constant	81·0 (at 1 atm, 17°C, and 60 MHz)
Electrical conductivity	Less than 10^{-8} Ω^{-1} cm^{-1} (at 25°C and 1 atm)

TABLE 2

Concentrations of the major ions in sea water of various salinities ($g\ kg^{-1}$)*

Salinity (‰)	Na^+	Mg^{2+}	Ca^{2+}	K^+	Sr^{2+}	B	Cl^-	SO_4^{2-}	Br^-	F^-	HCO_3^-
5	1·539	0·185	0·058	0·057	0·001	0·001	2·763	0·387	0·010	0·0002	0·020
10	3·078	0·370	0·118	0·114	0·002	0·001	5·527	0·775	0·019	0·0004	0·041
15	4·617	0·555	0·177	0·171	0·003	0·002	8·290	1·162	0·029	0·0005	0·061
20	6·156	0·739	0·235	0·228	0·005	0·003	11·054	1·550	0·038	0·0007	0·081
25	7·695	0·924	0·294	0·285	0·006	0·003	13·817	1·937	0·048	0·0009	0·101
30	9·234	1·109	0·353	0·342	0·007	0·004	16·581	2·325	0·058	0·0011	0·122
31	9·542	1·146	0·365	0·353	0·007	0·004	17·133	2·402	0·059	0·0011	0·126
32	9·850	1·183	0·377	0·365	0·007	0·004	17·685	2·480	0·062	0·0012	0·130
33	10·157	1·220	0·388	0·376	0·007	0·004	18·239	2·557	0·063	0·0012	0·134
34	10·465	1·257	0·400	0·388	0·008	0·004	18·791	2·635	0·065	0·0012	0·137
35	10·773	1·294	0·412	0·399	0·008	0·004	19·344	2·712	0·067	0·0013	0·142
36	11·081	1·331	0·424	0·410	0·008	0·005	19·897	2·789	0·069	0·0013	0·146
37	11·389	1·368	0·435	0·422	0·008	0·005	20·449	2·867	0·071	0·0013	0·150
38	11·696	1·405	0·447	0·433	0·009	0·005	21·002	2·944	0·073	0·0014	0·154
39	12·004	1·442	0·459	0·445	0·009	0·005	21·555	3·022	0·075	0·0014	0·158
40	12·312	1·479	0·471	0·456	0·009	0·005	22·107	3·099	0·077	0·0015	0·162
41	12·620	1·516	0·482	0·467	0·009	0·005	22·660	3·177	0·079	0·0015	0·166
42	12·928	1·553	0·494	0·479	0·009	0·005	23·213	3·254	0·081	0·0015	0·170

* Cations concentrations; averages of mean results of Cox and Culkin (1967) and Riley and Tongudai (1967). Sulphate and bromide concentration based on mean values from Morris and Riley (1966).

TABLE 3

Preparation of artificial sea water (S = 35·00‰)

Lyman and Fleming (1940)	(g.)	Kalle (1945)	(g.)
NaCl	23·939	NaCl	28·566
$MgCl_2$	5·079	$MgCl_2$	3·887
Na_2SO_4	3·994	$MgSO_4$	1·787
$CaCl_2$	1·123	$CaSO_4$	1·308
KCl	0·667	K_2SO_4	0·832
$NaHCO_3$	0·196	$CaCO_3$	0·124
KBr	0·098	KBr	0·103
H_3BO_3	0·027	$SrSO_4$	0·0288
$SrCl_2$	0·024	H_3BO_3	0·0282
NaF	0·003		
Water to	1 kg	Water to	1 kg

Kester *et al.* (1967)

A. Gravimetric salts

	$g\,kg^{-1}$
NaCl	23·926
Na_2SO_4	4·008
KCl	0·667
$NaHCO_3$	0·196
KBr	0·098
H_3BO_3	0·026
NaF	0·003

B. Volumetric salts (standardized by Mohr method)

	Approx. molarity	Use volume equivalent to
$MgCl_26H_2O$	1·0 M	1·297 g Mg kg^{-1}
$CaCl_22H_2O$	1·0 M	0·406 g Ca kg^{-1}
$SrCl_26H_2O$	0·1 M	0·0133g Sr kg^{-1}

C Water to 1 kg

Note: (i) Allowance must be made for water of crystallization of any of the salts used.
(ii) After aeration the pH should lie between 7·9 and 8·3.

TABLE 4

Collected conversion factors

Conversion	Factor	Reciprocal
μg NO_3^- ⟶ μg N	0·2259	4·427
μg NO_2^- ⟶ μg N	0·3045	3·286
μg NH_3 ⟶ μg N	0·8225	1·216
μg NH_4^+ ⟶ μg N	0·7764	1·287
μg PO_4^{3-} ⟶ μg P	0·3261	3·066
μg P_2O_5 ⟶ μg P	0·4364	2·291
μg SiO_2 ⟶ μg Si	0·4675	2·139
μg SiO_4^{4-} ⟶ μg Si	0·3050	3·278
μg N ⟶ μg-at N	0·07138	14·008
μg P ⟶ μg-at. P	0·03228	30·975
μg Si ⟶ μg-at. Si	0·03560	28·09

TABLE 5

Table for conversion of weights of nitrogen, phosphorus and silicon expressed in terms of μg into μg-at.

μg N, P, or Si^{-1}	μg-at Nl^{-1}	μg-at. P l^{-1}	μg-at. Si l^{-1}
1	0·071	0·032	0·036
2	0·143	0·065	0·071
3	0·214	0·097	0·107
4	0·286	0·129	0·142
5	0·357	0·161	0·178
6	0·428	0·194	0·214
7	0·500	0·226	0·249
8	0·571	0·258	0·284
9	0·643	0·291	0·320
10	0·714	0·323	0·356
20	1·428	0·646	0·712
30	2·142	0·968	1·068
40	2·856	1·291	1·424
50	3·569	1·614	1·780
60	4·283	1·937	2·136
70	4·997	2·260	2·492
80	5·711	2·582	2·848
90	6·425	2·905	3·204
100	7·139	3·228	3·560

TABLE 6

*Solubility of oxygen (C) in sea water ($cm^3\ dm^{-3}$) with respect to an atmosphere of 20·95% oxygen and 100% relative humidity at a total atmospheric pressure of 760 mm Hg. (UNESCO, 1973)**

T	Salinity (‰)														
(°C)	0	5	10	15	20	25	30	31	32	33	34	35	36	37	38
0	10·22	9·87	9·54	9·22	8·91	8·61	8·32	8·27	8·21	8·16	8·10	8·05	7·99	7·94	7·88
1	9·94	9·60	9·28	8·97	8·68	8·39	8·11	8·05	8·00	7·94	7·89	7·84	7·78	7·73	7·68
2	9·67	9·35	9·04	8·74	8·45	8·17	7·90	7·85	7·79	7·74	7·69	7·64	7·59	7·53	7·48
3	9·41	9·10	8·80	8·51	8·23	7·96	7·70	7·65	7·60	7·55	7·50	7·45	7·40	7·35	7·30
4	9·16	8·86	8·57	8·29	8·02	7·76	7·51	7·46	7·41	7·36	7·31	7·26	7·22	7·17	7·12
5	8·93	8·64	8·36	8·09	7·83	7·57	7·33	7·28	7·23	7·18	7·14	7·09	7·04	7·00	6·95
6	8·70	8·42	8·15	7·89	7·64	7·39	7·15	7·11	7·06	7·01	6·97	6·92	6·88	6·83	6·79
7	8·49	8·22	7·95	7·70	7·45	7·22	6·98	6·94	6·89	6·85	6·81	6·76	6·72	6·67	6·63
8	8·28	8·02	7·76	7·52	7·28	7·05	6·82	6·78	6·74	6·69	6·65	6·61	6·57	6·52	6·48
9	8·08	7·83	7·58	7·34	7·11	6·89	6·67	6·63	6·59	6·54	6·50	6·46	6·42	6·38	6·34
10	7·89	7·64	7·41	7·17	6·95	6·73	6·52	6·48	6·44	6·40	6·36	6·32	6·28	6·24	6·20
11	7·71	7·47	7·24	7·01	6·80	6·58	6·38	6·34	6·30	6·26	6·22	6·18	6·14	6·10	6·07
12	7·53	7·30	7·08	6·86	6·65	6·44	6·24	6·21	6·17	6·13	6·09	6·05	6·01	5·98	5·94
13	7·37	7·14	6·92	6·71	6·50	6·31	6·11	6·07	6·04	6·00	5·96	5·93	5·89	5·85	5·82
14	7·20	6·98	6·77	6·57	6·37	6·17	5·99	5·95	5·91	5·88	5·84	5·80	5·77	5·73	5·70
15	7·05	6·84	6·63	6·43	6·24	6·05	5·87	5·83	5·79	5·76	5·72	5·69	5·65	5·62	5·58
16	6·90	6·69	6·49	6·30	6·11	5·93	5·75	5·71	5·68	5·64	5·61	5·58	5·54	5·51	5·48
17	6·75	6·55	6·36	6·17	5·99	5·81	5·64	5·60	5·57	5·53	5·50	5·47	5·43	5·40	5·37
18	6·61	6·42	6·23	6·05	5·87	5·69	5·53	5·49	5·46	5·43	5·40	5·36	5·33	5·30	5·27
19	6·48	6·29	6·11	5·93	5·75	5·59	5·42	5·39	5·36	5·33	5·29	5·26	5·23	5·20	5·17
20	6·35	6·17	5·99	5·81	5·64	5·48	5·32	5·29	5·26	5·23	5·20	5·17	5·14	5·10	5·07
21	6·23	6·05	5·87	5·70	5·54	5·38	5·22	5·19	5·16	5·13	5·10	5·07	5·04	5·01	4·98
22	6·11	5·93	5·76	5·60	5·44	5·28	5·13	5·10	5·07	5·04	5·01	4·98	4·95	4·92	4·89
23	5·99	5·82	5·65	5·49	5·34	5·18	5·04	5·01	4·98	4·95	4·92	4·89	4·87	4·84	4·81

TABLE 6 *cont.*

*Solubility of oxygen (C) in sea water ($cm^3\ dm^{-3}$) with respect to an atmosphere of 20·95 % oxygen and 100 % relative humidity at a total atmospheric pressure of 860 mm Hg. (UNESCO, 1973)**

T	Salinity (‰)														
(°C)	0	5	10	15	20	25	30	31	32	33	34	35	36	37	38
24	5·88	5·71	5·55	5·39	5·24	5·09	4·95	4·92	4·89	4·86	4·84	4·81	4·78	4·75	4·73
25	5·77	5·61	5·45	5·30	5·15	5·00	4·86	4·84	4·81	4·78	4·75	4·73	4·70	4·67	4·65
26	5·66	5·51	5·35	5·20	5·06	4·92	4·78	4·75	4·73	4·70	4·67	4·65	4·62	4·59	4·57
27	5·56	5·41	5·26	5·11	4·97	4·83	4·70	4·67	4·65	4·62	4·60	4·57	4·54	4·52	4·49
28	5·46	5·31	5·17	5·03	4·89	4·75	4·62	4·60	4·57	4·55	4·52	4·50	4·47	4·45	4·42
29	5·37	5·22	5·08	4·94	4·81	4·67	4·55	4·52	4·50	4·47	4·45	4·42	4·40	4·37	4·35
30	5·28	5·13	4·99	4·86	4·73	4·60	4·47	4·45	4·43	4·40	4·38	4·35	4·33	4·31	4·28
31	5·19	5·05	4·91	4·78	4·65	4·53	4·40	4·38	4·36	4·33	4·31	4·28	4·26	4·24	4·22
32	5·10	4·96	4·83	4·70	4·58	4·45	4·33	4·31	4·29	4·26	4·24	4·22	4·20	4·17	4·15

* Based on measurements by Carpenter (1966) and Murray and Riley (1969a) fitted by Weiss (1970) to the thermodynamically consistent equation:

$\ln C = A_1 + A_2(100/T) + A_3 \ln(T/100) + A_4(T/100) + S‰[B_1 + B_2(T/100) + B_3(T/100)^2]$

where

A_1	A_2	A_3	A_4	B_1	B_2	B_3
−173·4292	249·6339	143·3483	−21·8492	−0·033096	0·014259	−0·0017000

and T and $S‰$ are the absolute temperature (K) and salinity in parts per mille respectively.

TABLE 7

Solubility of nitrogen in sea water ($cm^3\ dm^{-3}$) with respect to an atmosphere of 78·084 % nitrogen and 100 % relative humidity at a total pressure of 760 mm Hg (Weiss (1970) from data by Murray and Riley (1969b)).

T(°C)	Salinity ‰ 0	10	20	30	34	35	36	38	40
−1	—	—	16·28	15·10	14·65	14·54	14·44	14·22	14·01
0	18·42	17·10	15·87	14·73	14·30	14·19	14·09	13·88	13·67
1	17·95	16·67	15·48	14·38	13·96	13·86	13·75	13·55	13·35
2	17·50	16·26	15·11	14·04	13·64	13·54	13·44	13·24	13·05
3	17·07	15·87	14·75	13·72	13·32	13·23	13·13	12·94	12·76
4	16·65	15·49	14·41	13·41	13·03	12·93	12·84	12·66	12·47
5	16·26	15·13	14·09	13·11	12·74	12·65	12·56	12·38	12·21
6	15·88	14·79	13·77	12·83	12·47	12·38	12·29	12·12	11·95
8	15·16	14·14	13·18	12·29	11·95	11·87	11·79	11·62	11·46
10	14·51	13·54	12·64	11·80	11·48	11·40	11·32	11·17	11·01
12	13·90	12·99	12·14	11·34	11·04	10·96	10·89	10·74	10·60
14	13·34	12·48	11·67	10·92	10·63	10·56	10·49	10·35	10·21
16	12·83	12·01	11·24	10·53	10·25	10·19	10·12	9·99	9·86
18	12·35	11·57	10·84	10·16	9·90	9·84	9·77	9·65	9·52
20	11·90	11·16	10·47	9·82	9·57	9·51	9·45	9·33	9·21
22	11·48	10·78	10·12	9·50	9·26	9·21	9·15	9·03	8·92
24	11·09	10·42	9·79	9·20	8·98	8·92	8·87	8·76	8·65
26	10·73	10·09	9·49	8·92	8·71	8·65	8·60	8·50	8·39
28	10·38	9·77	9·20	8·66	8·45	8·40	8·35	8·25	8·15
30	10·06	9·48	8·93	8·41	8·21	8·16	8·12	8·02	7·92
32	9·76	9·20	8·67	8·18	7·99	7·94	7·89	7·80	7·71
34	9·48	8·94	8·43	7·96	7·77	7·73	7·68	7·59	7·51
36	9·21	8·69	8·20	7·75	7·57	7·53	7·48	7·40	7·31
38	8·95	8·46	7·99	7·55	7·38	7·33	7·29	7·21	7·13
40	8·71	8·23	7·78	7·36	7·19	7·15	7·11	7·03	6·95

The solubility at any value of salinity and temperature in the above range can be calculated if the following constants are substituted in the equation below (Table 6).

A_1	A_2	A_3	A_4	B_1	B_2	B_3
−172·4965	248·4262	143·0738	−21·7120	−0·049781	0·025018	−0·003486

TABLE 8

Solubility of argon in sea water (cm^{-3} dm^{-3}) *with respect to an atmosphere of 0·934% argon and 100% relative humidity at a total atmosphere pressure of 760 mm Hg* (*Weiss* (*1970*) *from data by Douglas* (*1964, 1965*)).

T (°C)	Salinity ‰ 0	10	20	30	34	35	36	38	40
−1	——	——	0·4456	0·4156	0·4042	0·4014	0·3986	0·3931	0·3877
0	0·4980	0·4647	0·4337	0·4048	0·3937	0·3910	0·3883	0·3830	0·3777
1	0·4845	0·4524	0·4224	0·3944	0·3837	0·3811	0·3785	0·3733	0·3682
2	0·4715	0·4405	0·4115	0·3845	0·3741	0·3716	0·3691	0·3641	0·3592
3	0·4592	0·4292	0·4012	0·3750	0·3650	0·3625	0·3601	0·3552	0·3505
4	0·4474	0·4184	0·3912	0·3659	0·3562	0·3538	0·3515	0·3468	0·3422
5	0·4360	0·4080	0·3817	0·3572	0·3478	0·3455	0·3432	0·3387	0·3342
6	0·4252	0·3980	0·3726	0·3488	0·3397	0·3375	0·3353	0·3309	0·3265
8	0·4049	0·3794	0·3555	0·3331	0·3246	0·3225	0·3204	0·3162	0·3121
10	0·3861	0·3622	0·3397	0·3186	0·3106	0·3086	0·3066	0·3027	0·2989
12	0·3688	0·3463	0·3251	0·3053	0·2977	0·2958	0·2939	0·2902	0·2866
14	0·3528	0·3316	0·3116	0·2929	0·2857	0·2839	0·2822	0·2787	0·2752
16	0·3380	0·3180	0·2991	0·2814	0·2746	0·2729	0·2712	0·2679	0·2647
18	0·3242	0·3053	0·2875	0·2707	0·2642	0·2626	0·2610	0·2579	0·2548
20	0·3114	0·2935	0·2766	0·2607	0·2546	0·2531	0·2516	0·2486	0·2457
22	0·2995	0·2825	0·2665	0·2514	0·2455	0·2441	0·2427	0·2399	0·2371
24	0·2883	0·2722	0·2570	0·2426	0·2371	0·2357	0·2344	0·2317	0·2291
26	0·2779	0·2626	0·2481	0·2344	0·2292	0·2279	0·2266	0·2241	0·2215
28	0·2681	0·2535	0·2398	0·2268	0·2217	0·2205	0·2193	0·2169	0·2144
30	0·2588	0·2450	0·2319	0·2195	0·2147	0·2136	0·2124	0·2101	0·2078
32	0·2502	0·2370	0·2245	0·2127	0·2081	0·2070	0·2059	0·2037	0·2015
34	0·2420	0·2294	0·2175	0·2062	0·2019	0·2008	0·1997	0·1976	0·1955
36	0·2342	0·2222	0·2109	0·2001	0·1959	0·1949	0·1939	0·1919	0·1899
38	0·2269	0·2154	0·2046	0·1943	0·1903	0·1893	0·1883	0·1864	0·1845
40	0·2199	0·2090	0·1986	0·1888	0·1849	0·1840	0·1831	0·1812	0·1794

The solubility at any value of salinity and temperature in the above range can be calculated if the following constants are substituted in the equation below (Table 6).

A_1	A_2	A_3	A_4	B_1	B_2	B_3
−173·5146	245·4510	141·8222	−21·8020	−0·034474	0·014934	−0·0017729

TABLE 9

Literature citations for solubilities of other gases in sea water

Gas	Reference
Carbon dioxide	Murray and Riley (1971); see also Chapter 9, Table 9.
Helium	Weiss (1971); see also Chapter 8, Table A8.5.
Neon	Weiss (1971); see also Chapter 8, Table A8.4
Krypton	Wood and Caputi (1966); see also Chapter 8, Table 8.5.
Xenon	Wood and Caputi (1966); see also Chapter 8, Table 8.5.
Carbon monoxide	Douglas (1967); see also Chapter 8, Table 8.12.
Hydrogen	Crozier and Yamamoto (1974).

TABLE 10

*The density of artificial sea water as a function of temperature and chlorinity**
(Millero and Lepple, 1973)

Cl (‰)	0°C	5°C	10°C	15°C	20°C	25°C	30°C	35°C	40°C
0	0·999868	0·999992	0·999728	0·999129	0·998234	0·997075	0·995678	0·994063	0·992247
$3{\cdot}42_6$	1·004944	1·004959	1·004599	1·003921	1·002962	1·001744	1·000295	0·998643	0·996783
$6{\cdot}05_5$	1·008665	1·008705	1·008292	1·007566	1·006575	1·005335	1·003868	1·002190	1·000307
$8{\cdot}17_4$	1·011851	1·011731	1·011265	1·010502	1·009472	1·008201	1·006707	1·005013	1·003113
$11{\cdot}69_5$	1·016982	1·016758	1·016208	1·015368	1·014275	1·012949	1·011407	1·009669	1·007745
$13{\cdot}67_3$	1·019835	1·019564	1·018970	1·018102	1·016986	1·015641	1·014087	1·012346	1·010406
$16{\cdot}33_3$	1·023703	1·023352	1·022695	1·021772	1·020611	1·019229	1·017642	1·015866	1·013920
$19{\cdot}05_6$	1·027648	1·027227	1·026511	1·025538	1·024335	1·022921	1·021311	1·019528	1·017564
$21{\cdot}53_7$	1·031240	1·030774	1·029989	1·028941	1·027731	1·026307	1·024658	1·022890	1·020925

* These densities are relative to those tabulated by Kell (1967) for pure water assuming the density of pure water is 1·000000 g ml^{-1} at 3·98°C.

TABLE 11

The expansibility of artificial sea water as a function of temperature and chlorinity, $\alpha \times 10^6$ (deg.$^{-1}$) (Millero and Lepple, 1973)

Cl‰	0°C	5°C	10°C	15°C	20°C	25°C	30°C	35°C	40°C
0·000	−68·1	16·0	87·9	150·7	206·6	257·0	303·1	345·7	385·4
3·426	−46·9	35·5	105·1	165·4	218·7	266·7	310·7	351·8	391·0
6·055	−28·0	49·4	115·2	172·7	224·1	271·0	314·8	356·4	396·7
8·174	−14·8	60·4	124·4	180·5	230·7	276·6	319·4	359·9	398·8
11·695	8·2	79·2	140·2	194·1	242·7	287·2	328·3	366·7	402·6
13·673	18·4	88·1	147·5	199·6	246·6	289·8	330·1	368·6	405·7
16·333	36·1	102·3	159·2	209·4	254·8	296·8	335·8	372·8	407·9
19·056	51·0	115·2	170·2	218·5	262·2	302·4	339·9	375·2	408·8
21·537	61·9	127·6	181·6	227·5	267·9	304·9	340·1	375·0	410·7

$\alpha = -1/d(\partial d/\partial t)$ where d is the density of the sea water.

TABLE 12

The isothermal compressibility of sea water at l atm as a function of salinity and temperature (Lepple and Millero, 1971)

				$\beta \times 10^{-6}$ (bar^{-1})					
S(‰)	0°C	5°C	10°C	15°C	20°C	25°C	30°C	35°C	40°C
0·00	50·886	49·171	47·811	46·736	45·895	45·250	44·774	44·444	44·243
6·14	50·07	48·42	47·10	46·09	45·31	44·71	44·26	43·92	43·75
11·80	49·25	47·70	46·43	45·41	44·66	44·13	43·68	43·34	43·19
14·75	48·84	47·30	46·11	45·15	44·38	43·83	43·43	43·13	43·01
21·01	48·14	46·71	45·59	44·63	43·92	43·45	42·96	42·71	42·63
24·52	47·63	46·25	45·17	44·29	43·61	42·98	42·68	42·33	42·23
29·38	47·01	45·62	44·62	43·74	43·17	42·56	42·24	41·96	41·86
34·25	46·49	45·17	44·15	43·32	42·69	42·18	41·88	41·69	41·55
35·00	46·32	45·03	44·02	43·19	42·58	42·11	41·78	41·49	41·48
39·00	45·84	44·62	43·63	42·80	42·30	41·73	41·53	41·23	41·15

TABLE 13

Observed values for the change in the specific volume of sea water from 0° to T°C at various pressures and salinities. Unit of specific volume = 10^{-6} cm^3 g^{-1}. (Cox et al., 1970)

	S = 35·00‰						S = 30·50‰				S = 39·50‰			
P, bars absolute→	8·3	201·3	401·2	601·0	800·9	1000·8	8·3	201·3	601·0	1000·9	8·3	201·3	601·0	1000·8
S, ‰→	35·000	35·004	35·005	35·002	35·002	35·002	30·502	30·504	30·506	30·510	39·503	39·502	39·504	39·507
pH (1 bar, 25°C)→	7·91	7·95	7·94	7·94	8·00	7·96	8·06	7·98	8·03	8·00	8·22	8·18	8·13	8·16
T(°C)														
−2·000	—	—	−277·1	−356·9	−424·3	−480·5		—	−341·9	−472·3	—	—	−370·6	−489·4
−1·000	—	−97·5	—	—	—	—	—	−86·9	—	—	—	−107·1	—	—
0·000	0	0	0	0	0	0	0	0	0	0	0	0	0	0
2·000	132·2	224·9	310·03	383·2	444·9	497·6	106·9	204·7	368·6	489·0	155·8	245·2	394·4	504·3
4·000	311·2	489·5	652·0	791·7	910·6	1012·5	262·7	450·5	766·4	998·5	355·8	527·9	815·7	1026·8
6·000	535·0	791·0	1023·0	1225·3	1396·8	1544·7	464·7	734·7	1189·8	1523·0	599·1	846·2	1259·1	1556·1
8·000	801·0	1127·3	1424·4	1683·3	1902·4	2094·3	712·1	1055·7	1637·2	2064·2	883·0	1198·6	1726·2	2117·5
10·000	1107·1	1498·0	1854·4	2163·3	2427·7	2660·1	1000·9	1412·3	2106·9	2623·2	1205·8	1582·7	2216·4	2686·7
12·000	1452·7	1901·6	2312·4	2668·5	2971·6	3243·6	1330·2	1802·6	2603·7	3197·3	1566·5	1999·4	2729·5	3274·6
14·000	1836·3	2336·9	2796·2	3198·3	3535·9	3827·4	1698·8	2226·3	3123·4	3790·3	1962·6	2446·4	3264·0	3872·9
16·000	2255·5	2804·0	3306·4	3745·3	4119·2	4448·9	2104·7	2682·5	3665·3	4398·0	2394·3	2923·1	3818·4	4489·2
18·000	2709·3	3299·8	3840·9	4315·0	4721·5	5075·4	2547·1	3169·6	4228·6	5021·7	2858·7	3428·0	4394·0	5119·5
20·000	3196·3	3823·9	4400·2	4906·4	5341·0	5719·9	3022·8	3685·2	4815·6	5661·0	3355·8	3961·8	4994·4	5764·6
22·000	3717·0	4376·5	4984·1	5516·1	5975·8	6378·6	3533·0	4230·0	5421·9	6315·4	3883·4	4522·1	5610·2	6423·2
24·000	4268·2	4957·5	5591·4	6151·3	6630·1	7051·0	4075·6	4804·1	6049·9	6985·9	4442·8	5108·6	6246·4	7100·4
26·000	4850·1	5564·4	6223·6	6803·5	7303·4	7738·2	4649·9	5403·9	6698·9	7673·4	5031·4	5721·8	6901·7	7789·7
28·000	5461·6	6197·0	6877·1	7472·2	7990·8	8439·8	5255·2	6032·2	7367·9	8374·8	5648·6	6358·8	7576·4	8493·4
30·000	6102·8	6855·3	7554·3	8165·9	8693·8	9159·8	5889·4	6687·1	8056·6	9091·8	6294·8	7021·5	8269·4	9211·1

TABLE 14

Specific gravity and percentage volume reduction of sea water under pressure (amended from Cox, 1965)*

Pressure (db)	Specific gravity	% decrease in volume
0	1·02813	0·000
100	1·02860	0·046
200	1·02908	0·093
500	1·03050	0·231
1,000	1·03285	0·460
2,000	1·03747	0·909
3,000	1·04199	1·349
4,000	1·04640	1·778
5,000	1·05071	2·197
6,000	1·05494	2·609
7,000	1·05908	3·011
8,000	1·06314	3·406
9,000	1·06713	3·794
10,000	1·07104	4·175

* Salinity, 35·00‰; Temperature 0°C.

TABLE 15

Percentage reduction in volume of sea water under a pressure of 1,000 db at various temperatures and salinities. (After Cox, 1965).

S‰	Temperature (°C)			
	0	10	20	30
0	0·500	0·470	0·451	0·440
10	0·486	0·459	0·442	0·432
20	0·474	0·448	0·432	0·423
30	0·462	0·438	0·424	0·415
35	0·457	0·433	0·419	0·411
40	0·450	0·428	0·415	0·407

TABLE 16

Thermal expansion of sea water under pressure (10^{-6} cm^3 (°C)$^{-1}$). (Bradshaw and Schleicher, 1970)

Pressure (bars)	Temperature (°C)			
	0	10	20	30
		$S = 30{\cdot}50‰$		
1	39	155	246	324
500	158	229	290	346
1000	240	284	323	362
		$S = 35{\cdot}00‰$		
1	52	162	251	327
500	166	234	293	347
1000	244	286	325	363
		$S = 39{\cdot}50‰$		
1	65	170	256	329
500	174	239	296	348
1000	248	289	326	363

TABLE 17a*

Velocity of sound in sea water†

Pressure (db)	Temperature (°C)						
	0	5	10	15	20	25	30
0	1449·3	1471·0	1490·4	1507·4	1522·1	1534·8	1545·8
1000	1465·8	1487·4	1506·7	1523·7	1538·5	1551·3	1562·5
2000	1482·4	1504·0	1523·2	1540·2	1555·0	1567·9	1579·2
3000	1499·4	1520·7	1538·6	1555·6			
4000	1516·5	1537·7	1555·2	1572·2			
5000	1533·9	1554·8	1571·9	1588·9			
6000	1551·5	1572·1					
7000	1569·3						
8000	1587·3						
9000	1605·4						
10000	1623·5						

* Reproduced by permission of U.S. Navy Oceanographic Office.

† Velocities in m s^{-1}; pressures in decibars above atmosphere. Salinity 35‰. For other salinities see Table 17b.

For detailed tables of the velocity of sound in sea water, see U.S. Naval Oceanographic Office (1962) and Bark *et al.* (1964).

TABLE 17b*

Effect of salinity on sound velocity†

S‰	Temperature (°C)						
	0	5	10	15	20	25	30
30	−7·0	−6·7	−6·5	−6·2	−5·9	−5·6	−5·3
32	−4·2	−4·0	−3·9	−3·7	−3·5	−3·4	−3·2
33	−2·8	−2·7	−2·6	−2·5	−2·4	−2·2	−2·1
34	−1·4	−1·3	−1·3	−1·2	−1·2	−1·1	−1·1
35	0	0	0	0	0	0	0
36	1·4	1·3	1·3	1·2	1·2	1·1	1·1
37	2·8	2·7	2·6	2·5	2·4	2·3	2·1
38	4·2	4·1	3·9	3·7	3·6	3·4	3·2
40	7·0	6·8	6·5	6·2	6·0	5·7	5·3

* Reproduced by permission of U.S. Navy Oceanographic Office.

Corrections to be applied to the values in Table 17a for salinities other than 35‰.

TABLE 18

Specific heat of sea water at constant pressure ($J g^{-1} {}^\circ C^{-1}$) at various salinities and temperatures (Millero et al., 1973).

Salinity, ‰	0°C	5°C	10°C	15°C	20°C	25°C	30°C	35°C	40°C
0	4·2174	4·2019	4·1919	4·1855	4·1816	4·1793	4·1782	4·1779	4·1783
5	4·1812	4·1679	4·1599	4·1553	4·1526	4·1513	4·1510	4·1511	4·1515
10	4·1466	4·1354	4·1292	4·1263	4·1247	4·1242	4·1248	4·1252	4·1256
15	4·1130	4·1038	4·0994	4·0982	4·0975	4·0977	4·0992	4·0999	4·1003
20	4·0804	4·0730	4·0702	4·0706	4·0709	4·0717	4·0740	4·0751	4·0754
25	4·0484	4·0428	4·0417	4·0437	4·0448	4·0462	4·0494	4·0508	4·0509
30	4·0172	4·0132	4·0136	4·0172	4·0190	4·0210	4·0251	4·0268	4·0268
35	3·9865	3·9842	3·9861	3·9912	3·9937	3·9962	4·0011	4·0031	4·0030
40	3·9564	3·9556	3·9590	3·9655	3·9688	3·9718	3·9775	3·9797	3·9795

TABLE 19

The relative partial equivalent heat capacity of sea salt (cal(eq deg)$^{-1}$) (Millero et al., 1973a)

Salinity (‰)	Temperature (°C) 0	5	10	15	20	25	30
0	0	0	0	0	0	0	0
5	2·8	3·1	3·4	3·7	4·0	4·3	4·6
10	4·7	5·0	5·4	5·7	5·9	6·3	6·6
15	6·9	7·1	7·3	7·5	7·7	7·9	8·1
20	9·5	9·4	9·4	9·4	9·3	9·3	9·2
25	12·2	11·9	11·5	11·2	10·8	10·5	10·2
30	15·3	14·6	13·8	13·1	12·3	11·6	10·8
35	18·6	17·4	16·2	15·0	13·8	12·6	11·3
40	22·2	20·5	18·7	17·0	15·2	13·4	11·7

TABLE 20

*Thermal conductivity (K in 10^{-5} W cm deg^{-1}) of sea water (S = 34·994‰) as a function of temperature and pressure. (After Castelli et al., 1974)**

Pressure (p) (bars)	Temperature (t°C) 1·82	10	20	30
200	563	578	594	605
400	570	585	601	613
600	578	592	609	619
800	585	599	615	627
1000	591	606	622	634
1200	596	613	628	641
1400	602	618	634	647

$K = 5{\cdot}5286 \times 10^{-3} + 3{\cdot}4025 \times 10^{-7}P + 1{\cdot}8364 \times 10^{-7}t - 3{\cdot}3058 \times 10^{-9}t$

* Other data have been published by Caldwell (1974).

TABLE 21

Freezing point of sea water (T_f) at atmospheric pressure based on the data of Doherty and Kester (1974).

S‰	T_f (°C)	S‰	T_f (°C)	S‰	T_f (°C)
5	−0·275	17	−0·918	29	−1·582
6	−0·328	18	−0·973	30	−1·638
7	−0·381	19	−1·028	31	−1·695
8	−0·434	20	−1·082	32	−1·751
9	−0·487	21	−1·137	33	−1·808
10	−0·541	22	−1·192	34	−1·865
11	−0·594	23	−1·248	35	−1·922
12	−0·648	24	−1·303	36	−1·979
13	−0·702	25	−1·359	37	−2·036
14	−0·756	26	−1·414	38	−2·094
15	−0·810	27	−1·470	39	−2·151
16	−0·864	28	−1·526	40	−2·209

The freezing point at *in situ* pressure is given by
$T_f(°C) = -0{\cdot}0137 - 0{\cdot}051990\,S‰ - 0{\cdot}00007225\,(S‰)^2 - 0{\cdot}000758z$ where z is the depth in metres.

TABLE 22

Boiling point elevation of sea water (S = 35·00 ‰) at various temperatures, (Stoughton and Lietzke, 1967)

Temp. (°C)	30	40	50	60	70	80	90	100
Vap. press. (atm)	0·042	0·073	0·122	0·197	0·309	0·469	0·694	1·003
Elevation of B.P (°C)	0·325	0·350	0·377	0·405	0·433	0·463	0·493	0·524
Temp. (°C)	120	140	160	180	200	220	240	260
Vap. press. (atm)	1·965	3·577	6·119	9·931	15·407	22·99	33·18	46·52
Elevation of B.P (°C)	0·590	0·660	0·735	0·817	0·906	1·003	1·111	1·232

TABLE 23

Osmotic pressure and vapour depression of sea water at 25°C (Robinson, 1954)

	Chlorinity									
	12	13	14	15	16	17	18	19	20	21
Osmotic pressure (atm)	15·51	16·85	18·19	19·55	20·91	22·28	23·366	25·06	26·47	27·89
Vap. press. lowering* $\times 10^2$	1·139	1·237	1·334	1·433	1·532	1·631	1·732	1·832	1·936	2·039

$(p^0 - p)/p^0$ where p and p^0 are the vapour pressures of sea water and pure water respectively ($p^0 = 23{\cdot}75$ mm at 25°C).

TABLE 24

Surface tension of clean sea water (in N *m*$^{-1}$*) at various salinities and temperatures (from data by Krümmel (1900)* and others (After Fleming and Revelle, 1939)*

S‰	Temperature (°C)			
	0	10	20	30
0	75·64 $\times 10^{-3}$	74·20 $\times 10^{-3}$	72·76 $\times 10^{-3}$	71·32 $\times 10^{-3}$
10	75·86	74·42	72·98	71·54
20	76·08	74·64	73·20	71·76
30	76·30	74·86	73·42	71·98
35	76·41	74·97	73·53	72·09
40	76·52	75·08	73·64	72·20

Surface tension (N m^{-1}) = 10^3 (75·64–0·144t + 0·0221 S‰)
* Measurements made on bubbles below the surface, they therefore take no account of the effects of surface contamination which may be very considerable (e.g. see Lumby and Folkard, 1956 and Vol. 2, pp. 233–4.

TABLE 25

The viscosity of sea water (η) at various salinities and temperatures (in centipoises) computed from values for distilled water (η_0) by Korson et al. (1969) using equations developed by Millero (1974)

Salinity ‰	Temperature °C															
	0	2	4	6	8	10	12	14	16	18	20	22	24	26	28	30
0	1·7916	1·6739	1·5681	1·4725	1·3857	1·3069	1·2349	1·1691	1·1087	1·0532	1·0020	0·9547	0·9109	0·8703	0·8326	0·7975
5	1·8049	1·6868	1·5808	1·4849	1·3979	1·3189	1·2466	1·1807	1·1200	1·0644	1·0129	0·9655	0·9215	0·8807	0·8428	0·8076
10	1·8180	1·6995	1·5930	1·4968	1·4095	1·3302	1·2576	1·1913	1·1304	1·0745	1·0228	0·9751	0·9309	0·8900	0·8519	0·8165
15	1·8312	1·7122	1·6054	1·5087	1·4210	1·3412	1·2685	1·2018	1·1407	1·0845	1·0327	0·9847	0·9402	0·8991	0·8608	0·8252
20	1·8445	1·7251	1·6178	1·5208	1·4325	1·3525	1·2794	1·2125	1·1513	1·0945	1·0424	0·9942	0·9495	0·9082	0·8697	0·8339
25	1·8579	1·7380	1·6302	1·5327	1·4442	1·3638	1·2903	1·2231	1·1614	1·1046	1·0522	1·0036	0·9588	0·9172	0·8786	0·8426
30	1·8713	1·7509	1·6427	1·5448	1·4560	1·3751	1·3012	1·2338	1·1717	1·1146	1·0619	1·0132	0·9682	0·9263	0·8875	0·8513
32	1·8767	1·7563	1·6478	1·5497	1·4607	1·3797	1·3057	1·2379	1·1758	1·1186	1·0658	1·0171	0·9719	0·9300	0·8910	0·8547
34	1·8823	1·7643	1·6528	1·5545	1·4652	1·3843	1·3101	1·2423	1·1800	1·1227	1·0698	1·0210	0·9757	0·9336	0·8945	0·8582
36	1·8876	1·7696	1·6578	1·5594	1·4701	1·3888	1·3146	1·2465	1·1841	1·1267	1·0737	1·0248	0·9793	0·9372	0·8981	0·8617
38	1·8932	1·7752	1·6630	1·5644	1·4748	1·3934	1·3189	1·2508	1·1883	1·1308	1·0778	1·0286	0·9831	0·9409	0·9017	0·8651
40	1·8986	1·7805	1·6680	1·5692	1·4795	1·3980	1·3233	1·2551	1·1925	1·1348	1·0817	1·0325	0·9869	0·9446	0·9053	0·8686
42	1·9041	1·7861	1·6732	1·5741	1·4842	1·4026	1·3278	1·2595	1·1967	1·1389	1·0857	1·0363	0·9906	0·9483	0·9089	0·8721

Viscosity of pure water η_t at temperature t°C is given by $\log \frac{\eta_t}{\eta_{20}} = \frac{1{\cdot}1709(20 - t) - 0{\cdot}001827(t - 20)^2}{t + 89{\cdot}93}$ where η_{20} is the viscosity at 20°C.

Viscosity of sea water calculated from ratio $\frac{\eta}{\eta_0} = 1 + A Cl_v^{\frac{1}{2}} + B Cl_v$

Where Cl_v is the volume chlorinity ($Cl_v = Cl‰ \times$ density) and A = 0·000366, 0·001403 and B = 0·002756, 0·003416 at 5° and 25°C; constants at other temperatures obtained by linear interpolation or extrapolation.

According to Matthäus (1972) the change in dynamic viscosity ($\Delta\eta_p$, centipoises) produced by increase in pressure (P, kg cm^{-2}) at temperature, T°C) can be calculated from the expression

$$\Delta\eta_p = -1{\cdot}7913 \times 10^{-4} P + 9{\cdot}5182 \times 10^{-8} P^2 + P(1{\cdot}3550 \times 10^{-5} T - 2{\cdot}5853 \times 10^{-7} T^2 - P^2(6{\cdot}0833 \times 10^{-9} T - 1{\cdot}1652 \times 10^{-10} T^2)$$

* The assistance of Miss J. Wolfe with the computations is gratefully acknowledged.

TABLE 26

Relative viscosity of Standard Sea Water (S = 35·00‰) at various temperatures and pressures. (Stanley and Batten, 1969)

Pressure, kg cm^{-2}	η_p/η_1 at −0·024°C	η_p/η_1 at 2·219°C	η_p/η_1 at 6·003°C	η_p/η_1 at 10·013°C	η_p/η_1 at 15·018°C	η_p/η_1 at 20·013°C	η_p/η_1 at 29·953°C
176	0·9828	0·9852	0·9891	0·9914	0·9949	0·9977	0·9997
352	0·9709	0·9742	0·9814	0·9876	0·9926	0·9972	0·0001
527	0·9620	0·9670	0·9766	0·9843	0·9900	0·9978	1·0031
703	0·9560	0·9626	0·9735	0·9821	0·9915	0·9998	1·0071
878	0·9533	0·9598	0·9733	0·9836	0·9932	1·0040	1·0131
1055	0·9526	0·9600	0·9750	0·9874	0·9964	1·0070	1·0179
1230	0·9533	0·9637	0·9767	0·9902	1·0014	1·0110	1·0244
1406	0·9559	0·9673	0·9821	0·9961	1·0073	1·0166	1·0313

Where η_p/η_1 is the ratio of the viscosity at pressure p (kg cm^{-2}) relative to that at 1 atm.

TABLE 27

*Specific conductivity of sea water** (*Weyl (1964). From data by Thomas et al., (1934)*)

$S‰$	Temperature (°C)					
	25	20	15	10	5	0
10	17·345	15·628	13·967	12·361	10·816	9·341
20	32·188	29·027	25·967	23·010	20·166	17·456
30	46·213	41·713	37·351	33·137	29·090	25·238
31	47·584	42·954	38·467	34·131	29·968	26·005
32	48·951	44·192	39·579	35·122	30·843	26·771
33	50·314	45·426	40·688	36·110	31·716	27·535
34	51·671	46·656	41·794	37·096	32·588	28·298
35	53·025	47·882	42·896	38·080	33·457	29·060
36	54·374	49·105	43·996	39·061	34·325	29·820
37	55·719	50·325	45·093	40·039	35·190	30·579
38	57·061	51·541	46·187	41·016	36·055	31·337
39	58·398	52·754	47·278	41·990	36·917	32·094

* Conductivity in millimho cm^{-1}.

TABLE 28

Effect of pressure on the conductivity of sea water (after Bradshaw and Schleicher, 1965)*

Temp.	Pressure (db)	S‰			Temp.	S‰		
		31	35	39		31	35	39
	1,000	1·599	1·556	1·512		1·032	1·008	0·985
	2,000	3·089	3·006	2·922		1·996	1·951	1·906
	3,000	4·475	4·345	4·233		2·895	2·830	2·764
	4,000	5·759	5·603	5·448		3·731	3·646	3·562
0°C	5,000	6·944	6·757	6·569	15°C	4·506	4·403	4·301
	6,000	8·034	7·817	7·599		5·221	5·102	4·984
	7,000	9·031	8·787	8·543		5·879	5·745	5·612
	8,000	9·939	9·670	9·401		6·481	6·334	6·187
	9,000	10·761	10·469	10·178		7·031	6·871	6·711
	10,000	11·499	11·188	10·877		7·529	7·358	7·187
	1,000	1·368	1·333	1·298		0·907	0·888	0·868
	2,000	2·646	2·578	2·510		1·755	1·718	1·680
	3,000	3·835	3·737	3·639		2·546	2·492	2·438
	4,000	4·939	4·813	4·686		3·282	3·212	3·142
5°C	5,000	5·960	5·807	5·655	20°C	3·964	3·879	3·795
	6,000	6·901	6·724	6·547		4·594	4·496	4·399
	7,000	7·764	7·565	7·366		5·174	5·064	4·954
	8,000	8·552	8·333	8·114		5·706	5·585	5·464
	9,000	9·269	9·031	8·794		6·192	6·060	5·929
	10,000	9·915	9·661	9·408		6·633	6·492	6·351
	1,000	1·183	1·154	1·125		0·799	0·783	0·767
	2,000	2·287	2·232	2·177		1·547	1·516	1·485
	3,000	3·317	3·237	3·157		2·245	2·200	2·156
	4,000	4·273	4·170	4·067		2·895	2·837	2·780
10°C	5,000	5·159	5·034	4·910	25°C	3·498	3·429	3·359
	6,000	5·976	5·832	5·688		4·056	3·976	3·896
	7,000	6·728	6·565	6·402		4·571	4·481	4·390
	8,000	7·415	7·236	7·057		5·045	4·945	4·845
	9,000	8·041	7·847	7·652		5·478	5·369	5·261
	10,000	8·608	8·400	8·192		5·872	5·756	5·640

* Percentage increase compared with the conductivity at one atmosphere.

TABLE 29

Conductivity ratio of sea water at 15°C (R_{15}) and 20°C (R_{20}) relative to sea water of salinity 35·000‰. (From data in UNESCO, 1966).

S‰	15°C	20°C	S‰	15°C	20°C
29·50	0·85795	0·8583	36·00	1·02545	1·0254
30·00	0·87101	0·8714	36·50	1·03814	1·0380
30·50	0·88404	0·8844	37·00	1·05079	1·0506
31·00	0·89705	0·8973	37·50	1·06341	1·0632
31·50	0·91002	0·9103	38·00	1·07601	1·0758
32·00	0·92296	0·9232	38·50	1·08858	1·0883
32·50	0·93588	0·9361	39·00	1·10112	1·1008
33·00	0·94876	0·9489	39·50	1·11364	1·1133
33·50	0·96160	0·9617	40·00	1·12613	1·1257
34·00	0·97444	0·9745	40·50	1·13849	1·1381
34·50	0·98724	0·9873	41·00	1·15103	1·1505
35·00	1·00000	1·0000	41·50	1·16344	1·1629
35·50	1·01275	1·0127	42·00	1·17583	1·1752

For 15°C $S‰ = -0{\cdot}08996 + 28{\cdot}29720\,R_{15} + 12{\cdot}80832\,R_{15}^2 - 10{\cdot}678969\,R_{15}^3 + 5{\cdot}98624\,R_{15}^4 - 1{\cdot}32311\,R_{15}^5$.

TABLE 30

Correction values (× 10^4) to be applied to conductivity ratios measured at temperatures differing from 20°C to correct them to ratios at 20°C (to be used only in conjunction with 20°C ratios in Table 29). (After UNESCO, 1966)

Measured ratio	Temperature (°C)								
	10	12	14	16	18	20	22	24	26
0·85	80	62	45	29	14	0	−14	−26	−38
0·90	56	44	32	21	10	0	−9	−18	−27
0·95	29	23	17	11	5	0	−5	−10	−14
1·00	0	0	0	0	0	0	0	0	0
1·05	−33	−25	−19	−12	−6	0	5	11	15
1·10	−69	−54	−39	−25	−12	0	11	22	32
1·15	−109	−85	−62	−40	−19	0	18	35	50

TABLE 31

Light absorption of typical sea waters. Extinction for 10 cm path length. (After Clarke and James, 1939)

Sample	Wavelength Å							
	3600	4000	5000	5200	6000	7000	7500	8000
Pure water	0·001	0·001	0·002	0·002	0·010	0·025	0·115	0·086
Artificial sea water	0·011	0·003	0·005	0·007	0·010	0·025	0·115	0·086
Ocean water, unfiltered	0·012	0·009	0·007	0·008	0·011	0·025	0·115	0·086
Continental slope waters, unfiltered	0·052	0·030	0·011	0·010	0·012	0·035	0·130	0·088
Continental slope waters, filtered	0·016	0·010	0·005	0·005	0·012	0·030	0·115	0·086
Inshore water unfiltered	0·055	0·042	0·028	0·026	0·035	0·052	0·140	0·100
Inshore water, filtered	0·015	0·010	0·005	0·005	0·010	0·025	0·110	0·086

TABLE 32

Differences between the extinctions of sea waters and pure water. (From data by Clarke and James, 1939)*

Sample	Wavelength Å							
	3600	4000	5000	5200	6000	7000	7500	8000
Artificial sea water	0·010	0·002	0·003	0·005	nil	nil	nil	nil
Ocean water unfiltered	0·011	0·008	0·005	0·006	0·001	nil	nil	nil
Continental slope water, unfiltered	0·051	0·029	0·009	0·008	0·002	0·010	0·015	0·002
Continental slope water, filtered	0·015	0·009	0·003	0·003	0·002	0·005	nil	nil
Inshore water, unfiltered	0·054	0·041	0·026	0·024	0·025	0·027	0·025	0·015
Inshore water, filtered	0·014	0·009	0·003	0·003	nil	nil	nil	nil

* $E_{SW(10\ cm)} - E_{PW(10\ cm)}$

Note: The values given in Tables 31 and 32 for unfiltered inshore waters should be taken as no more than a rough indication, since actual values vary widely with time and location.

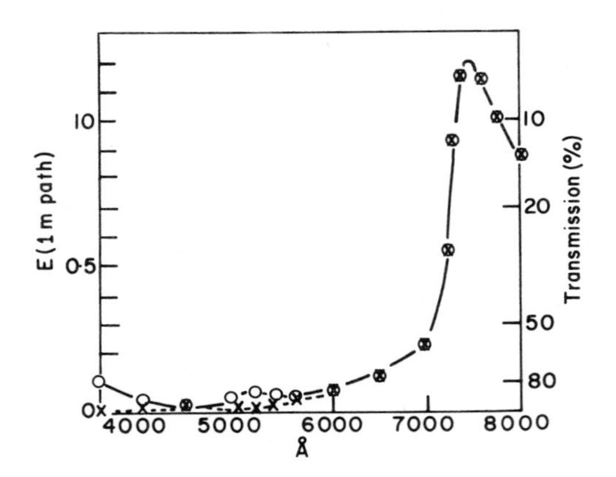

FIG. 1. Extinction (1 m path) against wavelength. Solid line—filtered ocean water. Broken line—pure water. (After Clarke and James, 1939).

TABLE 33

Refractive index differences (Δn) for sea water at a wavelength of 589·3 nm at various temperatures and salinities. ($\Delta n = (n-1{\cdot}30000) \cdot 10^5$). (Matthäus, 1974)

T[°C]	S [‰] 0	2	4	6	8	10	12	14	16	18	20	22	24	26	28	30	32	34	36	38	40
0	3402	3441	3481	3520	3559	3598	3637	3677	3716	3755	3794	3833	3873	3912	3951	3990	4029	4069	4108	4147	4186
1	3400	3439	3478	3517	3556	3595	3634	3674	3713	3752	3791	3830	3869	3908	3947	3986	4025	4064	4103	4142	4181
2	3398	3437	3476	3515	3553	3592	3631	3670	3709	3748	3787	3826	3865	3904	3942	3981	4020	4059	4098	4137	4176
3	3395	3434	3473	3511	3550	3589	3628	3666	3705	3744	3783	3821	3860	3899	3938	3976	4015	4054	4093	4131	4170
4	3392	3431	3469	3508	3547	3585	3624	3662	3701	3740	3778	3817	3855	3894	3933	3971	4010	4048	4087	4126	4164
5	3389	3427	3466	3504	3543	3581	3620	3658	3697	3735	3773	3812	3850	3889	3927	3966	4004	4043	4081	4120	4158
6	3385	3424	3462	3500	3538	3577	3615	3653	3692	3730	3768	3807	3845	3883	3922	3960	3998	4037	4075	4113	4152
7	3381	3419	3458	3496	3534	3572	3610	3648	3687	3725	3763	3801	3839	3878	3916	3954	3992	4030	4068	4107	4145
8	3377	3415	3453	3491	3529	3567	3605	3643	3681	3719	3757	3795	3833	3871	3909	3948	3986	4024	4062	4100	4138
9	3372	3410	3448	3486	3524	3562	3600	3638	3675	3713	3751	3789	3827	3865	3903	3941	3979	4017	4055	4093	4130
10	3367	3405	3443	3518	3518	3556	3594	3632	3669	3707	3745	3783	3821	3858	3896	3934	3972	4010	4047	4085	4123
11	3362	3399	3437	3475	3512	3550	3588	3625	3663	3701	3738	3776	3814	3851	3889	3927	3964	4002	4040	4077	4115
12	3356	3394	3431	3469	3506	3544	3581	3619	3656	3694	3732	3769	3807	3844	3882	3919	3957	3994	4032	4069	4107
13	3350	3387	3425	3462	3500	3537	3575	3612	3649	3687	3724	3762	3799	3837	3874	3911	3949	3986	4024	4061	4098
14	3344	3381	3418	3456	3493	3530	3568	3605	3642	3679	3717	3754	3791	3829	3866	3903	3941	3978	4015	4053	4090
15	3337	3374	3411	3449	3486	3523	3560	3597	3635	3672	3709	3746	3783	3821	3858	3895	3932	3969	4006	4044	4081
16	3330	3367	3404	3441	3478	3515	3552	3590	3627	3664	3701	3738	3775	3812	3849	3886	3923	3960	3997	4035	4072
17	3323	3360	3397	3434	3470	3507	3544	3581	3618	3655	3692	3729	3766	3803	3840	3877	3914	3951	3988	4025	4062
18	3315	3352	3389	3425	3462	3499	3536	3573	3610	3647	3684	3720	3757	3794	3831	3868	3905	3942	3979	4016	4052
19	3307	3344	3380	3417	3454	3491	3527	3564	3601	3638	3675	3711	3748	3785	3822	3858	3895	3932	3969	4006	4042
20	3298	3335	3372	3408	3445	3482	3518	3555	3592	3629	3665	3702	3739	3775	3812	3849	3885	3922	3959	3995	4032
21	3290	3326	3363	3399	3436	3473	3509	3546	3582	3619	3656	3692	3729	3765	3802	3838	3875	3912	3948	3985	4021
22	3281	3317	3354	3390	3427	3463	3500	3536	3573	3609	3646	3682	3719	3755	3792	3828	3865	3901	3938	3974	4011
23	3271	3308	3344	3380	3417	3453	3490	3526	3562	3599	3635	3672	3708	3745	3781	3817	3854	3890	3927	3963	3999
24	3261	3298	3334	3370	3407	3443	3479	3516	3552	3588	3625	3661	3697	3734	3770	3806	3843	3879	3915	3952	3988
25	3251	3288	3324	3360	3396	3433	3469	3505	3541	3578	3614	3650	3686	3723	3759	3795	3831	3868	3904	3940	3976
26	3241	3277	3331	3349	3386	3422	3358	3494	3530	3566	3603	3639	3675	3711	3747	3783	3820	3856	3892	3928	3964
27	3230	3266	3302	3338	3375	3411	3447	3483	3519	3555	3591	3627	3663	3699	3736	3772	3808	3844	3880	3916	3952
28	3219	3255	3291	3327	3363	3399	3435	3471	3507	3543	3579	3615	3651	3687	3723	3759	3796	3832	3868	3904	3940
29	3208	3244	3279	3315	3351	3387	3423	3459	3495	3531	3567	3603	3639	3675	3711	3747	3783	3819	3855	3891	3927
30	3196	3232	3268	3303	3339	3375	3411	3447	3483	3519	3555	3591	3627	3662	3698	3734	3770	3806	3842	3878	3914

TABLE 34

Refractive index differences (Δn) for sea water of salinity 35·00‰ at various temperatures and wavelengths ($\Delta n = (n - 1{\cdot}30000) . 10^5$). (After Matthäus, 1974)

T[°C]	Wavelength (nm) 404·7	435·8	457·9	467·8	480·0	488·0	501·7	508·5	514·5	546·1	577·0	579·1	589·3	632·8	643·8
0	5099	4840	4684	4621	4549	4504	4433	4400	4372	4240	4130	4124	4091	3961	3929
1	5094	4835	4679	4616	4544	4500	4428	4395	4367	4235	4126	4119	4086	3956	3925
2	5089	4830	4674	4611	4529	4495	4423	4390	4362	4230	4121	4114	4081	3951	3920
3	5084	4825	4669	4606	4534	4489	4418	4385	4357	4225	4115	4109	4076	3946	3914
4	5078	4819	4664	4601	4528	4484	4412	4379	4351	4219	4110	4103	4070	3941	3909
5	5072	4814	4658	4595	4522	4478	4407	4374	4345	4213	4104	4097	4065	3935	3903
6	5066	4807	4552	4589	4516	4472	4400	4367	4339	4207	4098	4091	4058	3929	3897
7	5060	4801	4645	4582	4510	4465	4394	4361	4333	4201	4091	4085	4052	3922	3890
8	5053	4794	4639	4576	4503	4459	4387	4354	4326	4194	4085	4078	4045	3916	3884
9	5046	4787	4632	4569	4496	4452	4380	4347	4319	4187	4078	4071	4038	3909	3877
10	5039	4780	4624	4561	4489	4444	4373	4340	4312	4180	4071	4064	4031	3901	3869
11	5031	4773	4617	4554	4481	4437	4366	4332	4304	4172	4063	4056	4023	3894	3862
12	5023	4765	4609	4546	4473	4429	4358	4325	4297	4164	4055	4048	4016	3886	3854
13	5015	4757	4601	4538	4465	4421	4350	4317	4288	4156	4047	4040	4008	3878	3846
14	5007	4748	4592	4529	4457	4412	4341	4308	4280	4148	4039	4032	3999	3869	3837
15	4998	4740	4584	4521	4448	4404	4333	4300	4271	4139	4030	4023	3991	3861	3829
16	4989	4731	4575	4512	4439	4395	4324	4291	4262	4130	4021	4014	3982	3852	3820
17	4980	4721	4566	4503	4430	4386	4314	4281	4253	4121	4012	4005	3972	3843	3811
18	4971	4712	4556	4493	4421	4376	4305	4272	4244	4111	4002	3995	3963	3833	3801
19	4961	4702	4546	4483	4411	4366	4295	4262	4234	4102	3993	3986	3953	3823	3791
20	4951	4692	4536	4473	4401	4356	4385	4252	4224	4092	3982	3976	3943	3813	3781

TABLE 34 *cont.*

T[°C]	0,4047	0,4358	0,4579	0,4678	0,4800	0,4880	0.5017	0.5085	0.5145	0,5461	0,5770	0,5791	0.5893	0,6328	0,6438
21	4940	4682	4526	4463	4390	4346	4275	4242	4214	4081	3972	3965	3933	3803	3771
22	4930	4671	4515	4452	4380	4335	4264	4231	4203	4071	3961	3955	3922	3792	3760
23	4919	4660	4504	4441	4369	4324	4253	4220	4192	4060	3951	3944	3911	3781	3749
24	4908	4649	4493	4430	4358	4313	4242	4209	4181	4048	3939	3932	3900	3770	3738
25	4896	4637	4482	4419	4346	4302	4230	4197	4169	4037	3928	3921	3888	3759	3727
26	4884	4626	4470	4407	4334	4290	4219	4186	4157	4025	3916	3909	3877	3747	3715
27	4872	4614	4458	4395	4322	4278	4207	4174	4145	4013	3904	3897	3865	3735	3703
28	4860	4601	4445	4382	4310	4265	4194	4161	4133	4001	3892	3885	3852	3722	3690
29	4847	4589	4433	4370	4297	4253	4182	4149	4120	3988	3879	3872	3840	3710	3678
30	4834	4576	4420	4357	4284	4240	4169	4136	4108	3975	3866	3859	3827	3697	3665

λ[nm]

TABLE 35

Absolute refractive index of sea water (S = 35·00‰) as a function of temperature, pressure and wavelength. (Stanley, 1971)

Pressure	Temperature (°C) 0·03	5·03	10·03	15·02	20·00	24·99	29·98
			6328 Å				
Atm.	1·34015	1·33977	1·33935	1·33899	1·33850	1·33795	1·33737
352 kg cm^{2}	1·34539	1·34487	1·34431	1·34388	1·34331	1·34270	1·34207
703 kg cm^{-2}	1·35025	1·34962	1·34896	1·34844	1·34780	1·34713	1·34647
1055 kg cm^{-2}	1·35481	1·35403	1·35380	1·35269	1·35200	1·35129	1·35059
1406 kg cm^{-2}	—	1·35813	1·35738	1·35668	1·35592	1·35519	1·35443
			5017 Å				
Atm	1·34455	1·34455	1·34422	1·34379	1·34327	1·34272	1·34215
352 kg cm^{-2}	1·35008	1·34969	1·34924	1·34873	1·34813	1·34757	1·34694
703 kg cm^{-2}	1·35507	1·35450	1·35394	1·35333	1·35269	1·35208	1·35137
1055 kg cm^{-2}	1·35953	1·35891	1·35834	1·35764	1·35695	1·35632	1·35561
1406 kg cm^{-2}	—	1·36314	1·36241	1·36166	1·36095	1·36019	1·35946

TABLE 36

Velocity of light (λ = 589·3 nm) in sea water at 1 atm ($km\ s^{-1}$) (Sager, 1974)

S‰	Temperature (°C) 0	5	10	15	20	25	30	35	40
0	224·732	224·749	224·785	224·837	224·904	224·985	225·080	225·185	225·305
2·5	224·650	224·668	224·705	224·759	224·827	224·909	225·004	225·110	225·230
5·0	224·567	224·588	224·626	224·681	224·749	224·832	224·928	225·035	225·156
7·5	224·485	224·507	224·547	224·603	224·672	224·756	224·852	224·960	225·081
10·0	224·402	224·426	224·468	224·524	224·595	224·679	224·776	224·885	225·006
12·5	224·319	224·346	224·388	224·446	224·518	224·603	224·700	224·810	224·931
15·0	224·236	224·265	224·309	224·368	224·441	224·527	224·625	224·735	224·857
17·5	224·154	224·185	224·230	224·290	224·364	224·450	224·549	224·660	224·782
20·0	224·072	224·104	224·151	224·212	224·287	224·374	224·473	224·585	224·707
22·5	223·990	224·024	224·072	224·134	224·210	224·297	224·398	224·510	224·633
25·0	223·907	223·943	223·994	224·057	224·133	224·221	224·322	224·435	224·559
27·5	223·825	223·863	223·915	223·979	224·056	224·145	224·247	224·360	224·485
30·0	223·743	223·783	223·836	223·901	223·979	224·069	224·171	224·285	224·411
32·5	223·661	223·703	223·758	223·823	223·903	223·993	224·096	224·211	224·336
35·0	223·579	223·623	223·679	223·746	223·826	223·917	224·020	224·136	224·262
37·5	223·498	223·543	223·600	223·669	223·749	223·841	223·945	224·061	224·188
40·0	223·416	223·463	223·521	223·591	223·673	223·765	223·870	223·986	224·114

TABLE 37

Osmotic pressures (atm) of sea salt solutions (Stoughton and Lietzke, 1965)

Sea salts weight %	Temperature °F 25	40	60	80	100
2·0	14·3	14·9	15·7	16·4	16·9
3·45	25·1	26·3	27·7	28·9	29·9
5	37·5	39·3	41·5	43·3	44·7
10	84	89	99	98	101
15	145	153	162	168	173
20	230	240	250	260	270

REFERENCES

Bark, L. S., Ganson, P. P. and Meister, N. A. (1964), "Tables of the Velocity of Sound in Sea Water" Pergamon, Oxford.
Bradshaw, A. and Schleicher, K. E. (1965). *Deep-Sea Res.* **12**, 151.
Bradshaw, A. and Schleicher, K. E. (1970). *Deep-Sea Res.* **16**, 691.
Caldwell, D. T. (1974). *Deep-Sea Res.* **21**, 131.
Carpenter, J. H. (1966). *Limnol. Oceanogr.* **11**, 264.
Castelli, V. J., Stanley, E. M. and Fischer, E. C. (1974). *Deep-Sea Res.* **21**, 311.
Clarke, G. L. and James, H. R. (1939). *J. Opt. Soc. Amer.* **29**, 43.
Cox, R. A. (1965). *In* "Chemical Oceanography" (J. P. Riley and G. Skirrow, eds), Vol. I. Academic Press, London.
Cox R. A. and Culkin, F. (1967). *Deep-Sea Res.* **13**, 789.
Cox, R. A., McCartney, M. J. and Culkin, F. (1970). *Deep-Sea Res.* **17**, 679.
Crozier, T. E. and Yamamoto, S. (1974). *J. Chem. Eng. Data,* **19**, 242.
Doherty, B. T. and Kester, D. R. (1974), *J. Mar. Res.* **32**, 285
Dorsey, N. E. (1940). "Properties of Ordinary Water-substance". Reinhold, New York.
Douglas, E. (1964). *J. Phys. Chem.* **68**, 169.
Douglas, E. (1965). *J. Phys. Chem.* **69**, 2608.
Douglas, E. (1967). *J. Phys. Chem.* **71**, 1931.
Fleming, R. H. and Revelle, R. R. (1939). "Recent Marine Sediments" (N. Trask ed.). Amer. Soc. Petrol. Geol., Tulsa, Oklahoma.
Kalle, K. (1945). In "Probleme der Kosmischen Physik" 2nd Edn., Vol. 23. Leipzig.
Korson, L., Drost-Hansen, W. and Millero, F. J. (1969). *J. Phys. Chem.* **73**, 34.
Kester, D. R., Duedall, I. W., Connor, D. N. and Pytkowicz, R. M. (1967). *Limnol. Oceanogr.* **12**, 176.
Krümmel, O. (1900). *Wiss. Meeresuntersuch.* **5**, 9.
Lepple, F. K. and Millero, F. J. (1971). *Deep-Sea Res.* **18**, 1233.
Lumby, J. R. and Folkard, A. R. (1956). *Bull. Inst. Océanogr. Monaco,* **1080**, 1.
Lyman, J. and Fleming, R. H. (1940). *J. Mar. Res.* **3**, 134.
Matthäus, W. (1974). *Beitr. Meeresk.* **29**, 93.

Matthäus, W. (1974). *Beitr. Meeresk.* **33**, 73.
Millero, F. J. (1974). *In* "The Sea" (E. D. Goldberg ed.), Vol. 5, Interscience, New York.
Millero, F. J. and Lepple, F. K. (1973). *Mar. Chem.* **1**, 89.
Millero, F. J., Hansen, L. D. and Hoff, E. V. (1973b). *J. Mar. Res.* **31**, 21.
Millero, F. J., Perron G. and Desnoyers, J. E. (1973). *J. Geophys. Res.* **78**, 4499.
Morris, A. W. and Riley, J. P. (1966). *Deep-Sea Res.* **13**, 689.
Murray, C. N. and Riley, J. P. (1969a). *Deep-Sea Res.* **16**, 311.
Murray, C. N. and Riley, J. P. (1969b). *Deep-Sea Res.* **16**, 297.
Murray, C. N. and Riley, J. P. (1971). *Deep-Sea Res.* **18**, 533.
Riley, J. P. and Tongudai, M. (1967). *Chem. Geol.* **2**, 263.
Robinson, R. A. (1954). *J. Mar. Biol. Ass. U.K.* **33**, 449.
Sager, G. (1974). *Beitr. Meeresk.* **33**, 68.
Stanley, E. M. (1971). *Deep-Sea Res.* **18**, 833.
Stanley, E. M. and Batten, R. C. (1969). *J. Geophys. Res.* **74**, 3415.
Stoughton, R. W. and Lietzke, M. H. (1965). *J. Chem. Eng. Data,* **10**, 254.
Stoughton, R. W. and Lietzke, M. H. (1967). *J. Chem. Eng. Data,* **12**, 101.
Thomas, B. D., Thompson, T. G. and Utterback, C. L. (1934). *J. Cons. Int. Explor. Mer,* **9**, 28.
UNESCO (1966). International Oceanographic Tables, Vol. 1. National Institute of Oceanography, Wormley, Surrey, England.
UNESCO (1973). International Oceanographic Tables. Vol. 2. National Institute of Oceanography, Wormley, Surrey, England.
U.S. Naval Oceanogr. Office (1962) Tables of sound speed in sea water. Publ. SP58 U.S. Naval Oceanographic Office, Washington, U.S.A.
U.S. Navy (1961). Tables for the velocity of sound in sea water. Bureau of ships reference NObsr 81564 S-7001-0307, Washington, U.S.A.
Weiss, R. F. (1970). *Deep-Sea Res.* **17**, 721.
Weiss, R. F. (1971). *J. Chem. Eng. Data,* **16**, 235.
Weyl, P. (1964). *Limnol. Oceanogr.* **9**, 75.
Wilson, W. D. (1960). *J. Acoust. Soc. Amer.* **32**, 1357.
Wood, D. and Caputi, R. (1966). "Technical Report No. 988" U.S. Naval Radiological Defense Laboratory, San Francisco.

Subject Index

(Numbers in bold type indicate the page on which a subject is treated most fully.)

D

E

F

G